Taste and Odour in Source and Drinking Water: Causes, Controls, and Consequences

Taste and Odour in Source and Drinking Water: Causes, Controls, and Consequences

Edited by
Tsair-Fuh Lin, Susan Watson,
Andrea M. Dietrich and I. H. (Mel) Suffet

IWA PUBLISHING

Published by **IWA Publishing**
Alliance House
12 Caxton Street
London SW1H 0QS, UK
Telephone: +44 (0)20 7654 5500
Fax: +44 (0)20 7654 5555
Email: publications@iwap.co.uk
Web: www.iwapublishing.com

First published 2019

British Library Cataloguing in Publication Data
A CIP catalogue record for this book is available from the British Library

ISBN: 9781780406657 (Print)
ISBN: 9781780406664 (eBook)
ISBN: 9781789060201 (ePUB)

Contents

Editors

Tsair-Fuh Lin, PhD
Department of Environmental Engineering
National Cheng Kung University
Tainan, Taiwan
Chair (2009–2011), International Water Association (IWA) Specialist Group (SG) on Off-Flavours in the Aquatic Environment

Dr. Lin has been affiliated with the Department of Environmental Engineering (DEnvE) at National Cheng Kung University (NCKU) in Taiwan since receiving his doctoral degree from the University of California, Berkeley in 1995. Dr. Lin's research interests include: (1) monitoring, treatment, and modeling of cyanobacteria, cyanotoxins, and taste and odour (T&O) compounds in drinking water sources and plants, and (2) adsorption and novel treatment techniques for contaminated ground water. Dr. Lin has published more than 100 refereed papers, co-authored and co-edited 3 books, and has owned 11 patents and 11 technology transfers. In the last 20+ years, Dr. Lin's team has helped many water agencies and water utilities in identifying and solving cyanotoxin and T&O relevant problems in the drinking water systems of Taiwan and Asian countries. Dr. Lin has received many outstanding teaching and research awards, including IWA Fellow (2012, 2018), International Honorary Member of American Academy of Environmental Engineers and Scientists (2015), Ho-Chin-Tsui Outstanding Award (2016), and the Outstanding Research Award from Taiwan Ministry of Science and Technology (2017).

Susan Watson, PhD
Department of Biology
University of Waterloo
Waterloo, ON, Canada
Chair (2011–2013), IWA SG on Off-Flavours in the Aquatic Environment

Dr Watson has wide experience and expertise in T&O and harmful algal/cyanobacterial blooms (HABs), with over 100 peer-reviewed publications as articles, book chapters, special journal issues and reports She established and led Canada's Federal research in these areas, incorporating multidisciplinary molecular, biochemical, physiological and ecosystem-based work. Her field and lab-based research addresses the causes, controls, ecophysiology and prediction of T&O and HABs, the species involved and the bioactive compounds that they produce (toxins, inhibitors, antioxidants, volatiles, grazer deterrents etc.) which affect drinking- and recreational water quality and safety, particularly in relation to nutrient management, climate change and source-water protection. She works with management, industry, academia, remedial and stewardship groups at international, national, and local levels, and specialty groups such as the International Water Association, the American Waterworks Association, the International Joint Commission and the International Association of Great Lakes Research. She has served on journal editorial boards, taught, trained and supervised students and staff, and given presentations and workshops at national and international symposia in countries such as Canada, UK, Switzerland, Finland, Germany, Italy, Denmark, Korea, China, the USA.

Andrea M. Dietrich, PhD
Department of Civil and Environmental Engineering
Virginia Polytechnic Institute and State University
Blacksburg, VA, USA
Chair (2014–2017), IWA SG on Tastes, Odours, and Algal Toxins in Drinking Water Resources and Aquaculture

Dr. Andrea M. Dietrich is a Professor of Civil and Environmental Engineering at Virginia Polytechnic Institute and State University. For over 30 years she has expanded knowledge related to drinking water though her teaching, research, outreach, and service. Professor Dietrich's extensive global expertise in T&O encompass understanding sensory, consumer, and chemistry aspects of drinking water that influence whether water is a desirable beverage or unsatisfactory product. The importance of consumer and societal issues of drinking water quality was emphasized during her leadership in both the International Water Association and the American Water Works Association. She conducts T&O workshops and sensory training sessions across the globe and is the author of over 150 peer-reviewed and widely read articles, book chapters, and reports. In addition to sensory and consumer issues, Professor Dietrich's expertise also includes human exposure to drinking water contaminants at the air-water interface and chemical spills that degrade drinking water quality and consumer confidence. At Virginia Polytechnic Institute and State University she co-directs the Interdisciplinary Water Interface program that unites students and professionals across the disciplines of sensory science, health, biology, chemistry, and engineering. She enjoys drinking safe and palatable water of many flavors with friends in many countries.

I. H. (Mel) Suffet, PhD
Department of Environmental Health Sciences
School of Public Health
University of California
Los Angeles, CA, USA
Chair (1995–2002), IWA SG on Off-Flavours in the Aquatic Environment

Dr. Suffet joined UCLA in 1991 after 21 years at Drexel University (Professor of Chemistry) and receiving his Ph.D. from Rutgers University. He has published over 225 papers and co-edited 9 treatises. He is a Board Certified Environmental Scientist of the American Academy of Environmental Engineers and Scientists, IWA certified, as a Fellow (2012, 2018) and received the outstanding publication award of the American Academy of Environmental Engineering and Scientists Professors (2010) for a seminal paper with C. W. Carter, on Binding of DDT to natural organic matter (NOM), Environmental Science and Technology 1982 and the A. P. Black Research Award from American Water Works Association in 2002 for "lifetime achievement in drinking water quality research". He has been studying throughout his career taste and odour problems in drinking water and odour nuisances in air using "odour wheels" as a template to characterize the odours, the effect of NOM on disinfection byproducts during water treatment from haloforms to now nitrosamines and the effects of NOM on the toxicity of organic chemicals in the aquatic environment. His proudest accomplishments are the 65 doctoral students and hundreds of Masters students that he has helped graduate.

Author List

Note: *indicating corresponding author of the chapter.

Chapter 1

Tsair-Fuh Lin*
National Cheng Kung University, Taiwan

Susan Watson
University of Waterloo, Canada

Andrea M. Dietrich
Virginia Polytechnic Institute and State University, USA

I. H. (Mel) Suffet
University of California, Los Angeles, USA

Chapter 2

I. H. (Mel) Suffet*
University of California, Los Angeles, USA

Scott Braithwaite
University of California, Los Angeles, USA

Yubin Zhou
University of California, Los Angeles, USA

Auguste Bruchet
CIRSEE, Suez Environment, Le Pecq, France

Chapter 3

Susan Watson*
University of Waterloo, Canada

Friedrich Jüttner
University of Zürich, Switzerland

Chapter 4

Auguste Bruchet
CIRSEE, Suez Environment, Le Pecq, France

Chapter 5

Andrea M. Dietrich*
Virginia Polytechnic Institute and State University, USA

Pinar Ömür-Özbek
Colorado State University, USA

Chapter 6

Chao Chen*
Tsinghua University, China

Arash Zamyadi
University of Montreal, Canada

Tsair-Fuh Lin
National Cheng Kung University, Taiwan

Daniel Gallagher
Virginia Polytechnic Institute and State University, USA

Chapter 7

Zhiquan Liu
Guangzhou University, China

Tsair-Fuh Lin*
National Cheng Kung University, Taiwan

Michael Burch
University of Adelaide, Australia

Chapter 8

Andrea M. Dietrich*
Virginia Polytechnic Institute and State University, USA

Ricard Devesa
Aigües de Barcelona, Spain

Chapter 9

Gary Burlingame*
Philadelphia Water Department, USA

Sarah Neiderer
Arlington, Virginia, USA

Lisa Ragain
Metropolitan Washington Council of Governments, USA

Chapter 1

Introduction

Tsair-Fuh Lin[1], *Susan Watson*[2], *Andrea M. Dietrich*[3] *and I. H. (Mel) Suffet*[4]

[1]*Department of Environmental Engineering, National Cheng Kung University, Tainan, Taiwan*
[2]*Department of Biology, University of Waterloo, ON, Canada*
[3]*Department of Civil and Environmental Engineering, Virginia Polytechnic Institute and State University, Blacksburg, VA, USA*
[4]*School of Public Health, University of California, Los Angeles, CA, USA*

Taste and odour are primary olfactory signals that can affect human behaviour, and play a key role in how we judge and react to our environment. Clean, safe drinking water is essential for human life and offensive tasting/smelling drinking water is instinctively rejected by consumers as an indication of contaminated and/or harmful supplies (Dietrich, 2006; McGuire, 1995). There has been considerable improvement in our ability to deliver safe drinking water in many areas of the world, yet despite this progress, poor taste and odour (T&O) in source and finished water supplies continues to be a major issue, costing the drinking water and other associated industries (aquaculture, tourism, recreation, fishing, food and beverage production) millions of dollars annually in treatment, management and lost income (e.g. Dodds *et al.*, 2009; Dunlap *et al.*, 2015; Engle *et al.*, 1995; Tucker, 2000). This book addresses this issue, providing an updated evaluation of the causes, characterization and management of T&O in source and

 Taste and Odour in Source and Drinking Water: Causes, Controls, and Consequences
Editor(s): Tsair-Fuh Lin, Susan Watson, Andrea M. Dietrich and I. H. (Mel) Suffet
doi: 10.2166/9781780406664_0001

drinking waters. Authored by international experts from the International Water Association (IWA) Specialist Group (SG) on Tastes, Odours, and Algal Toxins in Drinking Water Resources and Aquaculture (formerly known as IWA SG on Off-flavours in the Aquatic Environment), the book is an important synthesis of current knowledge on the origins, mitigation, and management of aquatic T&O problems. The material provides updated information on the causes and management of the increasing and widespread degradation of source waters and global demand for safe, high quality potable water. Key topics covered in the following nine chapters include early warning, detection and source tracking, chemical, sensory and molecular diagnosis, treatment options for common odourants and minerals, source management, and modelling and risk assessment.

The book is directed towards a wide readership of scientists, engineers, technical operators and managers. It presents both practical and theoretical material, including an updated version of the benchmark Drinking Water Taste and Odour Wheel and a new biological wheel to provide a practical and informative tool for the initial diagnosis of chemical and biological sources of aquatic T&O. This first chapter reviews the history of T&O issues in drinking water and the development of knowledge and research in this area, and provides a brief synopsis of each chapter.

1.1 T&O IN DRINKING WATER: HISTORY AND RESEARCH

The history of drinking water T&O research is in fact quite brief, only gaining momentum in the latter part of the last century, and pioneered by a relatively small number of scientists. Taste, odour, and appearance are three key criteria of drinking water acceptability (WHO, 2017), because for the general public the aesthetics and palatability of drinking water are often linked with their perception of its safety (McGuire, 1995). In essence, people avoid bad-tasting or malodorous water.

Although offensive T&O in drinking water was reported in the 1800s (Barker, 1981), peer-reviewed articles on this topic have mostly been published over roughly the last 100 years (Figure 1.1). In an extensive review of research on aquatic off-flavours in the 19th and early 20th centuries, Persson (1995) noted that the first report on adverse T&O in drinking water was as early as in 1855 (Horsford & Jackson, 1855), and was caused by a peculiar cucumber odour in the Cochituate water supply (Boston, USA) in 1854. Later, a few T&O cases in drinking water in Europe and north-eastern US between 1860 and 1910 were also reported (Persson, 1995).

A 1905 collection of laboratory methods for drinking water analysis, the *Standard Methods of Water Analysis* (APHA, 1905), emphasized that odours were caused by both chemicals and microorganisms, especially algae. The document stressed the importance of odour analysis for drinking water quality and stated: 'The observation of the odor of cold and hot samples of surface water is very important, as the odors are usually connected with some organic growths or with sewage contamination, or both. The odor of ground waters is often caused by the earthy constituents of the water-bearing strata. The odor of contaminated well water is often decisive evidence of pollution'.

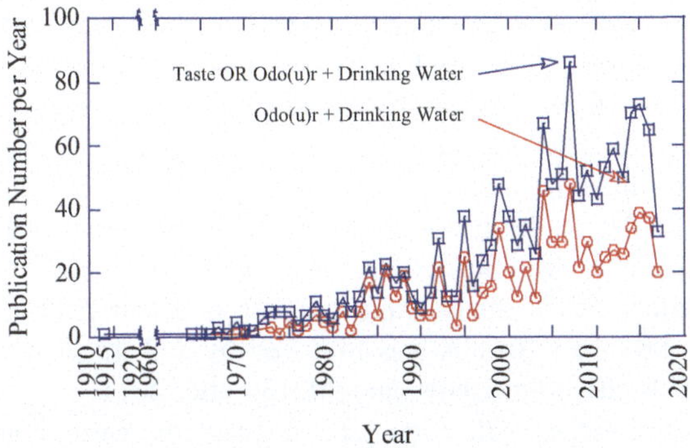

Figure 1.1 Publications on taste and odo(u)r and drinking water from 1913 to 2017 (based on a Scopus database search using keywords shown in legend).

In 1913, unpleasant odour in drinking water supplies from Hampstead, London, UK was documented (Anonymous, 1913) and although no data were provided, the T&O was attributed to an outbreak of an organism 'with no pathogenic significance'. Later in 1925, The British Waterworks Association discussed 'the taste factor in drinking water' in its annual meeting in London, focusing on two frequent T&O descriptors in the City's water, 'chlorinous' and 'iodoform' (Anonymous, 1925). At the same time, in Canada and the USA there were reports of increasing concerns and treatment strategies for odours caused both by microorganisms and inorganic sources (sulfur, chlorine, chlorinated phenols and iron; Howard, 1922). These reports demonstrate the importance of T&O issues as far back as the 19th and the early 20th centuries.

The first peer reviewed scientific papers dealing with T&O treatment issues appeared in the published literature around the mid 1900s. For example, a notable contribution at this time discussed the removal of salts from sea water using a chemical precipitation method (Spealman, 1944). This was followed in 1947 by a paper proposing the removal of T&O compounds using activated carbon (Sigworth, 1947), a treatment process which is still widely applied in many drinking water supplies. Procedures for the concentration, extraction, and separation of organic compounds in water using carbon filter as the media were described in 1951, and proposed as a method to measure objectionable tastes and odours in potable supplies (Braus *et al.*, 1951). These and other studies confirm that T&O treatment was already an issue between 1850s and the first half of 1900s.

Poor T&O had been linked to microorganisms for some time, but the chemistry of biological T&O was unresolved until the mid-1960s. Geber and co-workers

(Gerber, 1967, 1969; Gerber & Lechevalier, 1965) and Medsker *et al.* (1969) pioneered this work, identifying the structure and odour properties of geosmin and 2-methylisoborneol (2-MIB), two of the most commonly detected T&O compounds worldwide, thereby advancing our understanding on the chemistry behind microbial producers (e.g. Jüttner, 1995; Persson, 1995; Watson, 2010). Since then, numerous other compounds derived from both biotic and abiotic sources have been identified in drinking water systems (Figure 1.1).

In the past few decades, research into the different aspects of T&O (biotic and abiotic origins, detection/identification, treatment and management) has escalated, and resulted in tremendous advancement in the field. A search of several major literature databases yields a total of 3,086 papers published since 1913 that match with keywords 'taste or odo(u)r' and 'drinking water' (Figure 1.1). Before 1970, the average number of listed publications per year in the field was less than 10; this was followed by a sharp increase, jumping to 20–60 in the 1970s, 40–80 in 1990s, and 90–130 after 2011.

Exploring this in more depth, Figure 1.2 displays the word cloud representing the keywords provided in the articles represented in the previous figure (Figure 1.1). Of the major keywords relevant to the 1,440 papers published before 2000 (Figure 1.2(a)), 'water treatment' (188 papers) is the most common, followed by 'taste preference' (84), 'ozone' (65), 'water analysis' (60), 'odor control' (58), 'geosmin' (57), 'disinfection' (52), 'water pollution' (51), 'instrument conditions' (51), 'activated carbon' (44), 'chlorides' (42), 'flavour' (40), 'gas chromatography' (4), 'smell' (39), 'mass spectrometry' (36), 'algae' (34), 'taste threshold' (34), 'chlorination' (32), '2-methylisoborneol' (28), 'distribution system' (24), and 'filtration' (24). These keywords represent the major research topics (water treatment and analysis), analytical methods (instrumental conditions, gas chromatography and mass spectrometry), treatment methods (ozonation, activated carbon, chlorination and filtration), chemicals (chloride, chlorine, geosmin and 2-methylisoborneol (2-MIB)) and sources of T&O issues (algae, disinfection, and distribution systems) that were addressed. For the 1,636 papers published after 2001, similar, but an increased number of, keywords are found (Figure 1.2(b)), demonstrating the increasing recognition of the complexity and ubiquity of this issue, from source to tap. Thus, new keywords such as 'water quality' (432), 'cyanobacteria' (99), 'environmental monitoring' (83), 'genetics' (49), 'reservoirs' (49), 'risk assessment' (44), and 'biofilm' (41) appear, while others are referred to more frequently, likely indicative of the relative importance of these particular issues ('2-MIB', 'gas chromatograph', and 'oxidation').

Co-word analysis was further conducted to find the research trends, using the papers published with keywords of 'taste or odo(u)r' and 'drinking water' searched from Web of Science in August 2018. In the analysis, an open source tool, *SciMAT* (Cobo *et al.*, 2012), was used and the papers were categorized into three groups: before year 2000 (the number of papers = 193); between year 2001 and 2010 (388); and between 2011 and 2018 (523).

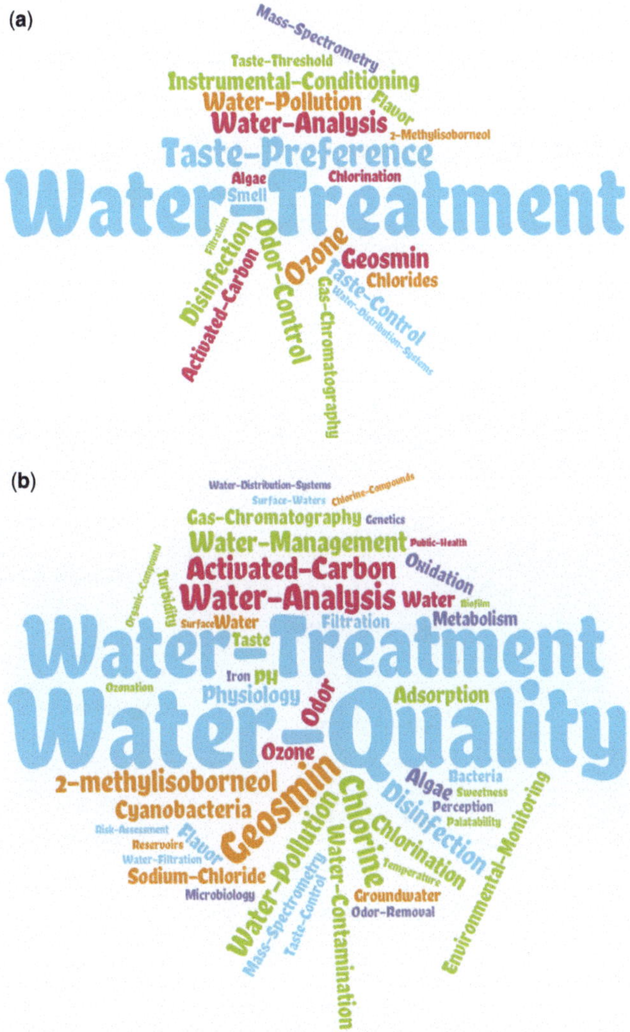

Figure 1.2 Word cloud of the keywords in papers published (a) before 2000, and (b) after 2001 to May 2018 (based on a Scopus search using the terms ('taste' and/or 'odo(u)r') and 'drinking water').

Figure 1.3 shows the evolution map for the keywords of T&O relevant studies. In Figure 1.3(a), the solid lines and dotted lines are for the linked clusters that share the main and non-main items, respectively, while the size of the spheres represents the number of papers relevant to the cluster (Cobo *et al.*, 2012). Figure 1.3(a) shows the major keywords in the published papers. As shown in Figure 1.3(b), the number of keywords for the three periods are 515, 949, 1430,

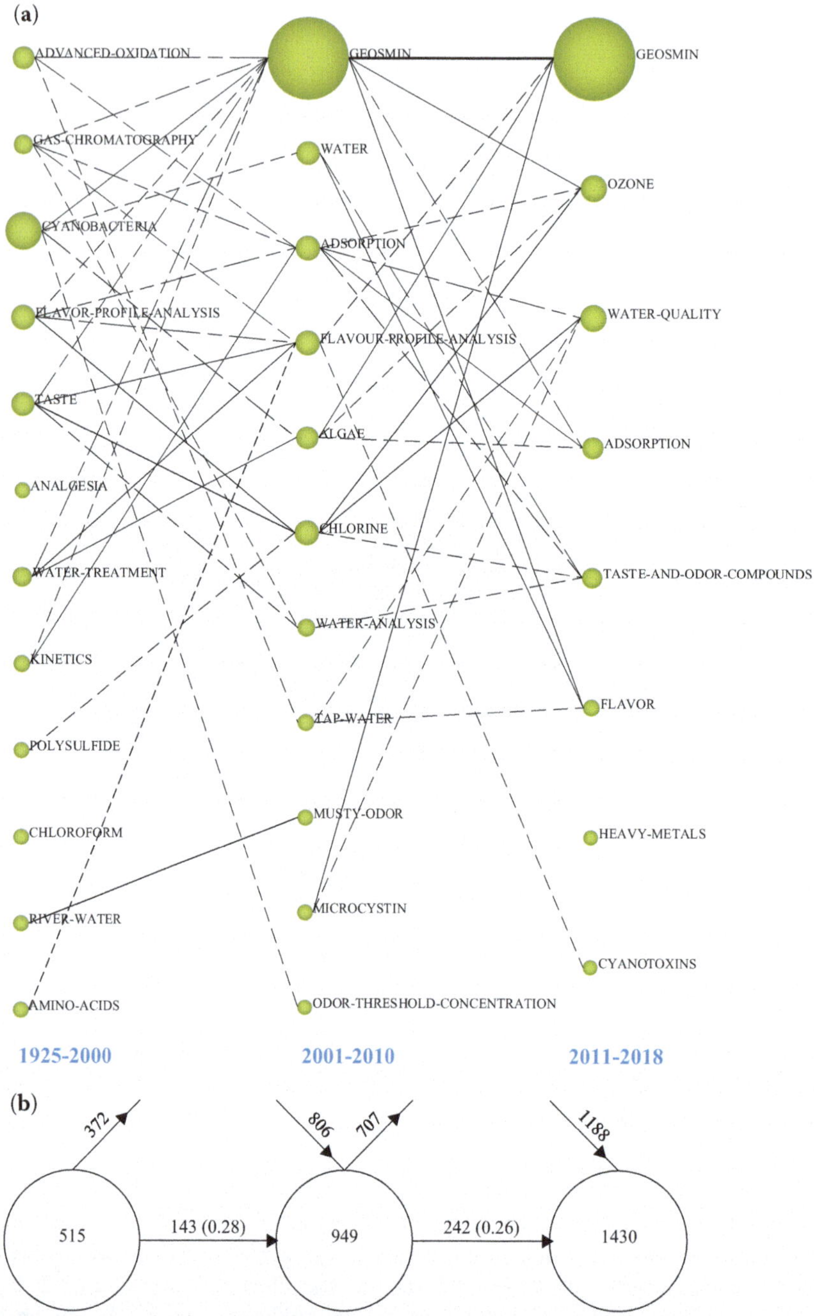

Figure 1.3 The evolution map for the keywords of T&O relevant studies.

respectively. Only 28% of keywords before year 2000 were carried to the next period (2001–2010). For the second period, 806 new keywords were brought in, with only 26% of keywords being retained in the period after 2011. It is clear to see that cyanobacteria are strongly linked with geosmin in the three studied periods, with geosmin appearing in many published papers. Taste was linked with flavour profile analysis and chlorine, while chlorine was linked with both ozone and water quality. In addition, water treatment (before 2000) was linked with algae (between year 2001 and 2010) and then with geosmin (after 2011).

1.2 INTRODUCTION OF BOOK CHAPTERS

The Taste and Odour Wheel (Mallevialle & Suffet, 1987) has provided an essential tool to the drinking water industry for over 30 years. Over recent years it has undergone improvements, but the fundamental idea has remained – providing a valuable front-line diagnosis of the likely cause of a T&O event, which can then be followed up by further investigation. Importantly, minimal training is required and can be used by operators and managers at all levels. Chapter 2, '*The Drinking Water Taste-and-Odour Wheel After 30 Years*', presents an updated version that can be applied as a frame of reference for identifying taste-and-odour problems in drinking water supplies. As with earlier versions, the wheel is based on thirteen primary categories: eight classes of odours, four tastes, and a mouth-feel/nose-feel category. Each of these shows a secondary ring of specific tastes and odours that are defined by trained sensory panels. Specific chemicals, either confirmed odourants or representative taste and odour standards, surround the outside of the wheel. Discussion and tables provide foundational and updated information on each of the taste-and-odour categories. In addition, a taste section is included which provides the user with, for the first time, the value of the wheel as an illustrated approach to help identify odourants from chemical spills.

The majority of all reported T&O events is derived from biological activity, and these can be generated throughout the origin, treatment and distribution of potable water. Chapter 3 ('*Biological Production of Taste and Odour Compounds*') presents an in-depth review of the source organisms and the odourants they produce. The chapter reviews the major genetic and biochemical pathways involved in the production of these metabolites, and the environmental and sampling/analytical challenges which can impede source tracking and management. It describes the chemistry and odour qualities of the odourants (terpenoids such as geosmin and 2-MIB, sulfides, pigment/fatty acid derivatives (oxylipins), amines, fermentation products, etc.), and the factors moderating their production and fate. The taxonomy and ecology of the organisms – bacteria (including two major groups, actinomycetes and Cyanobacteria), fungi, micro and macroalgae, etc. – their habitats and the growth conditions conducive to their development are covered, providing insight into the seasonal and geographical patterns of T&O occurrence. The tenuous, but often assumed, relationship between T&O and cyanobacterial toxins which are derived from entirely different cellular pathways is also critically examined, with links to further reading. The chapter advocates a multifaceted approach to T&O, combining traditional approaches with molecular probes, biosensors, 'omics' (e.g. metagenomics, proteomics), behavioural/toxicity assays and state-of-the art analytical methods. The chapter concludes with a discussion of future research and the need to delineate the function of the compounds (i.e. their chemical ecology) and their intra/extracellular control mechanisms in order to better predict and manage this issue in the face of global eutrophication, climate change and rising demand.

The purpose of Chapter 4 is to show the principles and areas of applications of analytical methods in the T&O field. Target odorous chemicals and their detection limits are listed. Methods based on static and dynamic head-space extraction with and without an adsorption or absorption phase are first described. These methods include closed-loop-stripping extraction (CLSA), purge-and-trap (P&T), solid-phase microextraction (SPME), and stir-bar-sorptive extraction (SBSE). Detailed analytical conditions for important odourous chemicals such as geosmin and 2-MIB, methyl *tert*-butyl ether and ethyl *tert*-butyl ether, halogenated anisoles and halogenated phenols, are provided. Large volume injection in GC/MS as a method to enhance sensitivity for eight earthy-musty odorous compounds including halogenated anisoles is discussed. The design of an olfactory-GC/MS system that can be homemade is presented. Other extraction techniques that have found limited application in the T&O field such as solid-phase extraction, GC-atomic emission detection (AED), CLSA/SPME, hollow-fiber stripping analysis (HSFA), solid-phase dynamic extraction (SPDE), are mentioned briefly. As a conclusion, the number of samples that could be processed per day with the most common methods are compared.

Consumers judge water quality based on its sensory and aesthetic characteristics, and Chapter 5, '*Advances in Sensory Measurement Determinations*', describes traditional and new sensory methods used to collect information from professionals and consumers on the assessment and acceptability of (finished) drinking water. The chapter first provides a comprehensive description of the different sensory methods used to evaluate preference, acceptance, discrimination, description, and thresholds for foods and beverages. This is followed by a discussion of tracking and assessing consumer feedback and applying consumer complaint surveillance techniques.

Chapter 6 ('*Removal of Odourants from Drinking Water*') focusses on the development and application of methods for the removal of T&O compounds from drinking water, including the chemicals, sources and treatment processes involved. Additionally, removal technologies for the two most commonly detected T&O compounds in drinking water, geosmin and 2-MIB, are specifically discussed. The chapter also reviews two common treatment methods for odour compounds, activated carbon adsorption and oxidation, presenting both the fundamental theories and applications of these methods. Finally, five full-scale water treatment plant (WTP) case studies for the removal of odour compounds are discussed, including two cases for routine treatment and three for emergency conditions.

The management and mitigation of T&O issues in source water is discussed in Chapter 7, '*Management of T&O in Source Water*'. The fate of T&O compounds in lakes and reservoirs are first systematically analyzed, including the processes involved in both the source production and elimination/loss of T&O compounds. Then appropriate treatment methods for the management of T&O, including inhibition and control of the growth of T&O compound producers, are considered. Finally, the efficiency and cost of different management methods and the safety of these methods are reviewed.

An informative and timely assessment of the effects of total dissolved solids (TDS) as an aggregate measure of mineral taste, and the effects of individual and combinations of cations (e.g. Fe^{2+}, Na^{+}) and anions (e.g. Cl^{-}, SO_4^{2-}) is provided in Chapter 8 ('*Characterization and Removal of Minerals that Cause Taste*'). The chapter compares global water quality standards for tastes and provides key information on the salinization of freshwaters, an increasing and widespread water quality challenge which adversely affects drinking water taste, and addresses its treatment with desalinization to improve palatability.

Chapter 9, '*Risk Management of Public Perception*', emphasizes that drinking water providers must become trusted sources of information for their consumers and that communication with consumers about water is essential. The information in this chapter enlightens the drinking water community concerning proactive and reactive strategies for effectively communicating with consumers. Effective communication is not only needed to address risks and outbreaks, but also to have a dialogue with consumers about their perceptions and satisfaction with their drinking water. Specific tools for communicating within the utility and to consumers are presented.

REFERENCES

American Public Health Association (1905). Standard Methods of Water Analysis, 1st edn, American Public Health Association, Washington, DC.

Anonymous (1913). Unpleasant odour and taste in drinking water. *The Lancet*, **181**(4674), 912.

Anonymous (1925). The taste factor in drinking water. *The Lancet*, **206**(5315), 76–77.

Barker M. N. (1981). The Quest for Pure Water. American Water Works Association, Denver, CO, USA, Vol. **1**, 527 pp.

Braus H., Middleton F. M. and Walton G. (1951). Organic chemical compounds in raw and filtered surface waters. *Analytical Chemistry*, **23**(8), 1160–1164.

Cobo M. J., López-Herrera A. G., Herrera-Viedma E. and Herrera F. (2012). *SciMAT*: a new science mapping analysis software tool. *Journal of the American Society for Information Science and Technology*, **63**(8), 1609–1630.

Dietrich A. M. (2006). Aesthetic issues for drinking water. *Journal of Water and Health*, **4**(Supplement 1), 11–16.

Dodds W. K., Bouska W. W., Eitzmann J. L., Pilger T. J., Pitts K. L., Riley A. J., Schloesser J. T. and Thornbrugh D. J. (2009). Eutrophication of U.S. freshwaters: analysis of potential economic damages. *Environmental Science & Technology*, **43**(1), 12–19.

Dunlap C. R., Sklenar K. S. and Blake L. J. (2015). A costly endeavor: addressing algae problems in a water supply. *Journal of the American Waterworks Association*, **107**(5), E255–E262.

Engle C. R., Pounds G. L. and Van Der Ploeg M. (1995). The cost of off-flavor. *Journal of the World Aquaculture Society*, **26**(3), 297–306.

Gerber N. N. (1967). Geosmin, an earthy smelling substance isolated from actinomycetes. *Biotechnology and Bioengineering*, **9**(3), 321–327.

Gerber N. N. (1969). A volatile metabolite of actinomycetes, 2-methylisoborneol. *Journal of Antibiotics*, **22**(10), 508–509.

Gerber N. N. and Lechevalier H. A. (1965). Geosmin, an earthy smelling substance isolated from actinomycetes. *Applied Microbiology*, **13**(6), 935–938.

Horsford E. N. and Jackson C. T. (1855). Reports on recent impurities in the water. Report on Cochituate. City Document 9 32-58. Water Board, City Council, Boston, USA.

Howard J. N. (1922). Modern practice in the removal of taste and odor. *Journal of the American Water Works Association*, **9**(5), 766–782.

Jüttner F. (1995). Physiology and biochemistry of odorous compounds from freshwater cyanobacteria and algae. *Water Science and Technology*, **31**(11), 69–78.

Mallevialle J. and Suffet I. H. (1987). Identification and Treatment of Tastes and Odors in Drinking Water. AWWA Research Foundation and Lyonnaise Des Eaux Cooperative Research Report. American Water Works Association, Denver, CO, USA.

McGuire M. (1995). Off-flavor as the consumer's measure of drinking water safety. *Water Science and Technology*, **31**(11), 1–8.

Medsker L. L., Jenkins D., Thomas J. F. and Koch C. (1969). Odorous compounds in natural waters. 2-exo-hydroxy-2-methylbornate, the major odorous compound produced by actinomycetes. *Environmental Science and Technology*, **3**(5), 476–477.

Persson P. (1995). 19th Century and early 20th century studies on aquatic off-flavours – a historical review. *Water Science and Technology*, **31**(11), 9–13.

Sigworth E. A. (1947). Taste and odor control with activated carbon. *Water and Sewage Works*, **94**(7), 135.

Spealman C. R. (1944). The chemical removal of salts from sea water to produce potable water. *Science*, **99**(2566), 184–185.

Tucker C. S. (2000). Off-flavor problems in aquaculture. *Reviews in Fisheries Science*, **8**(1), 45–88.

Watson S. B. (2010). Algal Taste-Odour. Ch. 15. In: Algae: Source to Treatment, American Water Works Association, Denver, CO, USA, Vol. **M57**.

World Health Organization (2017). Chapter 10: acceptability aspects: taste, odour and appearance. In: Guidelines for Drinking Water Quality, WHO, Geneva, Switzerland, pp. 219–230.

Chapter 2

The drinking water taste-and-odour wheel after 30 years

I. H. (Mel) Suffet[1], Scott Braithwaite[2], Yubin Zhou[3] and Auguste Bruchet[4]

[1]Department of Environmental Health Sciences and Environmental Science and Engineering, UCLA Schools of Public Health and Arts and Sciences, Los Angeles, CA 90095, USA
[2]Environmental Science and Engineering Program, UCLA Institute of the Environment and Sustainability, School of Arts and Sciences, Los Angeles, CA 90095, USA
[3]Department of Environmental Health Sciences, UCLA School of Public Health, Los Angeles, CA 90095, USA
[4]CIRSEE, Suez, 38 rue du Président Wilson, 78230 Le Pecq, France

2.1 INTRODUCTION

This chapter presents the frame of reference of our knowledge of taste-and-odour problems in drinking water supplies by updating the 'Taste-and-Odour Wheel'. The wheel organizes the relationships between specific taste-and-odour problems and their causes. The 'Year 2016 Taste-and-Odour Wheel' is shown in Figure 2.1 as an upgrade of the 'Year 2000 and 2006 Taste-and-Odour Wheel' (Suffet *et al.*, 1999, 2004; Suffet & Rosenfeld, 2007). The wheel describes primary taste-and-odour categories (inner circle) and those common tastes and odours from each primary category that are defined by trained sensory panels (outer circle) by the flavor-profile analysis (FPA) (APHA *et al.*, 2012; Method, 2170).

Editor(s): Tsair-Fuh Lin, Susan Watson, Andrea M. Dietrich and I. H. (Mel) Suffet
doi: 10.2166/9781780406664_0011

The primary odour categories are akin to the primary colors of the rainbow. The outer circle indicates variations within each primary category, i.e. secondary colors. Outside the circle are chemicals marked with an asterisk (*) that have been confirmed as odourants in drinking waters while the others have been used by sensory panels as 'representative' standards for the particular organoleptic responses (Burlingame *et al.*, 1991a). Over the years, the drinking water 'Taste-and-Odour Wheel' has been joined by similar wheels for water distribution systems (Khiari *et al.*, 2002), wastewater (Burlingame *et al.*, 2004), landfill (Decottignies *et al.*, 2009), compost (Suffet *et al.*, 2009), urban odours (Suffet & Rosenfeld, 2007) and others (Curren, 2012).

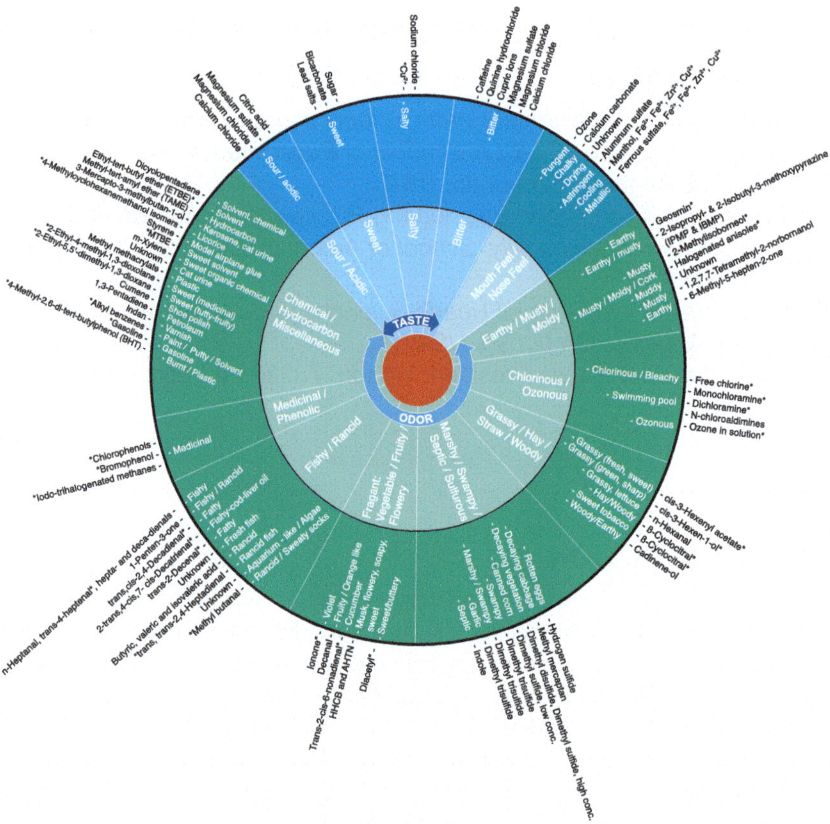

Figure 2.1 Year 2016 drinking water taste-and-odour wheel.

Three different philosophies guide the usage of the 'Taste-and-Odour Wheel'. First, the wheel has been successfully used to help develop a common language for taste-and-odour sensory panels and drinking water practitioners around the world. Second, the wheel has been devised to present the water industry with the current knowledge about the identification of the 'common' organoleptic characteristics found in drinking water. Finally, the wheel has helped develop an understanding of the chemical causes of tastes and odours in drinking water supplies. This has helped the drinking water industry determine what chemical analysis to use to evaluate a taste-and-odour problem and thereby identify the chemical causes of the particular problem. More importantly, the control of taste-and-odour problems in water supplies has been developed based upon the chemical causes of the taste-and-odour problems. The presence of certain items in drinking water, however, is necessary for pleasant taste (e.g. minerals) or often tolerated (e.g. slight chlorinous odour).

The 'Taste-and-Odour Wheel', developed over the past 31 years, includes eight classes of odours, four tastes, and a mouth-feel/nose-feel category (Brady *et al.*, 1988; Burlingame *et al.*, 1991a; Mallevialle & Suffet, 1987; Suffet *et al.*, 1988, 1995a, 1999, 2004; Suffet & Rosenfeld, 2007). The identification of odourants is well understood for two groups of chemicals: chlorinous/ozonous and earthy/musty/mouldy. Of the latter category, *trans*-1,10-dimethyl-*trans*-9-decalol (geosmin) (earthy odour) and 2-methylisoborneol (MIB) (musty odour) are especially well known odourants. Over the past 25 years, however, researchers have begun to identify other types of odours by utilizing odour panels with the FPA (APHA *et al.*, 2012; Method, 2170) and gas chromatography/mass spectrometry (GC/MS) together with sensory GC analyses (Khiari *et al.*, 1992).

In addition to primary drinking water standards that protect human health (USEPA, 2010), secondary standards that evaluate aesthetic qualities are important for ensuring the palatability of drinking water (USEPA, 2007). Secondary standards are in place for the taste and odour of drinking water and thus tests have been developed to quantify the degree of odour in potable water. Responsible drinking water utilities meet secondary standards that are mandatory only in a few US states, including California; moreover, these water utilities feel accountable for maintaining the palatability of their drinking water when the quality changes. Some highly soluble, off-odour or off-taste chemicals, which are not a toxic threat, may enter the water supply. In such cases, it is helpful to know both the concentration at which the consumers can detect a difference (i.e. odour-threshold concentrations (OTC) or flavor-threshold concentrations (FTC)) in their drinking water and the concentration at which consumers can detect specific odours or flavors (i.e. odour-recognition concentrations (ORC) or flavor recognition concentrations).

The Threshold Odour Number (TON) (APHA *et al.*, 2012; Method, 2170) is the secondary drinking water standard used in the United States (USEPA, 2007) and in many parts of the world (e.g. WHO, 2011). It can be used to determine the strength of a general odour problem by making dilutions of a particular sample and determining how many dilutions it takes to produce 'odour-free' water. A TON of 3 is the standard in the United States as well as in many countries.

The development of the wheel was based on the FPA panels assigning odour descriptors and intensities rather than the dilution-to-threshold approach of the TON (APHA *et al.*, 2012; Method, 2170). The present TON method does not provide any odour description, so it cannot be used to identify the source or cause of the taste-and-odour event as described by Mallevialle and Suffet (1987). Furthermore, the TON method uses dilution, which may alter the type of odour. The FPA method, instead, determines odour descriptors as well as intensities of each odour descriptor on a 7-point scale: threshold (1), slight (2), weak (4), medium (6), medium strong (8), strong (10), and very strong (12). FPA and GC/MS identification must be confirmed by sensory GC analysis using the appropriate standards (Khiari *et al.*, 1992). Only by a combination of approaches can sources of contaminants be determined that cause taste-and-odour problems in drinking water.

Chemical identification of the odour-causing substance(s) during a taste-and-odour problem is critical. Odour production is really a result of biological and chemical reactions that produce odourants. A decision tree was developed by McGuire *et al.* (2005) to help describe an off-flavor, determine the origin, quickly learn how to proceed with an investigation of the event, and identify potential solutions. Unfortunately, there remains a general lack of identification of compounds that cause taste-and-odour problems, especially when compounds are present in complex mixtures at or below the specific substance's OTC. Some chemicals can mask other odours (serving as antagonists) or act synergistically to make odours stronger than when a single odourant is present. Also, the chemical's odour characteristics may change as a function of concentration or for other reasons. For example, the odours produced by algal growth progress from cucumber to fishy as their growth cycle and abundance change (Rashash *et al.*, 1993). In addition, odour-causing compounds may hydrolyze, photolyze, or biodegrade during a taste-and-odour episode or unexpectedly during the sampling and analysis steps (Khiari *et al.*, 1999b).

2.2 BACKGROUND

Consumers expect their drinking water to taste good and often associate off-flavors with potential health risks; however, there are only a few enforceable regulations of taste-and-odour compounds in drinking water worldwide. Although the secondary Maximum Contaminant Levels for taste and odour are listed as '3' on the TON scale in use by the USEPA (2007), in reality the drinking water industry has used the following as the benchmarks for managing taste-and-odour problems: Consumer complaints; FPA action levels; GC/MS chemical analysis levels of the causative odour observed by FPA; Relative rankings of odourants by dividing the measured concentration by the OTC or ORC (i.e. C/OTC or C/ORC); OTC – defined as the level at which 50% of an odour panel detects a difference in the odour of the water during a series of triangle tests with ascending concentrations.

OTC is completed by ASTM Method 679-04 (2011), while ORC is a new addition to the benchmarks (see McGuire *et al.*, 2014b; Suffet *et al.*, 2013). ORC is defined as the concentration at which 50% of the consumers will recognize specific odours in their drinking water and is completed by ASTM Method 679-04 (2011). Odour Objection Concentration (OOC) is another developing benchmark (see McGuire *et al.*, 2014b; Suffet *et al.*, 2013). It is defined as the concentration at which 50% of the consumers will object to specific odours in their drinking water.

Site-specific drinking water standards for drinking-water facilities, especially for earthy and musty odours, have been developed (e.g. Mackey *et al.*, 2008; Suffet *et al.*, 2008; Yu *et al.*, 2014). These standards have been set due to consumer complaints and employ quantitation by FPA and associated GC/MS chemical analysis of the causative agents. In addition, certain places in the world (e.g. the City of Philadelphia and the Metropolitan Water District of Southern California) have set a limit for the earthy and musty odourants, MIB and geosmin, at 10 ng/L (Mackey *et al.*, 2008; Suffet *et al.*, 2008). China has varied the limits by watershed and region for MIB and geosmin (Yu *et al.*, 2014).

The ASTM Standard Method E679-04 (ASTM, 2011) is generally respected as the 'gold standard' for the objective determination of OTCs. ASTM E679-04 utilizes a geometric series of three samples, where only one is spiked with the chemical of interest. It is an ascending concentration, forced-choice test, meaning the panelist must choose the sample deemed different and mark it on the answer sheet for the OTC determination. To further build on the use of these tests, the panelist may be asked to describe the odour and thus determine the ORC. While a panelist may be able to detect an odour or taste at a certain concentration (OTC), they may not be able to recognize that odour correctly until a higher concentration is administered (ORC).

In addition, a subjective (hedonistic) test can be added to determine the OOC by the panel using different methods as described by Suffet *et al.* (2013) and McGuire *et al.* (2014b). Knowing the concentrations at which consumers can detect (OTC), recognize (ORC) and object (OOC) to an odour in their drinking water is important for utilities that wish to reduce the number of consumer complaints and allows operators to adjust treatment regimens to reduce the concentrations of odourants to acceptable levels.

ASTM E679-04 has been used to determine the OTCs for a range of odourants (Young *et al.*, 1996). Details on the determination the OTC of a chemical are illustrated for methyl *tert*-butylether (MTBE) by Stocking *et al.* (2001) and the textbook by Lawless and Heymann (2010), which includes the Stocking *et al.* (2001) study as a case study on how to complete consumer odour panels. The more recent OOC can be determined as described by others (McGuire *et al.*, 2014b; Suffet *et al.*, 2013) and added after the OTC and ORC determinations.

Water utilities have primarily used the OTC as a guide for any known chemical causing an odour. If the concentration of the odourant is near or below the OTC, consumer complaints rarely occur. If the concentration of the odourant is at or above the ORC, consumer complaints can be expected. Only the OTC, however, is widely available for chemicals. ORCs and OOCs would be very helpful tools for water utilities. For example, a site-specific drinking water standard for geosmin and MIB is 10 ng/L, which is above the OTC of 4–9 ng/L for geosmin and MIB (Mallevialle & Suffet, 1987) and probably closer to the yet-unknown ORC. As another example, the OTC, ORC and OOC have been determined by a large-scale consumer panel study for licorice-odour chemical spill in 2014 in West Virginia to enable an understanding of the effects on the population (McGuire *et al.*, 2014b).

2.3 CAUSE-AND-EFFECT RELATIONSHIPS IN DRINKING WATER TASTE-AND-ODOUR PROBLEMS

The literature is filled with presumptive statements about the causes of taste-and-odour problems that often mislead the water industry. For example, the mere number or density of algae or actinomycetes may not actually create a taste-and-odour problem, the cause of the earthy/musty/mouldy odour in water may not be due to geosmin or MIB, and odours from chemical spills may be caused by mixtures of chemicals.

Rules of evidence must be satisfied before a cause-and-effect relationship can be established between the presence of an organism or chemical and a taste-and-odour problem. These rules follow the 'scientific method' by defining presumptive and subsequent confirmatory testing procedures to validate the true cause of a taste-and-odour event in drinking water (Mallevialle & Suffet, 1987; Persson, 1992).

Microbiological causes of taste-and-odour problems require ecological (concurrence of the organism and the odour in the field), sensory (the isolated organism must be able to produce the odour in pure culture), and chemical (the odorous compound must be isolated from the original water and chemically identified) evidence. Accordingly, a presumptive test (identifying the organism believed to be the cause of the taste-and-odour problem and quantifying its population density) must be confirmed by either: isolating the organism by culture techniques and determining whether the odour produced by the culture matches the original problem odour in the water, as determined by a sensory panel; isolating microbial products associated with the microbial growth in the culture and confirming it by GC/MS or other detection method for the metabolite(s); isolating from natural water samples the microbial product(s) that cause that odour, with confirmation by a sensory panel; or developing a relationship between the chemical concentration in actual water samples and the odour intensity determined by a sensory panel.

Chemical causes of taste-and-odour problems rely on a presumptive statistical correlation between the chemical compound(s) in the water sample and the intensity of tastes or odours observed by the sensory panel as well as the ability to identify individual compounds from among those that have the same sensory characteristics as described by a sensory panel, ultimately by sensory GC analysis (Khiari *et al.*, 1992). Final confirmation is by sensory panels evaluating the presumptive chemical identified by chemical analysis and sometimes for mixtures after sensory GC analysis. Sensory methods are often more sensitive than analytical instrument-based methods, or in other words the reason many chemicals are of taste-and-odour concern is due to their extrememly low odour thresholds. Through further study, the relationship between the chemical concentration(s) in actual water samples and odour intensity is determined by the sensory panel studying the identified individual compounds and, possibly, mixtures of the compounds chemically identified.

2.4 FPA EVALUATION

The 'Taste-and-Odour Wheel' (Figure 2.1) presents reference standards for odours as 'known' (marked with *) and 'representative' compounds. However, odour descriptions vary from individual to individual due to life experiences. As a further complication, odours in the earthy/musty/mouldy category can be produced by several different compounds, such as geosmin, MIB, and 2,4,6-trichloroanisole (TCA). Trained panelists using known compounds as standards at different concentrations learn to define the difference between musty and earthy. A novice panel may describe both geosmin and MIB as earthy/musty/mouldy using the inner circle of the wheel, where as a highly trained panel may describe a standard of geosmin in distilled water by its refined descriptors in the second band of the wheel, namely earthy.

A third category of reference standards besides the 'known' and 'representative' compounds (Burlingame *et al.*, 1991a) on the outside of the odour wheel is the 'substitutes'. These are standards that are made from natural materials that produce an odour found in water yet can be prepared in the laboratory to give a predictable, consistent odour in water. Examples include aged solutions of grass that yield decayed vegetation and septic odours (APHA *et al.*, 2012; Method, 2170). The goal is to develop an odour reference library that links odours to culprit compounds and to move from representative odourants to identified compounds. For example, the representative compound for the cucumber odour was determined to be *trans,cis*-2,6-nonadienal, which, rather than cucumber itself, is the actual culprit for cucumber odour in water (Burlingame *et al.*, 1991b).

Known, naturally present, internal reference standards found in untreated or finished drinking water need to be evaluated for both OTC and dose-response relationships. An understanding of these reference relationships can help elucidate the contribution of odourous compounds of targeted interest to the overall sensory quality of drinking water. For each odourous constituent – either reference or target – the relationship between odour intensity and odourant concentration in water can be described by either of the following models:

Weber-Fechner model (Thiemer, 1982)

$$\text{Intensity} = m \log(\text{concentration}) + b \tag{2.1}$$

Stevens' Power Law model (Moskowitz *et al.*, 1974)

$$\text{Intensity} = k(\text{concentration})^n \tag{2.2}$$

where m = gradient, b = y-axis intercept, and k = gradient. For the Stevens' Power Law model, an exponent (n) of less than 1.0 indicates that a change in intensity results from a greater change in odourant concentration (i.e. the sensory ratio is smaller than the physical ratio), which generally holds true for odours. For example, an exponent of 0.5 would mean that an increase of 1 to 10 in concentration corresponds to an increase of 1 to 3.2 in odour intensity. Butanol, a frequently used standard, has an exponent of 0.66, so its odour intensity increases roughly 1.5 times for each doubling of concentration (Thiemer, 1982). To compare the two models, three reference standards (geosmin, MIB, and *n*-hexanal) were evaluated at a variety of concentrations to obtain dose-response relationships (Burlingame *et al.*, 1991a). The expected concentrations agreed fairly well at low-odour intensities for both models, while at higher intensities the Stevens' Power Law model produced less deviation.

2.5 SPECIFIC TASTE-AND-ODOUR CAUSING COMPOUNDS

Both the 'Taste-and-Odour Wheel' (Figure 2.1) and Tables 2.1 and 2.2 below list confirmed causes of taste-and-odour problems in drinking water. The most common causes are summarized next.

2.5.1 Earthy/musty/mouldy

Geosmin, MIB, and TCA have been identified in water supplies as key odourants of the earthy/musty/mouldy group, with OTCs of <10 ng/L at 20°C (Suffet *et al.*, 1995b). Gerber and co-workers (see Table 2.1) were the first to give an accurate description of the major microbial metabolite, geosmin, that is responsible for producing the earthy odour in water supplies. Rosen and co-workers (1970) first reported the isolation of geosmin from natural water and went on to isolate MIB not only from pure actinomycete cultures but also from a natural water. A year earlier, Medsker *et al.* (1969) and Gerber (1969) were the first to identify the musty odour of MIB during taste-and-odour episodes in natural waters by pure culture techniques.

Geosmin and MIB both have two optical isomers, typically referred to as the (−) and (+) forms (see Figure 2.2), although in nature only the (−) form exists (Polak &

Table 2.1 Compounds causing odours in drinking water.

Compound	Source	Odour	References
Earthy/Musty/Mouldy			
(1) Geosmin*	Actinomycetes, cyanobacteria	Earthy	Gerber and LeChevalier (1965), Gerber (1968, 1979), Safferman *et al.* (1967), Medsker *et al.* (1968), Rosen *et al.* (1970), Henley (1970), Collins *et al.* (1970), Piet *et al.* (1972), Kikuchi *et al.* (1973a, b), Narayan and Nunez (1974), Tabachek and Yurkowski (1976), Tsuchiya *et al.* (1978, 1981), Persson (1979a, b), Izaguirre *et al.* (1982), Berglind *et al.* (1983), Wood *et al.* (1983), Burlingame *et al.* (1986), Wu and Juttner (1988a, b), Matsumoto and Tsuchiya (1988), Izaguirre (1992), van Breemen *et al.* (1992), Jensen *et al.* (1994), Chen *et al.* (2010)
(2) 2-Methylisoborneol*	Actinomycetes, cyanobacteria	Musty	Rosen *et al.* (1968), Gerber (1969), Medsker *et al.* (1969), Collins (1971), Tsuchiya *et al.* (1978), Izaguirre *et al.* (1982, 1983), Negoro *et al.* (1988), Wu and Juttner (1988b), Martin *et al.* (1991), Izaguirre (1992), Jensen *et al.* (1994), Lin *et al.* (2002), Chen *et al.* (2010)
(3) Cadinene-ol	Actinomycetes	Woody/earthy	Collins (1971), Gerber (1971)
(4) 2,4,6-Trichloroanisole*	Biochemical methylation of chlorophenol	Musty	Montiel (1991), Nystrom *et al.* (1992), Welte and Montiel (1999)
	Algae	Earthy/Musty	Li *et al.* (2016), Malleret *et al.* (2001)

(*Continued*)

Table 2.1 Compounds causing odours in drinking water (*Continued*).

Compound	Source	Odour	References
(5) 1,2,7,7-Tetramethyl-2-norbornanol	Streptomyces	Musty	Gerber (1979), Scholler *et al.* (2002)
(6) 6-Methyl-5-hepten-2-one	Thermoactinomycetes	Earthy	Wilkins (1996)
(7) 2,4,6-Tribromoanisole (TBA)*	Algae	Earthy/Musty	Malleret *et al.* (2001)
(8) 2-Isopropyl-3-methoxypyrazine (IPMP)*	Algae	Potato bin, Earthy/Musty	Buttery and Ling (1973), Li *et al.* (2016) (vegetable to decaying odour according to Chen *et al.*, 2013; Khiari *et al.*, 1997; Peter *et al.*, 2009)
(9) 2-Isobutyl-3-methoxypyrazine (IBMP)*	Algae	Earthy/Musty	Li *et al.* (2016) (vegetable odour according to Chen *et al.*, 2013; Khiari *et al.*, 1997; Peter *et al.*, 2009)
(10) 1,3,5-trimethylbenzene	Unknown	Earthy/Musty	Bruchet *et al.* (1991)
Fragrant/Vegetable/Fruity/Flowery			
(1) *trans,cis*-2,6-Nonadienal*	Algae	Cucumber	Hayes and Burch (1989), Burlingame *et al.* (1991b)
(2) Aldehydes (>C5 molecular weight)	Ozonation	Fruity/fragrant	Anselme *et al.* (1985a, 1988), Suffet *et al.* (1986), Rashash *et al.* (1997), Richardson *et al.* (1999)
(3) Ionone*	Oxidation	Violet	Cotsaris *et al.* (1995), Peter *et al.* (2009)
(4) Musks	Fragrance in consumer products	Musk, Flowery, soapy, sweet	Stackelberg *et al.* (2004), Young *et al.* (1996)
(A) Galaxolide: 1,3,4,6,7,8-hexahydro-4,6,6,7,8,8-hexamethyl cyclopenta-γ-2-benzopyran (HHCB)			

(B) Tonalide: 7-acetyl-1,1,3,4,4, 6-hexamethyl tetrahydronaphthalene (AHTN)			
(5) 2,3-Butanedione (diacetyl)*	Bacteria	Sweet, buttery	Diaz *et al.* (2004)
Grassy/Hay/Straw/Woody			
(1) *cis*-3-Hexenyl-1-ol acetate*	Leaching of chemicals from grass into drinking water	Sweet, grassy	Khiari *et al.* (1995a, 1999a)
(2) *cis*-3-Hexen-1-ol*	Leaching of chemicals from grass into drinking water	Grassy (green, sharp)	Khiari *et al.* (1995a, 1999a)
(3) *n*-Hexanal	Algae	Grassy, lettuce	Ömür-Özbek and Dietrich (2008), Zhang *et al.* (2010)
(4) (a) β-Cyclocitral*	Cyanobacteria in a water supply	Tobacco-like	Slater and Blok (1983)
(b)	Microcystis cyanobacteria culture	1° Sweet-pipe tobacco	Juttner *et al.* (1986)
(c)		2° Grassy/fruity	Cotsaris *et al.* (1995)
(d)	Lake water (algal bloom) and drinking water intake	Fresh grass, <1 μg/L; Hay/woody, 2–20 μg/L; Tobacco-like, >10 μg/L	Young *et al.* (1999)

(*Continued*)

Table 2.1 Compounds causing odours in drinking water (*Continued*).

Compound	Source	Odour	References
Fishy/Rancid			
(1) *n*-Heptanal	Flagellated algae diatoms	Fishy	Collins and Kalnins (1965a, b, 1966, 1967), Kikuchi *et al.* (1974, 1986)
(2) *trans*, *cis*-2,4-Decadienal*	Flagellated algae	Cod-liver oil	Juttner (1981, 1983)
		Fatty	Watson *et al.* (2000)
(3) 2-*trans*,4-*cis*,7-*cis*-Decatrienal*	Algae – *Synura petersenii* and *Dinobryon cylindricum* cultures	Fishy/cod-liver oil	Rashash *et al.* (1993), Watson *et al.* (2000)
(4) Hepta- and deca-dienals	*Dinobryon* algae	Fishy	Juttner *et al.* (1986), Bayliss (1951), Yano *et al.* (1988)
(5) *trans*-2-Decenal*	Unknown	Fatty	Chen *et al.* (2013)
(6) Butyric acid, valeric acid, isovaleric acid	Fatty acids	Rancid	Karlsson *et al.* (1995)
(7) *trans,trans*-2,4-Heptadienal*	Drinking water supplies	Fishy/swampy	Khiari *et al.* (1995b)
(8) *trans*,4-Heptenal*	Drinking water supplies	Fishy	Khiari *et al.* (1995b)
(9) 1-Penten-3-one	Drinking water supplies	Fishy/rancid	Khiari *et al.* (1995b), Peter and von Gunten (2007)
Marshy/Swampy/Septic/Sulfurous			
(1) Mercaptans*	Decomposed or living blue-green algae	Odourous sulfur, rotten onion	Jenkins *et al.* (1967), Slater and Blok (1983), Zhang *et al.* (2010)
(2) Dimethyl polysulfides*	Bacteria	Swampy/fishy	Giger and Schaffner (1981), Wajon *et al.* (1985a, b, c), Krasner *et al.* (1986)

(3) Hydrogen sulfide	Anaerobic bacteria (reduce SO_4^{2-} to S^-)	Rotten egg	Hack (1981)
(4) Dimethyl disulfide*, dimethyl trisulfide (mixture)	Biochemical decay of grass and drinking water supplies	Decaying vegetation, swampy, rotten garlic	MacKenthun and Keup (1970), Khiari *et al.* (1997), Wajon *et al.* (1985b), Franzmann *et al.* (2001)
(5) Volatile sulfides, including: methyl thiols, dimethyl sulfide, dimethyl disulfide, and dimethyl trisulfide (mixture)	Lake Taihu, Wuxi, China (algal bloom death)	Swampy/ rotten vegetables	Zhang *et al.* (2010)
(6) Dimethyl disulfide, dimethyl trisulfide, 2-isobutyl-3-methoxypyrazine and 2-isopropyl-3-methoxypyrazine	Biochemical decay of grass and drinking water supplies	Decaying vegetation	Khiari *et al.* (1997)
(7) Indole*	Decomposition of vegetation, feces	Septic	Khiari *et al.* (1997), Zhang *et al.* (2010)
Medicinal/Phenolic			
(1) Chlorophenols: 2-CP, 4-CP, 2,4-DCP, 2,6-DCP, 2,4,6-TCP	Phenol chlorination/ chloramination	Medicinal	Burtschell *et al.* (1959), Lee (1967), Bryan *et al.* (1973), White (1980), Acero *et al.* (2005)
(2) Bromophenols	Phenol chlorination/presence of bromine ion	Medicinal	Suffet *et al.* (1995a), Whitfield *et al.* (1988), Al-Samarrai *et al.* (2004), Acero *et al.* (2005), Piriou *et al.* (2007)
(3) Iodo-trihalogenated methanes* (Iodinated Haloforms)	Chloramination	Medicinal	Bruchet *et al.* (1989), Gittelman and Yohe (1989), Richardson (2003)

(Continued)

Table 2.1 Compounds causing odours in drinking water (*Continued*).

Compound	Source	Odour	References
Chemical/Hydrocarbons/Miscellaneous			
(1) Unknown	Chlorine dioxide and new carpet	Cat urine	Hoehn *et al.* (1990), Dietrich *et al.* (1992)
(2) 5,5′-Dimethyl-l,3-dioxane (2-EDD)	Mixing waste: aldehyde and glycol	Sweet (tutti-frutti), olive	Pretti *et al.* (1993), Ventura *et al.* (1995), Schweitzer *et al.* (1999b), Bruchet *et al.* (2007)
(3) 2-Ethyl-4-methyl-1,3-dioxoane (2-EMD)	Mixing waste: aldehyde and glycol	Sweet (medicinal), olive	Schweitzer *et al.* (1999a), Bruchet *et al.* (2007)
(4) Methyl-*tert*-butyl ether (MTBE)*	Gasoline additive	Sweet solvent	Shen *et al.* (1997), Dale *et al.* (1997), Williams *et al.* (2002), van Wezel *et al.* (2009), Suffet *et al.* (2013)
(5) Dicyclopentadiene	Industrial chemical	Solvent, chemical	Ventura *et al.* (1997)
(6) Ethyl-tert-butyl ether (ETBE)*	Gasoline additive	Chemical	van Wezel *et al.* (2009), Leavey *et al.* (2013)
(7) Methyl-tert-amyl ether (TAME)	Gasoline additive	Hydrocarbon	van Wezel *et al.* (2009)
(8) 3-Mercapto-3-methylbutan-1-ol	Chlorine dioxide reacting with volatile organics in home	Kerosene, cat urine	Dietrich *et al.* (1992)
(9) 4-Methylcyclohexane-methanol (crude mixture)	Tank spill	Liquorice	McGuire *et al.* (2014a, b)
(10) 4-Methylcyclohexane-methanol (*cis* and *trans* isomers)	Tank spill	Liquorice	Dietrich *et al.* (2015)

(11) Styrene		Model airplane glue	Reference standard
(12) m-Xylene		Sweet organic	Reference standard
(13) Methyl methacrylate		Plastic	Reference standard
(14) Cumene		Shoe polish	Reference standard
(15) 1,3-Pentadiene		Petroleum	Reference standard
(16) Indan		Varnish	Reference standard
(17) Gasoline		Gasoline	Reference standard
(18) Alkyl benzenes		Paint/putty, solventy	Reference standard
(19) 4-Methyl-2,6-di-tert-butylphenol	HDPE pipes	Burnt plastic	Anselme *et al.* (1985b)
Chlorinous/Ozonous			
(1) Chlorine (free)*	Disinfection of water	Chlorinous	Krasner and Barrett (1984)
(2) Monochloramine*	Disinfection of water	Chlorinous	Krasner *et al.* (1986)
(3) Dichloramine*	Disinfection of water	Swimming pool	Krasner *et al.* (1986)
(4) Ozone*	Disinfection of water	Ozonous, very pungent	NIOSH (2005), Suffet *et al.* (2011)
(5) N-chloroaldimines	Disinfection of water	Swimming pool	Freuze *et al.* (2004), Brosillon *et al.* (2009)

Notes: *Confirmed in drinking water. Update of table from Ibrahim *et al.* (1990), Suffet *et al.* (1993, 1996, 1999, 2004).

Table 2.2 Compounds causing taste and mouth feel/nose feel in drinking water.

Compound	Source	Flavor or Feel	References
Bitter			
Caffeine	Reference Std. FPA Method	Bitter	APHA *et al.* (2012), Method 2170; Matia (1995)
Quinine hydrochloride dihydrate	Reference Std. FPA Method	Bitter	APHA *et al.* (2012), Method 2170; Matia (1995)
Cupric ions*	Corrosion	Bitter	Whelton and Dietrich (2004), Dietrich *et al.* (2004), Ömür-Özbek and Dietrich (2011)
Inorganic compounds (magnesium choride, calcium chloride, and Mohn salt)	Natural minerals	Bitter	Matia (1995)
Salty			
Sodium chloride*	Reference Std. FPA Method	Salty	APHA *et al.* (2012), Method 2170; Matia (1995)
Cupric ions*	Corrosion	Salty	Whelton *et al.* (2004), Dietrich *et al.* (2004), Ömür-Özbek and Dietrich (2011)
Sweet			
Sugar	Reference Std. FPA Method	Sweet	APHA *et al.* (2012), Method 2170
Bicarbonate	Natural minerals	Sweet	Mallevialle and Suffet (1987)
Lead salts	Unknown	Sweet	Burlingame *et al.* (2007)
Sour/Acidic			
Citric acid	Reference Std. FPA Method	Sour, acidic	APHA *et al.* (2012), Method 2170
Inorganic compounds (magnesium sulfate, magnesium chloride, and calcium chloride)	Natural minerals	Acidic	Matia (1995)
Mouth Feel/Nose Feel			
Aluminum sulfate	Reference standard	Astringent	Smith and Noble (1998)
Menthol	Reference standard	Cooling	Eccles (1994)
Ozone	Disinfection of water	Very pungent	NIOSH (2005)
Calcium carbonate	Cement	Chalky	Tomboulian *et al.* (2004)
Ferrous sulfate	Corrosion	Metallic	

(*Continued*)

Table 2.2 Compounds causing taste and mouth feel/nose feel in drinking water (*Continued*).

Compound	Source	Flavor or Feel	References
			Whelton *et al.* (2004), Dietrich *et al.* (2004), Ömür-Özbek and Dietrich (2011)
Iron, zinc and copper ions	Corrosion	Metallic	Mallevialle and Suffet (1987)
Iron, zinc and copper ions	Corrosion	Cooling, burning	Mallevialle and Suffet (1987)

*Confirmed in drinking water.

Provasi, 1992). The OTC for (−)geosmin has been reported as 10 to 15 ng/L in water at room temperature (Persson, 1979a; Polak & Provasi, 1992; Sugiura *et al.*, 1983). Polak and Provasi (1992) reported that (−)geosmin had an average OTC 11 times lower than the (+) form at 9.5 ± 1.3 ng/L and 78 ± 12 ng/L, respectively. This 11-fold difference was confirmed by European panels (Piriou *et al.*, 2009), although much lower OTCs of 0.6 ± 0.4 ng/L and 6.7 ± 0.4 ng/L, respectively, were determined at a higher temperature of 45°C. The chemically synthesized form of geosmin used in laboratory studies has a 50% mixture of both forms, and thus a higher measured OTC than the natural (−) material. Although Piriou *et al.* (2009) examined geosmin isomers individually and mixed, they and others have not done the same for MIB. The optical isomer form of geosmin and MIB must be reported, as is done in Table 2.3 for geosmin. MIB still requires isomer-specific study.

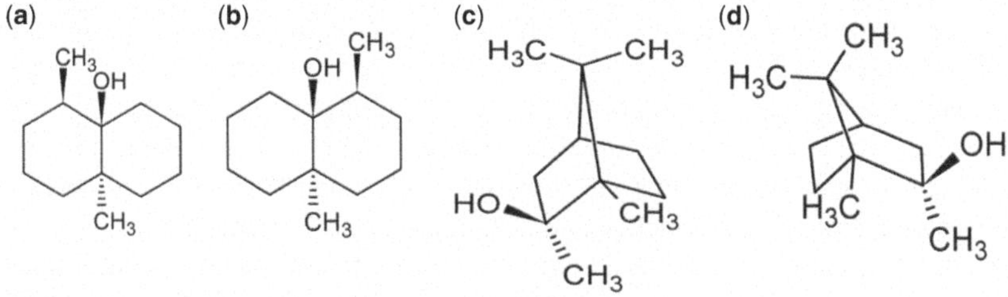

Figure 2.2 (a) (+)geosmin and (b) (−)geosmin (Brookes *et al.*, 2009); (c) (+)2-methyl isoborneol and (d) (−)2-methyl isoborneol (Bett *et al.*, 2000; Blank & Grosch, 2002; Leffingwell, 2016).

Suffet *et al.* (1995a) reported that the lowest OTCs of geosmin and MIB reported were 15 and 20 ng/L, respectively, while McGuire *et al.* (1981) reported OTC values of geosmin and MIB of 4 and 9 ng/L, respectively. The OTC of geosmin (mixed isomers at 40°C) and MIB were reported to be 3.8 ng/L and

15 ng/L, respectively, by Young *et al.* (1996). The OTC was determined by FPA to be 2.1 ng/L for MIB and 3.5 ng/L for geosmin at the experimental temperature of 45°C with unknown isomers (Mackey *et al.*, 2008). At about 20°C, the OTC of these chemicals ranges from 5 to 10 ng/L as compiled by Mallevialle and Suffet (1987) and Suffet *et al.* (1995a). The temperature is critical to report with the OTC (Table 2.3).

Table 2.3 Odour threshold concentration (OTC) for geosmin as a function of isomer and temperature.

Geosmin Optical Isomer	Temperature (°C)	OTC (ng/L)	References
(±)	45	0.8	Piriou *et al.* (2009)
(−)	45	0.6	Piriou *et al.* (2009)
Not reported	45	3.5	Mackey *et al.* (2008)
(±)	40	3.8	Young *et al.* (1996)
(+)	45	6.7	Piriou *et al.* (2009)
(−)	17–19	9.5 ± 1.3	Polak and Provasi (1992)
(−)	25	10	Sagiura *et al.* (1983)
(−)	Not reported	15	Persson (1979a)
(+)	17–19	78 ± 12	Polak and Provasi (1992)

A recent study (Yu *et al.*, 2014) found wide variation in MIB odour sensitivity between eight different communities using FPA-trained odour panels in China. The OTCs ranged from 7 to 20 ng/L for samples prepared in the laboratory using reagent water at 45°C. The isomer (+ or −) was not specified. Local sensitivities and preferences appear to affect OTCs. It must be noted that the effects of the presence of disinfection agents (e.g. chlorine or chloramines) were not included in these studies.

Many drinking water suppliers around the world have accepted a local standard for their drinking water for geosmin and MIB of 10 ng/L that is based primarily upon consumer complaints that are sometimes correlated with FPA and chemical analysis (Joe *et al.*, 2007; Mackey *et al.*, 2008; Yu *et al.*, 2014).

Under neutral pH and alkaline conditions, no significant change in concentration of geosmin and MIB was observed (Hsieh *et al.*, 2012). Under acidic condition, however, dehydration of MIB occurs and results in products with different odour profiles than that of the parent compound. Using ^{1}H and ^{13}C nuclear magnetic resonance and GC-MS analysis, Schumann and Pendleton (1997) found that under acidic conditions, MIB is converted to a 1:1 mixture of 2-methylenebornane and 1-methylcamphene. No odour, however, was associated with these compounds in a study of catfish fillets by Mills *et al.* (1993), contrary to an earlier study (Martin *et al.*, 1988).

Geosmin and MIB are found within the cyanobacteria, so care must be taken during bacterial removal. For example, chlorination of water containing *Anabaena circinalis* ruptures the cyanobacteria's cell walls and rapidly releases geosmin (Lin *et al.*, 2009). Multiple barriers are required to remove both the cells (by filtration) and any released compounds (by powdered and/or granular activated carbon) according to a study at a full-scale treatment facility (Zamyadi *et al.*, 2015). Two years after implementing the barriers, the geosmin was reduced 80% and MIB reduced 60% in finished water.

Two decades after finding geosmin and MIB, a third compound, 2,4,6-trichloroanisole (TCA), was discovered in water-distribution systems (Montiel, 1991; Nystrom *et al.*, 1992). TCA has an OTC of about 20–80 pg/L (UK Water Industry Research, 1996). In water distribution systems, trichlorophenols, which are chlorination byproducts, can be transformed to TCA via a methylation process (Montiel, 1991). Montiel (1991) showed that a phenol concentration of 0.1 μg/L was sufficient to produce a musty taste. Chemicals beyond geosmin, MIB, and TCA may also produce earthy/musty/mouldy odours, which is supported by FPA panels using these descriptors even when none of the three chemicals are identified by analytical methods (Bruchet, 1999).

2,4,6-Tribromoanisole (TBA) has an odour descriptor and odour threshold similar to that from TCA. It was first reported in a distribution system after a storage tank leaked 2,4,6-tribromophenol (Malleret & Bruchet, 2002). TBA was recently identified as the odour-causing agent during a musty odour episode that occurred at a plant producing drinking water from the river Lot, France. TBA and its tribromophenol precursor were both present in the river and drinking water. The treatment plant included ozonation but no activated carbon adsorption, and ozonation was not efficient enough at removing the musty odour from TBA, which resulted in many consumer complaints (Bruchet, 2015).

Two additional earthy/musty odourants with even lower OTCs than geosmin and MIB have been studied. One is 2-isopropyl-3-methoxypyrazine (IPMP) (Buttery & Ling, 1973; Gerber, 1971; Li *et al.*, 2016) and the other 2-isobutyl-3-methoxypyrazine (IBMP) (Li *et al.*, 2016). The OTCs of both compounds are in the lower ng/L range. Some have characterized the odours from IPMP and IBMP as vegetable or decaying vegetable (Chen *et al.*, 2013; Peter *et al.*, 2009) instead of earthy/musty/mouldy.

The Lyonnaise de Eaux (today Suez Environnement) FPA panel assigned a distinctive 'earthy-musty' descriptor to a 10 μg/L solution of 1,3,5-trimethylbenzene (Bruchet *et al.*, 1991). Testing of raw and treated drinking water near Paris found 1,3,5-trimethylbenzene up to 1.8 μg/L as well as about 150 other compounds (out of 436 different chromatographic peaks) (Bruchet *et al.*, 1991). This is a tentative identification.

For multi-component odours in drinking water, antagonism between odours can occur, such as chlorine masking the earthy/musty/mouldy odours of geosmin and MIB (Worley *et al.*, 2003). Curren *et al.* (2009) found that removing chlorine and chloramines from drinking water increased the intensity of the earthy odour of geosmin but not the geosmin-MIB mixture, nor MIB alone. Regarding taste, Zhang *et al.* (1992) found that the musty taste of MIB at concentrations of 10 and 20 ng/L decreased as the free chlorine concentration increased and that the musty taste of MIB in sand-filtered water increased during decay of the chlorine residual. Free chlorine appeared to mask MIB. Piriou *et al.* (2009) also found that the presence of chlorine partially masked the earthy-musty flavor of geosmin or MIB solutions for European panelists. In the Philadelphia water supply with 2 mg/L monochloramine residual, MIB was detected near its odour threshold suggesting that monochloramine did not mask MIB (Burlingame, 2003, personal communication, reported in Curren *et al.*, 2009).

Occasionally a reduction in aqueous concentration does not consistently reduce odour intensity. A study by Oestman *et al.* (2004) showed that chlorine 'did not strongly mask either geosmin or MIB'. This study, however, used a very wide range of concentrations of earthy/musty odourants (5 to 1,200 ng/L MIB or geosmin) and chlorination agents (0.5–20 mg/L free chlorine and 3–24 mg/L chloramine). Thus, the large range of MIB concentrations could have missed the masking effect of chlorine at low MIB concentrations. Some of the odour data may have been misinterpreted due to adaptation (panelists becoming accustomed to the odour, reducing their sensitivity to the odourant), which has been found at concentrations >400 ng/L of geosmin and >100 ng/L MIB (Whelton & Dietrich, 2004). That study evaluated actual drinking water with low levels of geosmin and MIB of <30 ng/L with a few samples between 30 and 60 ng/L.

An evaluation of the effect of background water quality on human sensitivity found that it did not affect the odour characteristic perceived by an FPA odour panel for geosmin or MIB, but it did affect the odour intensity (Mackey *et al.*, 2004). Dechlorination increased the earthy odour of geosmin and the earthy odour of the geosmin/MIB combination. This confirms the chlorine odour intensity effect – chlorine masks geosmin (earthy) odour; however, it did not seem to affect the musty odour of MIB at levels found in drinking water. Curren *et al.* (2009) found similar results.

Metal ions can also result in a type of musty, metallic odour. See the mouth-feel/nose-feel section for further details.

2.5.2 Chlorinous/ozonous

French consumers and panelists can detect chlorine at levels as low as 0.05 mg/L (Bruchet *et al.*, 2004). Their US counterparts have a much higher limit of 0.8 mg/L using the forced-choice triangle method (Mackey *et al.*, 2004). Trained panelists, however, overestimate the average consumer's ability to perceive chlorine and seem to be influenced by the chlorination practices of their home country (Piriou *et al.*, 2004). France has a low residual in the distribution system (around 0.1 mg/L) while many utilities in the United States rely on a comparatively high residual (0.3–1 mg/L).

Hypochlorous acid and hypochlorite ions share the same odour descriptor, namely 'bleach'. Hypochlorous acid has an acid-dissociation constant (pK_a) of 7.6 and ionizes to hypochlorite ion and hydronium ion. Krasner and Barrett (1984) reported in the USA, hypochlorous acid (at $pH < 6$) has an OTC of 0.28 mg/L, whereas the dissociated hypochlorite ion (at $pH > 9$) has an OTC of 0.36 mg/L.

The breakpoint chlorination curve describes the reaction of chlorine and ammonia. The dominant chlorine species in water prior to the breakpoint are monochloramine and dichloramine, with an odour descriptor of 'swimming pool'. The OTC of monochloramine is 0.65 mg/L (Krasner & Barrett, 1984), although American panels gave an average sensitivity limit of 5.4 mg/L (3.7 mg/L as Cl_2) using the forced-choice triangle method (Mackey *et al.*, 2004). According to data from the Metropolitan Water District of Southern California, monochloramine at concentrations of 0.5 to 1.5 mg/L has an odour-intensity level of 2 (very slight) on the seven-level FPA scale (Krasner & Barrett, 1984). Monochloramine, however, rarely causes taste-and-odour problems in drinking water unless its concentration exceeds 5 mg/L.

In the USA, most people perceive the unpleasant chlorinous smell once the concentration of dichloramine is above 0.5 mg/L (Krasner & Barrett, 1984). Dichloramine in concentrations of 0.1 to 0.5 mg/L has an odour-intensity level of 4 (slight) to 8 (moderate) on the seven-level FPA scale, yet when the concentration of dichloramine reaches 0.9 to 1.3 mg/L, the odour is described as offensive and not acceptable. When the Metropolitan Water District of Southern California switched from free chlorine to chloramines as the disinfectant, no odour problems were caused by chloramine species themselves; however, the chloramines were unable to effectively oxidize odourous compounds released by certain microorganisms that continued to impart garlic, decaying vegetable, or fishy-swampy odours (Krasner *et al.*, 1989). Other treatment approaches were used to remove these odours.

While researching a drinking water odour event at the Méry Sur Oise Plant near Paris, France, Freuze *et al.* (2004) found that N-chlorophenylacetaldimine was likely to blame for the 'swimming pool' odour. The OTC is about 3 μg/L, and the odour can persist for more than one week at 18°C. N-chloroisobutaldimine, N-chloromethyl-3-butaldimine and N-chloromethyl-2-butaldimine were subsequently detected in three different drinking water distribution systems in the Paris area at individual levels in the range 1–11 μg/L. These compounds are formed during the chlorination of amino acids that can permeate through nanofiltration membranes and generate 'chlorinated' odours in water (Brosillon *et al.*, 2009).

Investigation of transient odour episodes in the drinking water from the city of Edmonton, Canada, showed that these episodes were associated with high levels of low-molecular-weight aldehydes usually present above their individual OTCs and known to arise from the chlorination of amino acids (Hrudey *et al.*, 1988). Some people have found the combined aldehyde odour to be reminiscent of the characteristic odour in swimming pools (Hrudey *et al.*, 1989). This, plus the fact that amino acids are also present at high levels in swimming pools, led Froese *et al.* (1999) to state that 'the resulting aldehydes are believed to contribute to the characteristic "swimming pool" odour'. As chlorination of amino acids yields a mixture of aldehydes, nitriles and N-chloroaldimines, the relative contribution of these byproducts to swimming pool odours needs to be established more precisely.

Chlorine dioxide, an alternative disinfectant to chlorine, is not detected as an odour in water up to 0.2 mg/L (Bruchet *et al.*, 2004). Further, the reactions of chlorine dioxide are much more specific than those of chlorine, so they do not lead to the development of many chlorinous tastes and odours. Due to its volatility, however, when cold tap water (less than 8–10°C) exits the tap within a home, chlorine dioxide odour can be perceived as the water warms to room temperature (Bruchet *et al.*, 2004).

It is also noted that investigations by Welte and Montiel (1999) and Bruchet (1999) described intense chlorinous odours that occurred in the absence of free or combined chlorine. The cause of these odours has not been identified.

FPA panels have identified an odour associated with ozonated waters as 'oxidant' or 'ozonous' (Suffet *et al.*, 2011). Although an ozone residual may not have been present at the time of the sensory panel evaluations, the oxidant odour seemed to persist. Additional odours depend on the ozone dosage. Figure 2.3 shows that at low to medium doses, a sweet taste and odour appears due to straight-chain aldehyde formation (Anselme *et al.*, 1988). At higher doses and especially at high pH, soapy and rancid odours begin to appear (Crozes *et al.*, 1997).

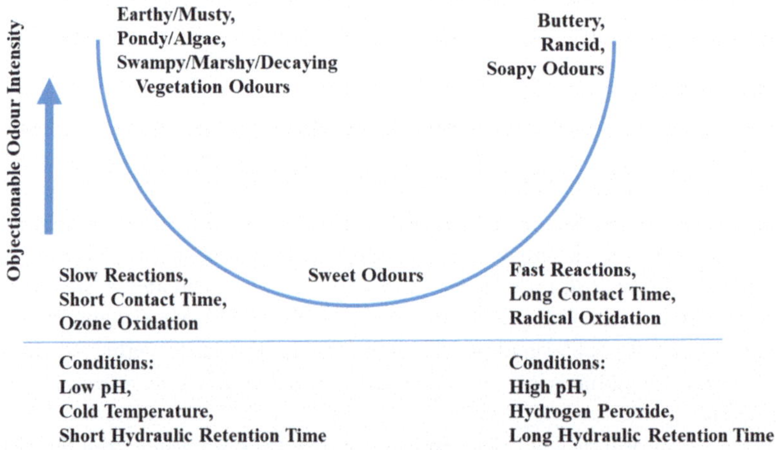

Figure 2.3 Trends in odour reduction and formation during ozonation (modified from Crozes *et al.*, 1997).

As each drinking water source during each season of the year is unique, the parameters for the disinfection process must be adjusted to achieve disinfection, minimize the formation of byproducts, and avoid taste-and-odour issues. For example, a statistically designed testing approach for optimized parameters for the removal of earthy (geosmin) and musty (MIB) odourants was used for water from the Detroit River, USA (Atasi *et al.*, 1999). In this case, because of the background water properties, peroxide addition was not as helpful as has been reported at other locations. Additional information on the factors impacting the oxidant or ozonous odours are needed.

2.5.3 Fragrant/vegetable/fruity/flowery

The high molecular weight aldehydes produced during ozonation have been linked to fruity, fragrant, and orange-like odours (Anselme *et al.*, 1988). Specifically, aldehydes with eight or more carbons (i.e. octanal and beyond) showed a positive statistical correlation with fruity odours produced during the ozonation of drinking water at the Morsang Treatment Plant near Paris, France. Such is an example of a presumptive test of a chemical cause of an odour (Anselme *et al.*, 1988). In an independent sensory panel study, a Weber-Fechner model relationship was found between the fruity odour intensity levels measured by an FPA sensory panel and the logarithm of the total aldehyde concentration in a mixture of seven straight-chain aldehydes: hexanal (C6) to dodecanal (C12). This is an example of sensory confirmation of a chemical cause of an odour. The fruity/orange-like odour of decanal (C10) can be used to represent this group of aldehydes.

Microbiological causes of taste-and-odour problems require: (1) concurrence of the organism and the odour in the field; (2) the isolated organism must be able to produce the odour in pure culture; and (3) odourous compound must be isolated from the original water and chemically identified. For example, the cucumber odour in water supplies is caused by the microbial metabolite *trans, cis*-2,6-nonadienal, and Burlingame *et al.* (1991b) found a mixed bloom that included the responsible microbe in the source water.

Musk compounds have associated odours (perfume, floral, soapy, musk, and sweet) and are used as fragrance in consumer products. Two specific musks – galaxolide (1,3,4,6,7,8-hexahydro-4,6,6,7,8,8-hexamethylcyclopenta-γ-2-benzopyran; HHCB) and tonalide (7-acetyl-1,1,3,4,4,6-hexamethyltetra hydronaphthalene; AHTN) – have been detected in raw, settled, filtered and finished drinking water (Stackelberg *et al.*, 2004). Concentrations throughout the treatment steps remained the same, approximately 0.08 μg/L for HHCB and 0.3 μg/L for AHTN, indicating little or no effect of treatment on these compounds. For HHCB, Young *et al.* (1996) determined an OTC of 5 μg/L at 40°C and 50 μg/L at 25°C.

Industrial sources can also add fragrant odours to drinking water sources. Near Barcelona, Spain, 2,3-butanedione (diacetyl), the smell of popcorn, from an unidentified industrial source caused sweet, buttery odours in drinking water (Diaz *et al.*, 2004).

Trichloramine was removed due to instability and lack of proof of a flowery (geranium) odour (Suffet *et al.*, 1999). Further study has shown that a chlorinous odour from trichloramine is also minor, with an OTC of 26 μg/L as $NaCl_3$ (or 15 μg/L as Cl_2) (Phattarapattamawong *et al.*, 2011).

2.5.4 Medicinal/phenolic

When bromide and iodide exist in water supplies, ozonation can convert these compounds to HOBr and HOI (Suffet *et al.*, 2011). The HOBr and HOI can then react with naturally occurring organic matter to produce various brominated and iodinated compounds. The possible reactions that could occur between HOBr or HOI and trace amounts of organic compounds present in water have been classified into three major groups: phenol odours, haloform odours, and unknown odours.

Chlorophenols and bromophenols share the same odour category, medicinal/phenolic (Whitfield *et al.*, 1988). The chlorine-to-phenol ratio is the major determinant for the formation of the phenolic byproducts resulting from the chlorination process (Lee, 1967). The highest taste intensity is obtained at a 2:1 chlorine:phenol ratio because odourous 2,6-dichlorophenol is the dominant chlorination byproduct with a FTC of 6 ng/L (Piriou *et al.*, 2007). When the ratio increases to 4:1 and the chlorine concentration increases to 10 mg/L, the medicinal/phenolic odour is no longer detected. When ammonia is present in the phenol solution, it consumes free chlorine thereby lowering the chlorine:phenol ratio, which may increase medicinal/phenolic odour (Lee, 1967). The formation of odourous chlorophenols is also highly pH dependent. There is no significant development of chlorophenolic odours at pH < 7. The optimum range for chlorophenolic odours is pH 8 to 9 (Lee, 1967).

Kinetics experiments and modeling studies show bromophenols are formed more than chlorophenols (by a factor of 100 to 1,000) yet dissipate within 10–20 min under typical drinking water chlorination conditions (Acero *et al.*, 2005; Al-Samarrai *et al.*, 2004; Piriou *et al.*, 2007). Similar to the chlorophenols, 2,6-bromophenol has the lowest OTC of 0.5 ng/L (Whitfield, 1988), with a FTC of 4 ng/L (Piriou *et al.*, 2007). For comparison, 2-bromophenol (a monophenol) has a higher OTC of 30 ng/L (Whitfield, 1988) and FTC of 12 ng/L (Piriou *et al.*, 2007). The formation of bromophenols is also a function of pH and follows the same pattern as chlorophenols. Bromophenols have been identified in water distribution systems where phenols were leaching from a reservoir-coating material and reacting with chlorine in the presence of bromide ion in the water (Khiari *et al.*, 1999a).

The iodomethanes formed depend on the organic content of the raw water and the chlorination process. Free chlorine (as HOCl) oxidizes both organic and inorganic compounds in water. Trace amounts of bromide and iodide ions as low as 0.1 mg/L in the water supplies can be converted into bromine as HOBr and iodine as HOI by chlorination. The natural organic matter in water reacts via the haloform reaction to produce noxious brominated and iodinated haloforms. The OTC of iodoform is 0.3 µg/L (Burlingame & Anselme, 1995; Khiari *et al.*, 2002). As a result, the presence of iodinated haloforms at concentrations between 0.3 and 10 µg/L will cause medicinal taste-and-odour problems in drinking water (Bruchet *et al.*, 1989; Gittelman & Yohe, 1989). The medicinal characteristics for iodinated trihalomethanes increase with the number of iodine atoms in the structure, with iodoform presenting the most characteristic medicinal descriptor (Cancho *et al.*, 2000).

Brominated disinfection byproducts from substitution or oxidation reactions have been observed (e.g. dibromoacetonitrile, cyanogen bromide, bromoacetic acids (mono-, di-, and tri-), bromopicrin, and bromate) as early as 1993 (e.g. Daniel & Meyerhofer, 1993; Glaze *et al.*, 1993; Krasner *et al.*, 1993). Furthermore, Khiari *et al.* (2002) developed odour descriptors and OTCs for halogenated disinfection byproducts. As of yet none has been found at concentrations that would be detected in a drinking water except for iodinated haloforms, as discussed above.

2.5.5 Grassy/hay/straw/woody

Grassy tastes and odours have frequently been reported in drinking water supplies, and the identification of two grassy compounds was confirmed in drinking water supplies: *cis*-3-Hexen-1-ol and *cis*-3-Hexenyl acetate (Khiari *et al.*, 1999b). Both were observed when fresh grass was mixed with water for less than a day (Khiari *et al.*, 1995a). The acetate, however, was released first when grass was allowed to decay in water (Khiari *et al.*, 1997), while the alcohol was released during later stages of decay. In water, Cotsaris *et al.* (1995) reported *cis*-3-Hexen-1-ol as a product of a green alga, *Scenedesmus subspicatus*. Both compounds were identified in a drinking water system from the Metropolitan Water District of Southern California's Oxidation Demonstration Plant (Khiari *et al.*, 1995a). When odour-threshold tests were performed, the acetate exhibited a low OTC of 1–2 µg/L, which is only 2–4% of the OTC of the alcohol (Khiari *et al.*, 1995a). Hydrolysis experiments indicated that degradation of the chemicals occurred rapidly at pH > 6.1 (Khiari *et al.*, 1999b).

The compound β-cyclocitral was first identified with a tobacco-like odour by Slater and Blok (1983) during an algae bloom and by Juttner *et al.* (1986) in a culture of cyanobacterium *Microcystis aeruginosa*. Subsequently, Young *et al.* (1999) identified it as producing a hay/woody odour in lake and treated water during a *Microcystis aeruginosa* lake algae bloom. On further evaluation,

Young *et al.* (1999) found that only concentrations between 2 and 20 μg/L give a hay/woody odour in distilled water while higher concentrations give a tobacco-like odour. This work demonstrates the importance of knowing the relationship of odour type and concentration, as provided by the Weber-Fechner model, because β-cyclocitral changes odour with concentration. Later, Lin *et al.* (2009) found that the concentration of β-cyclocitral quickly decreased upon chlorination of water containing the cyanobacterium *M. aeruginosa*, which typically produces the compound.

2.5.6 Fishy/rancid

In 1999, 'fishy' category was changed to 'fishy/rancid' because of studies in drinking water supplies (Khiari *et al.*, 1995b; Young & Suffet, 1999). Fishy odours occur naturally, as shown in algal cultures by many authors (see Table 2.1). *trans,trans*-2,4-Heptadienal, a metabolite of the algae *Uroglena americana*, contributed to the fishy odour in the Nunobiki reservoir in Japan (Yano *et al.*, 1988). The biological origin of *trans,trans*-2,4-heptadienal was confirmed during the study of natural odour standards (Khiari *et al.*, 1995b) where it was identified in a sample of decaying grass in water. Rashash *et al.* (1993) identified 2-*trans*-4-*cis*-7-*cis*-decatrienal, an alkyl trienal. Sensory GC analysis indicated that *trans*-4-heptenal is associated with fishy odours, and 1-pentene-3-one is associated with rancid odours (Khiari *et al.*, 1995a); however, these identifications need further confirmation.

A trend occurs when ozonating drinking waters at low to medium doses – many odourants are removed by ozonation yet a sweet taste and odour appears due to straight chain aldehyde formation (Figure 2.3). However, Figure 2.3 shows that at higher doses and especially at high pH, soapy, buttery and rancid odours can begin to appear (Crozes *et al.*, 1997, 1999; Suffet *et al.*, 2011). These odours have been observed during sensory GC analysis (Khiari *et al.*, 1995b) yet have unknown compounds associated with them on the wheel (Figure 2.1). The identification of specific rancid, buttery, and soapy odours needs further study.

The disinfection byproduct 3-methylbutanal has a rancid odour along with various other odour descriptors (sickening, heavy, sharp/pungent acid, sweaty, putrid/foul/decayed, etc.) (Dravnieks, 1985) and was linked in theory, along with other aldehydes, to drinking water odour complaints in Edmonton, Canada (Hrudey *et al.*, 1988). As further evidence emerges, 3-methylbutanal may be added to a future version of the wheel.

Rancid odours were also observed after slow sand filtration during unusually warm weather in southern Sweden (Karlsson *et al.*, 1995) and were attributed to the formation of short chain fatty acids.

Sensory GC analysis combined with GC/MS allowed the identification of *trans, trans*-2,4-heptadienal in raw water supplies (Khiari *et al.*, 1995b). The odour at the olfactory port of the GC indicated at low concentrations in one case a rancid odour and in another case a fishy odour. The Weber-Fechner model for *trans, trans*-2,4-heptadienal gives an OTC of about 5 μg/L with a rancid fishy odour (Young & Suffet, 1999). This confirms the identification of this compound in raw water supplies. There is a pH dependence on the stability of the *trans, trans*-2,4-heptadienal. This compound is more stable near pH 9, so samples should be collected and preserved at high pH and refrigerated prior to analysis (Suffet *et al.*, 1999). This contrasts, however, with the stability of grassy compounds at low pH (Khiari *et al.*, 1999b).

2.5.7 Marshy/swampy/septic/sulfurous

Khiari *et al.* (1997) followed the decay of grass in water with less and less oxygen present and found odours changing with time. Methyl mercaptan, dimethyl sulfide, and dimethyl disulfide generally contributed to the decaying vegetation odour that the FPA panel observed, and GC sensory analysis confirmed it. Accompanying compounds identified by GC sensory analysis included 2-isobutyl-3-methoxpyrazine, identified as a vegetable odour, and 2-isopropyl-3-methoxypyrazine, as a rotten vegetable odour (Khiari *et al.*, 1997). The literature, however, indicates that the odours associated with the pure 2-isobutyl-3-methoxypyrazine compound are earthy/musty and 'bell pepper' (Buttery & Ling, 1973; Gerber, 1971; Li *et al.*, 2016), and for the pure 2-isopropyl-3-methoxypyrazine compound they are earthy/musty and 'potato bin' (Li *et al.*, 2016). The effect of concentration on the combination of these chemicals appears to require further investigation in the presence of each other.

Another compound, dimethyl trisulfide, may be associated only with the swampy odour (Khiari *et al.*, 1997). It should be noted that sometimes swampy and septic odours are not easy to differentiate by FPA panelists. Both dimethyl trisulfide and indole present odour characteristics similar to the odours found in a sample analyzed by FPA, but they were not sufficiently extracted to produce odours at the olfactory port of the sensory GC. It is suspected that dimethyl trisulfide and indole may both contribute to the septic odour (Khiari *et al.*, 1997). Dimethyl trisulfide also has been described as a garlicy foul odour (Wajon *et al.*, 1988).

A taste and odour episode from the death of algae at Wuxi, China, in late May 2007 (Zhang *et al.*, 2010) mirrored the progression of odourous sulfur compounds that Khiari *et al.* (1997) found during a grass decay experiment. Residents of Wuxi complained of septic/marshy odours in their drinking water. Initially, GC/MS studies by Zhang *et al.* (2010) showed methyl mercaptan and dimethyl sulfide as the primary odourants (decaying vegetation odours), which shifted 3 days later to dimethyl disulfide (also a decaying vegetation odour) and 5 days later to dimethyl trisulfide (a more garlic odour).

Figure 2.4 shows that once reduced sulfur compounds have been produced in a water supply they may be further transformed by microrganisms into other reduced sulfur compounds. Methyl mercaptan and hydrogen sulfide are methylated by organisms using methoxylated aromatic compounds present in humic materials to dimethyl sulfide and methyl mercaptan, respectively (Higgins *et al.*, 2006). Dimethyl sulfide can be further acted on biochemically to form dimethyl disulfide (Lomans *et al.*, 2002). Methyl mercaptan can also be oxidized to form dimethyl disulfide (Adams *et al.*, 2003). The aerobic or anaerobic state of this reaction is unclear in the literature. Figure 2.4 details the formation and transformation of odourous sulfur compounds (Higgins *et al.*, 2006). Once microorganisms produce these odourous compounds, they can be eliminated by degradation by other types of microorganisms in the system.

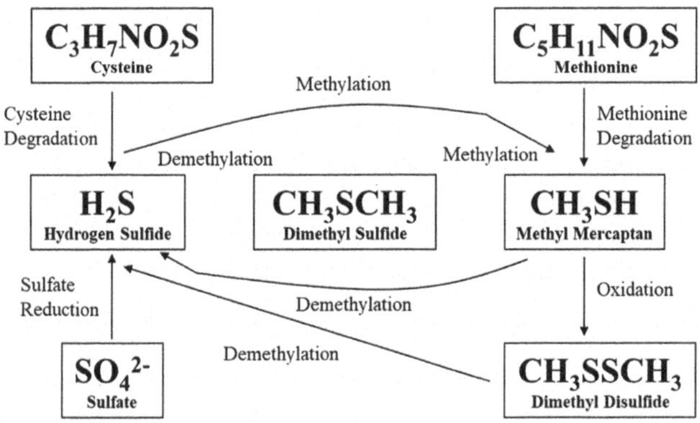

Figure 2.4 Production and transformations of volatile sulfur compounds in wastewater treatment plants.

As sulfur compounds are commonly found as mixtures, which complicates the associated odours, those included on the 2016 Drinking Water Taste and Odour Wheel (Figure 2.1) are limited to known odours associated with single compounds: hydrogen sulfide – rotten eggs (Mallevialle & Suffet, 1987; Suffet & Rosenfeld, 2007); methyl mercaptan – decaying cabbage or vegetation (Khiari *et al.*, 1997; Mallevialle & Suffet, 1987); dimethyl sulfide (high conc.) – decaying vegetation (Suffet & Rosenfeld, 2007); dimethyl sulfide (low conc.) – canned corn (Suffet & Rosenfeld, 2007); dimethyl disulfide – decaying vegetation (Suffet & Rosenfeld, 2007); dimethyl trisulfide – swampy (Mallevialle & Suffet, 1987); dimethyl trisulfide – marshy/swampy (Suffet & Rosenfeld, 2007); dimethyl trisulfide – garlic; indole – septic (Khiari *et al.*, 1997).

The odours from multiple odourants may blend or produce an integrated odour (Lawless, 1999), so the real-world challenges of multiple-odour management needs investigation. Further, odours mask each other and can be viewed as 'peeling an onion', i.e. when one high concentration odourant is removed, another takes its place and then dominates. In other words, after the primary odour is removed, other odours may become apparent or the odour intensity of other odours increases and subsequently need to be controlled. An example of this is when dechlorination increased the earthy odour intensity of geosmin (Mackey *et al.*, 2004). Predicting the nuisance level of each layer of the 'onion' is the broader challenge of odour control.

2.5.8 Chemical/hydrocarbon/miscellaneous

Of the many compounds in this group, odour episodes in drinking water have been due to byproducts of resin manufacturing and oxygenated fuel additives used for smog reduction. Byproducts of resin manufacturing processes have been reported to cause at least five different taste-and-odour episodes in drinking water around the world (Bruchet *et al.*, 2007; Noblet *et al.*, 1999; Pretti *et al.*, 1993; Schweitzer *et al.*, 1999a, b; Ventura *et al.*, 1995). The compounds are products of simple reactions of aldehydes and glycols. Of particular concern are the OTCs (<10 ng/L) of 2-ethyl-5,5-dimethyl-1,3-dioxane (2-EDD) from propionaldehyde and neopentyl glycol and 2-ethyl-4-methyl-1,3-dioxolane (2-EMD) from propionaldehyde and propylene glycol. 2-EDD has a sweet 'tutti-frutti' (described as 'olive juice' in Spain by Ventura *et al.*, 1995). Incidents have been confirmed in the Ohio River (Noblet *et al.*, 1999), in the Neshaminy Creek in Pennsylvania (Pretti *et al.*, 1993), in groundwater in an aquifer near Barcelona, Spain (Ventura *et al.*, 1995), and in South America (Bruchet *et al.*, 2007). These compounds are formed readily at room temperature.

MTBE and ethyl *tertiary*-butylether (ETBE) are octane boosters used in gasoline. They are very water-soluble compounds and thus of particular concern from leaky gasoline underground storage tanks, which can affect groundwater supplies, and in fuel usage from two-cycle, outboard motors used on reservoirs and lakes that provide drinking water. MTBE and ETBE are unique industrial pollutants in that it is their odour levels that set drinking water standards because the chemicals' odour is a problem well before any toxic effect. The MTBE odour is described as a sweet, solvent odour, and the ETBE odour is described as a harsher solvent odour, similar to low molecular weight alkylbenzenes such as the so-called BTEX (benzene, toluene, ethylbenzene, and *ortho*/*meta*/*para*-xylene). MTBE has been identified in many ground water and drinking water supplies around the world as exemplified by the data compilations for the Netherlands (van Wezel *et al.*, 2009) and California, USA (Shih *et al.*, 2004).

The OTC for MTBE is near 15 μg/L. The mean results and 95% confidence interval (CI) of two large-scale consumer studies (>50 consumers) were: 15 μg/L (9.0–23 μg/L CI) (Stocking *et al.*, 2001) and 22 μg/L (13–35 μg/L CI) (Suffet, 2007). In a small-panel studies, van Wezel *et al.* (2009) and Suffet (2007) found means of 7 μg/L (5–10 μg/L CI) and 18 μg/L, respectively, for the OTC of MTBE. The Stocking *et al.* (2001) consumer study has been used as a textbook example of how to complete an OTC study of consumer groups (Lawless & Heymann, 2010) as described previously.

During the course of a four-day study, untrained consumer panelists gained expertise for the sweet, solvent odour of MTBE. More than an order-of-magnitude decrease in the OTC was observed over the four days. In effect, the study trained the panelists to become experts in the same way an expert panel becomes more precise and detects lower concentrations than consumers of drinking water (Suffet *et al.*, 2013). Such a pattern has been documented previously for other chemicals (Engen, 1960; Pangborn *et al.*, 1964; Pfaffmann *et al.*, 1954; Powers & Shinholser, 1988; Rabin & Cain, 1986). Powers and Shinholser (1988) state that panelist training can lower the threshold concentration by around four orders of magnitude.

Suffet *et al.* (2013) provides the first determination of the ORC for MTBE, which was 3–4 times higher than the OTC for all waters tested. This pattern is consistent with other laboratory studies (Rabin & Cain, 1986). The presence of chlorine made the detection and recognition of MTBE more difficult. In other words, the OTC and ORC were much higher for the chlorinated water tested. The consumer odour panel data have indicated that 0.3 mg/L total chlorine does mask the odour threshold and recognition of MTBE's sweet, solvent odour.

ETBE has been identified in ground water and drinking water supplies as exemplified by the data compilations of van Wezel *et al.* (2009) in the Netherlands. The OTC of the ETBE solvent odour was reported to be near 1 μg/L (van Wezel *et al.*, 2009). Leavey *et al.* (2013) reported the OTC at 0.3 ug/L. The Leavey *et al.* (2013) study compared a consumer laboratory panel of 12 members and a large consumer panel of 57 consumers, each panel with about 50% of both sexes and even distribution across age groups (18–29, 30–48 and 49–65). Both panels led to OTCs of 0.3 μg/L, although the laboratory panel had a wider 95% confidence interval (CI) due to its smaller size. The same two panels also determined the ORC for ETBE. The consumer-panel ORC was 3.3 μg/L (1.6–6.7 μg/L 95% CI) and the laboratory-panel ORC was 1.8 μg/L (0.41–8.1 μg/L 95% CI). Although the laboratory panel followed a more relaxed protocol, the result was within the CI of the consumer panel.

During the 1980s, owing to an increase in the use of oil products, water utilities had to face many spills on both surface and groundwaters. Mallevialle *et al.* (1982) reported the analytical methods used by Lyonnaise des Eaux company (now Suez Environnmental) to deal with such cases. Table 2.4 summarizes the known detection thresholds of oil products, including gasoline. It is clear from Table 2.4 that oil products will be usually detected at very low levels during spills in source waters.

Some of the lighter compounds in oil products can also penetrate into distribution systems by permeation through plastic pipes. As an example, 1,3-pentadiene, which has a characteristic gasoline/petrol/kerosene-like odour (Stratford *et al.*, 2007), can arise in drinking water as a result of permeation of petrol, diesel and similar products through plastic pipes (Environment Agency, 2004).

Table 2.4 Known odour threshold concentrations of oil products, including gasoline.

Oil Product (Source)	Odour Threshold Concentration (mg/L)
Normal gasoline (ESSO)	0.01
Extra gasoline (ESSO)	0.01
Diesel fuel (ESSO)	0.0005
Light domestic oil (Shell)	0.005
Medium domestic oil (Shell)	0.5
Fuel oil 400 (Shell)	0.002
Fuel oil 800 and 3,500	0.1

During an odour episode that occurred immediately after the installation of 3 × 100 m sections of high density polyethylene (HDPE) pipes in a distribution system south of Paris, many consumers complained about plastic/burnt plastic odours. Investigation of this case by Anselme *et al.* (1985b), who used steam-distillation extraction followed by sensory GC, led to the conclusion that the burnt plastic odour was mainly caused by 4-methyl-2,6-di-*tert*-butylphenol (BHT).

Polyvinyl chloride (PVC) pipe and its joint solvent (primer and cement) leaching studies were completed at room temperature and showed glue/varnish odours (Wiesenthal & Suffet, 2007). A sweet/phenolic odour replaced the glue/varnish odour after the leaching media were diluted with Milli-Q water to threshold odour concentrations. The solvent/primer bonding of the pipe led to the glue/varnish odour due to the ingredients – acetone, tetrahydrofuran, methyl ethyl ketone, and cyclohexanone. Leaching studies of the pipe itself did not produce an odour. Similarly, Durand and Dietrich (2007) showed a chemical/solvent odour in drinking water from leaching of a silane crosslinked PEX pipe.

Cat-urine-like and kerosene-like odours have been associated with drinking water disinfection. A survey of chlorine dioxide use in water treatment in the USA found that almost half the complaints of such odours came from homes that had recently installed new carpeting (Dietrich *et al.*, 1992), likely due to reactions between escaping chlorine dioxide and off-gassing organic compounds from the new carpeting (Hoehn *et al.*, 1990). Cat excretion breaks down into mercaptans, such as 3-mercapto-3-methylbutanol, which may be the cause of the odour (Miyazaki *et al.*, 2006).

2.6 TASTE AND MOUTH FEEL/NOSE FEEL

Although smell is far more sensitive than taste in most cases, taste still can be the basis for objections to drinking water quality. Minerals are the primary source of the water's taste but do not typically contribute to its odour, while organic substances are typically at levels below their taste thresholds (Matia, 1995). The World Health Organization has guidelines for acceptable levels of minerals in drinking water (WHO, 1985). Similarly, the USEPA secondary drinking water standards (USEPA, 2007) provide guidance for 15 items with organoleptic effects (aluminum, chloride, colour, copper, corrosivity, fluoride, foaming agents, iron, manganese, odour, pH, silver, sulfate, total dissolved solids, and zinc), which has been reviewed (Dietrich & Burlingame, 2015). The role of minerals in drinking water taste (and consumer health) has also been reviewed (Whelton *et al.*, 2007).

The fundamentals of tap water taste have been reviewed (Burlingame *et al.*, 2007). The tastes of sweet, sour, bitter or salty as described below and in Table 2.2 are used as reference points for the taste-and-odour intensity within FPA (APHA *et al.* 2012; Method, 2170). Table 2.2 and the 'Taste-and-Odour Wheel' (Figure 2.1) show the primary tastes and feelings that are observed with drinking water.

2.6.1 Sour/acidic

A number of inorganic compounds – magnesium sulfate, magnesium chloride and calcium chloride – can cause an acid taste (Matia, 1995). An organic compound, trimethylcyclohexane, was responsible for a sour taste after ozone treatment (Anselme *et al.*, 1988).

2.6.2 Sweet

Bicarbonate alkalinity may be described as sweet (Mallevialle & Suffet, 1987) but may be masked by the presence of chlorine. The taste threshold concentrations of organic compounds that lead to sweet tastes are fairly high, in mg/L range (Alexander *et al.*, 1982). Lead salts taste sweet (Burlingame *et al.*, 2007).

2.6.3 Bitter

Many organic compounds give water a bitter taste, and among these quinine and caffeine are the most common (Matia, 1995). Many inorganic compounds can also lead to a bitter taste, and the intensity levels are especially important for magnesium choride, calcium chloride, and Mohr salt $[(NH_4)_2Fe(SO_4)_2 \cdot 6H_2O]$ (Matia, 1995). Copper in drinking water may produce unpleasant (bitter, salty) sensations that have been observed in the concentration range 1–5 mg/L (Dietrich *et al.*, 2004). Finally, it should be noted that an intense and persistent bitter taste can be generated by a continuous flow of very pure water (distilled water) over the tongue (Mallevialle & Suffet, 1987), which might explain why waters treated by reverse osmosis are considered as bitter by some consumers.

2.6.4 Salty

Salty taste is one of the major consumer complaints to water utilities. Sodium chloride is the most common cause of salty taste, with minor contributions from magnesium, potassium, calcium and lithium chlorides (Matia, 1995). The FPA taste panel method has been used to determine the average taste threshold concentration for sodium chloride in Milli-Q water, which was 640 $\pm$ 3 mg/L at pH 8 (Wiesenthal *et al.*, 2007). Although saliva contains sodium chloride above 500 mg/L, preference testing revealed that water containing NaCl $<$ 290 mg/L is acceptable while $>$465 mg/L is disliked (Whelton, 2009). The secondary drinking water standard (USEPA, 2007) for total dissolved solids is 500 mg/L, which addresses salty taste.

2.6.5 Mouth feel/nose feel

Mouth feel and nose feel consist of descriptors beyond traditional tastes yet are detectable by the mouth or nose. Examples include metallic, astringent, powdery, chalky, cooling, and burning sensations (Matia, 1995). Iron, zinc and copper ions can lead to these sensations, along with other mineral salts. Drying sensations have also been observed by FPA panelists, although the cause is unknown.

Iron has often been described as having a metallic taste (Hettinger *et al.*, 1990; Lawless *et al.*, 2004). Iron in drinking water occurs either naturally in groundwater or is from corrosion of older pipes (Volk *et al.*, 1967). The mean taste threshold for iron ions ranges from 3 mg/L (Mallevialle & Suffet, 1987) to 0.17 mg/L (Mirlohi *et al.*, 2011) to 0.03 mg/L (Ömür-Özbek & Dietrich, 2011). The secondary maximum contaminant level in the USA (USEPA, 2007) is 0.3 mg/L and in the United Kingdom is 0.2 mg/L to avoid colour, taste, and odour problems. The WHO (2003) recommends a dietary intake of 2 mg/L for iron in drinking water, which may be at odds with the lower taste and odour thresholds.

2.6.6 Flavour-by-mouth

A recent phenomenon has been observed: flavor-by-mouth. Flavor by mouth includes taste and retronasal odour, which occurs via the nasopharynx located at the back of the throat to the olfactory bulb while tasting water (Shepherd, 2006). Tucker *et al.* (2007) have found that iron taken into the mouth has both a taste and an odour component. Ferrous iron (Fe^{2+}) has been reported by Lawless *et al.* (2004) to produce a retronasal smell. A study by Hettinger *et al.* (1990) reported that the subject could taste a metallic flavor with both the nose and mouth unobstructed but could not taste the metallic flavor with the nose occluded; therefore, the taste of iron as metallic is due to olfaction. However, the odour of iron was unable to be smelled in the headspace over a water sample containing iron (Lawless *et al.*, 2004). This further supports the finding that the taste of iron in water may occur retronasally.

The type of iron found in water is dependent upon the pH and oxygen, which dictate whether it exists in the ferrous (Fe^{2+}) or ferric (Fe^{3+}) form (Tucker *et al.*, 2007). On the skin, ferrous iron reacts to produce a metallic odour comprised of *n*-aldehydes (e.g. *n*-hexanal) and ketones (e.g. 1-octene-3-one) (Tucker *et al.*, 2007). The metallic odour resulting from iron coming into contact with the skin is also described as a musty, metallic odour (Glindemann *et al.*, 2006). Moisture on the skin forms Fe^{2+} ions that are quickly oxidized to Fe^{3+} ions, which occurs concurrently with reduction and decomposition of existing skin lipidperoxides to odourous carbonyl hydrocarbons, which are perceived as metallic in odour (Glindemann *et al.*, 2006).

Complaints of metallic taste in drinking water are fairly common. In a survey of North American water utilities, 14–20% of taste-related complaints were attributed to a metallic flavor in the water (Suffet *et al.*, 1996). As iron is often found in drinking water, it may prove important as a retronasal odour for water providers. Fe^{2+} may mask or enhance other odourants, such as MTBE in drinking water.

For Cu^{2+}, the metallic flavour is a combination of a weak bitter taste and a strong metallic odour that is produced as a biochemical reaction in the mouth. Cu^{2+} in the presence of oxygen causes lipid oxidation of organic acids in the mouth (Dietrich, 2009). The taste threshold for copper ions is generally 3 mg/L (Mallevialle & Suffet, 1987), with slightly lower values found for tap water (2.6 mg/L), and somewhat higher values for uncarbonated mineral water (3.5–3.8 mg/L) as determined by Zacarias *et al.* (2001). The latter study did not find a significant difference when the nose was clamped closed. Ömür-Özbek and Dietrich (2011), however, calculated a much lower population threshold of 0.6 mg/L for Cu^{2+} by both logistic regression and geometric mean methods. Their results also show that metallic sensation has a significant odour component and should be treated as a flavour (i.e. taste and odour) instead of just a taste.

Magnesium or manganese in drinking water can impart an astringent taste. Although magnesium is detectable at 100–500 mg/L (Lockhart *et al.*, 1955), consumers dislike water containing $MgCl_2$ > 47 mg/L and $Mg(HCO_3)_2$ > 58 mg/L. Manganese has been found to impart an astringent taste at a threshold of 0.05 mg/L (Whelton, 2009).

2.7 RECENT CASE STUDIES OF ODOUR INCIDENCES

A recent case study demonstrates how the drinking water 'Taste-and-Odour Wheel' links odours to odiferous compounds and expedites identifying treatments. In late May 2007, residents of Wuxi, China, complained of septic/marshy odours in their drinking water. Analysis of samples from 30th May identified methyl mercaptan (204 μg/L) and dimethyl sulfide (94 μg/L) as the primary odourants, which shifted to dimethyl disulfide (46 μg/L) and dimethyl trisulfide (17 μg/L) in samples from 2nd June (Zhang *et al.*, 2010). On 4th June, another research group found dimethyl trisulfide to be the major constituent (Yu *et al.*, 2007).

Figure 2.4 probably best describes the biochemical transformations that occurred during this event as many of the volatile sulfur compounds were observed successively with time.

The most common compounds in this odour category – geosmin and MIB – were not found at sufficiently high levels to contribute. Instead, the death and decay of an enormous cyanobacterial bloom in the nearby water source, Lake Taihu, was the apparent source (Guo, 2007; Yang *et al.*, 2008). Fortunately, treatment by potassium permanganate oxidation at the water intake and powdered activated carbon adsorption at the treatment plant resolved the problem, although panic in the city of over 2 million led to a run on bottled water (Zhang *et al.*, 2010) and normal drinking water service was not fully restored for a week (Qin *et al.*, 2010). Further restrictions to avoid eutrophication (industrial discharges and agricultural runoff) and algal bloom rapid die-off (chemical company discharges) were mandated. Increasing global temperatures are expected to foster progressively earlier algal blooms in this and other water bodies (Qin *et al.*, 2010).

Even for substances not found on the Taste and Odour Wheel (Figure 2.1), the basic process of investigating an odour problem is the same. For the 2014 spill into the Elk River in West Virginia, USA, the liquid was easily identified: an approximately 80% 4-methylcyclohexanemethanol (MCHM) industrial mixture. As the capability of the down-river drinking-water treatment plant was overwhelmed (McGuire *et al.*, 2014a), the MCHM mixture gave a characteristic liquorice odour to the drinking water supply. A 60-member panel was used to determine the OTC, which was 0.55 µg/L (McGuire *et al.*, 2014b). The analytical reporting level at the time of the spill was much higher (10 µg/L) so it is not surprising that consumer complaints continued even though analytical reports came back 'non-detectable'.

Crude MCHM exists as a mixture of *cis* and *trans* isomers (ratio 1:2), which have OTCs that differ by a factor of 2000 (Gallagher *et al.*, 2015). The *trans*-MCHM isomer is the major odourant with the lower OTC of 1.2 µg/L in liquid (0.06 ppb in air) and descriptors of 'liquorice' and 'sweet'. The *cis*-MCHM isomer was described as 'fermented fruit'. Although individual responses varied by a factor of over 100, those with prior exposure were 5-fold more sensitive than those with none. Another study (Dietrich *et al.*, 2015) found a similar 1:1.75 *cis*:*trans* ratio for crude MCHM and also derived octanol-water parition coefficients for the isomers (225 for *cis*-MCHM and 291 for *trans*-MCHM), which indicate less water solubility and greater uptake by activated carbon of the more-odiferous *trans* isomer.

Gallagher *et al.* (2015) used a word-cloud technique to visually display the odour descriptions used by the panelists. Larger font size and weight indicated greater frequency of the same word being used. When describing the odour of *trans*-MCHM, the words 'liquorice' and 'sweet' were very prominent for those previously exposed, while a wider variety of descriptors was used by naïve subjects. Just as OTCs often vary greatly from one panelist to another, which is captured by the range and standard deviation of values, a word cloud is a useful tool that helps capture the variation in odour descriptors used by panelists.

REFERENCES

Acero J. L., Piriou P. and von Gunten U. (2005). Kinetics and mechanisms of formation of bromophenols during drinking water chlorination: assessment of taste and odour development. *Water Research*, **39**(13), 2979–2993.

Adams G. A., Witherspoon J., Card T., Erdal Z., Forbes B., McEwen D., Geselbracht J., Glindemann D., Hargreaves R., Hentz L., Higgins M. and Murthy S. (2003). Identifying and Control Odour in the Municipal Wastewater Environment Phase II: Impacts of In-plant Parameters on Biosolids Odour Quality. Water Environment Research Foundation. Alexandria, Virginia, USA.

Alexander H. C., Howard C., Alexander H. C., McCarty W. M., Bartlett E. A. and Syverud A. N. (1982). Aqueous odour and taste threshold values of industrial chemicals. *Journal American Water Works Association*, **74**(11), 595–599.

Al-Samarrai H., Matud J., Wiesenthal K., Atiyah P., Bruchet A. and Suffet I. H. (2004). Sensory evaluation of the odours produced during bromophenol formation using a multi-level statistical model. *Water Science and Technology*, **49**(9), 241–248.

American Public Health Association (APHA), American Water Works Association, and Water Environment Federation (2012). Standard Methods for the Examination of Water and Wastewater, 22nd edn. APHA, Washington, D.C., USA.

American Society for Testing and Materials (ASTM) (2011). E679-04 Standard practice for determination of odour and taste thresholds by a forced-choice ascending concentration series method of limits. ASTM, West Conshohocken, PA, USA, pp. 1–7.

Anselme C., Mallevialle J. and Suffet I. H. (1985a). Removal of Tastes and Odours by the Ozone-Granular Activated Carbon Water Treatment Processes. Proceedings of the 7th Ozone World Congress, Intl. Ozone Assoc., Tokyo, Japan.

Anselme C., N'Guen K., Bruchet A. and Mallevialle J. (1985b). Can polyethylene pipes impart odours in drinking water? *Environmental Technology Letters*, **6**, 477–488.

Anselme C., Suffet I. H. and Mallevialle J. (1988). Effects of ozonation on tastes and odours. *Journal American Water Works Association*, **80**(10), 45–51.

Atasi K. Z., Chen T., Huddleston J. I., Young C. C. and Suffet I. H. (1999). Factor screening for ozonating the taste- and odour-causing compounds in source water at Detroit, USA. *Water Science and Technology*, **40**(6), 115–122.

Bayliss J. R. (1951). Fishy odour in water caused by Dinobryon. *Pure Water*, **3**, 128.

Berglind L., Johnsen I. J., Onnerod K. and Skulberg O. M. (1983). *Oscillatoria brevis* (Kutz.) Gorn. and some other especially odouriferous benthic cyanophytes in Norwegian inland waters. *Water Science and Technology*, **15**(6–7), 241–246.

Bett K. L., Ingram D. A., Grimm C. C., Vinyard B. T., Boyette K. D. C. and Dionigi C. P. (2000). Alteration of the sensory perception of the muddy/earthy odourant 2-methylisoborneol in channel catfish *ictalurus punctatus* fillet tissues by addition of seasonings. *Journal of Sensory Studies*, **15**, 459–472.

Blank I. and Grosch W. (2002). On the role of (−)-2-methylisoborneol for the aroma of robusta coffee. *Journal of Agricultural and Food Chemistry*, **50**, 4653–4656.

Brady B., Bartels J. H. M., Mallevialle J. and Suffet I. H. (1988). Sensory analysis of drinking water. *Water Quality Bulletin*, **13**(2/3), 61.

Brookes J. C., Horsfield A. P. and Stoneham A. M. (2009). Odour character differences for enantiomers correlate with molecular flexibility. *Journal of The Royal Society Interface*, **6**, 75–86.

Brosillon S., Lemasle M., Renault E., Tozza D., Heim V. and Laplanche A. (2009). Analysis and occurrence of odourous disinfection by-products from chlorination of amino-acids in three different drinking water treatment plants and corresponding distribution networks. *Chemosphere*, **77**, 1035–1042.

Bruchet A. (1999). Solved and unsolved cases of taste and odour episodes in the files of Inspector Clouseau. *Water Science and Technology*, **40**(6), 15–22.

Bruchet A. (2015). Personal communication.

Bruchet A., N'Guyen K., Mallevialle J. and Anselme C. (1989). Identification and Behaviour of Iodinated Haloform Medicinal Odour. Proceedings of the AWWA Annual Conference, Los Angeles, CA, USA.

Bruchet A., N'Guyen K., Legrand M. F. and Mallevialle J. (1991). Variations in organic and organoleptic water quality during treatment of artificially recharged groundwater. *Journal American Water Works Association*, **83**(3), 63–70.

Bruchet A., Duguet J. P. and Suffet I. H. (2004). Role of oxidants and disinfectants on the removal, masking and generation of tastes and odours. *Reviews in Environmental Science & Bio/Technology*, **3**, 33–41.

Bruchet A., Hochereau C. and Campos C. (2007). An acute taste and odour episode solved by olfactory GC-MS. *Water Science and Technology*, **55**(5), 223–230.

Bryan P., Kuzcinski L., Sawyer M. and Feng R. (1973). Taste threshold of halogens in water. *Journal American Water Works Association*, **65**(5), 363–368.

Burlingame G. A. (2003). Personal communication, reported in Curren *et al.* (2009).

Burlingame G. A. and Anselme C. (1995). Distribution system tastes and odours. In: Advances in Taste-and-Odour Treatment and Control, I. H. Suffet, J. Mallevialle and E. Kawczynski (eds), AWWA Research Foundation and Lyonnaise Des Eaux Cooperative Research Report. American Water Works Assoc., Denver, CO, USA, pp. 281–319.

Burlingame G. A., Dann R. M. and Brock G. L. (1986). A case study of geosmin in Philadelphia's waters. *Journal American Water Works Association*, **78**(3), 56–61.

Burlingame G. A., Khiari D. and Suffet I. H. (1991a). Odour Reference Standards: The Universal Language. Proceedings of the AWWA Water Qualality and Technology Conference, Orlando, FL, USA, pp. 7–39.

Burlingame G. A., Muldowney J. J., Taylor I. and Maddrey R. E. (1991b). Cucumber flavor in Philadelphia drinking water. *Journal American Water Works Association*, **84**(8), 92–97.

Burlingame G. A., Suffet I. H., Khiari D. and Bruchet A. (2004). Development of an odour wheel classification scheme for wastewater. *Water Science and Technology*, **49**(9), 201–209.

Burlingame G. A., Dietrich A. M. and Whelton A. J. (2007). Understanding the basics of tap water taste. *Journal American Water Works Association*, **99**(5), 100–111.

Burtschell R. M., Rosen A. A., Middleton R. M. and Ettinger M. B. (1959). Chlorine derivatives of phenol causing taste and odour. *Journal American Water Works Association*, **51**(2), 205–214.

Buttery R. G. and Ling L. C. (1973). Earthy aroma of potatoes. *Journal of Agricultural and Food Chemistry*, **21**(4), 745–746.

Cancho B., Ventura F., Galceran M., Diaz A. and Ricart S. (2000). Determination, synthesis and survey of iodinated trihalomethanes in water treatment processes. *Water Research*, **34**(13), 3380–3390.

Chen X., Luo Q., Yuan S., Wei Z., Song H., Wang D. and Wang Z. (2013). Simultaneous determination of ten taste and odour compounds in drinking water by solid-phase microextraction combined with gas chromatography-mass spectrometry. *Journal Environmental Science*, **25**(11), 2313–2323.

Chen Y. M., Hobson P., Burch M. D. and Lin T. F. (2010). *In situ* measurement of odour compound production by benthic cyanobacteria. *Journal of Environmental Monitoring*, **12**(3), 769–775.

Collins R. P. (1971). Characterization of Tastes and Odours in Water Supplies. Wat. Poll. Control Res. Series, Project 16 040-DGM- 08/7 l. Office of Research Monitoring, USEPA, Washington, D.C., USA.

Collins R. P. and Kalnins K. (1965a). Volatile constituents produced by the alga *Synura pretersenii*. I. The carbonyl fraction. *Lloydia*, **28**, 48.

Collins R. P. and Kalnins K. (1965b). Volatile constituents produced by the alga *Synura petersenii*. II. Alcohols, esters, and acids. *International Journal of Air and Water Pollution*, **9**, 501–504.

Collins R. P. and Kalnins K. (1966). Carbonyl compounds produced by *Cryptomonas ovata* var. *palustris*. *Journal of Eukaryotic Microbiology*, **13**(3), 435–437.

Collins R. P. and Kalnins K. (1967). The fatty acids of *Synura petersenii*. *Lloydia*, **30**, 437–440.

Collins R. P., Knaak L. E. and Soboslai J. W. (1970). Production of geosmin and 2-exo-hydroxy-2-methylbornane by *Streptomyces odourifer*. *Lloydia*, **33**(1), 199–200.

Cotsaris E., Bruchet A., Mallevialle J. and Bursill D. B. (1995). The identification of odourous metabolites produced from algal monocultures. *Water Science and Technology*, **31**(11), 251–258.

Crozes G., Hagstrom J. and White P. (1997). Approach to Design of Ozone Facilities for Cryptosporadium Inactivation on Lake Michigan. Proceedings of AWWA Water Quality Technology Conference, Denver, CO, USA.

Crozes G., Hangstrom J., Suffet I. H. and Young C. (1999). Bench-scale evaluation of adsorptive processes for taste and odour control using rapid small scale column tests and flavor profile analysis. *Water Science and Technology*, **40**(6), 39–44.

Curren J. (2012). Characterization of Odour Nuisance. *Doctorial dissertation*, Environmental Science & Engineering Program, UCLA, Los Angeles, CA.

Curren J., Wang Z., Matud J., Mackey E. D. and Suffet I. H. (Mel). (2009). The effect of water source andchlorine and chloramine odourants in drinking water on earthy and musty odour intensity. *Journal of Water Supply: Research and Technology – AQUA*, **58**(8), 521–531.

Dale M. S., Moylan M. S., Koch B. and Davis M. K. (1997). MTBE: Taste-and-Odour Threshold Determinations Using the Flavor Profile Method. Proceedings of the AWWA Water Quality Technology Conference, Denver, CO, USA.

Daniel P. A. and Meyerhofer P. F. (1993). Ozone and water treatment. In: Proceedings of the 9th International Ozone Congress, L. J. Bolloky (ed.), International Ozone Association, Zurich, Switzerland, Vol. **1**.

Decottignies V., Bruchet A. and Suffet I. H. (2009). Landfill Odour Wheel: A New Approach to Characterize Odour Emissions at Landfill Sites. Twelfth International Waste

Management and Landfill Symposium, S. Margherita di Pula (Cagliari), Sardinia, Italy, 5–9 October.

Diaz A., Ventura F. and Galceran M. T. (2004). Identification of 2,3-butanedione (diacetyl) as the compound causing odour events at trace levels in the Llobregat River and Barcelona's treated water (Spain). *Journal of Chromatography A*, **1034**, 175–182.

Dietrich A. M. (2009). The sense of smell: contributions of orthonasal and retronasal perception applied to metallic flavor of drinking water. *Journal of Water Supply: Research and Technology – AQUA*, **58**(8), 562–570.

Dietrich A. M. and Burlingame G. A. (2015). Critical review and rethinking of USEPA secondary standards for maintaining organoleptic quality of drinking water. *Environmental Science & Technology*, **49**(2), 708–720.

Dietrich A. M., Orr M. P., Gallagher D. L. and Hoehn R. C. (1992). Tastes and odours associated with chlorine dioxide. *Journal American Water Works Association*, **84**(6), 82–88.

Dietrich A. M., Glindemann D., Pizarro F., Gidi V., Olivares M., Araya M., Camper A., Duncan S., Dwyer S., Whelton A. J., Younos T., Subramanian S., Burlingame G. A., Khiari D. and Edwards M. (2004). Health and aesthetic impacts of copper corrosion on drinking water. *Water Science and Technology*, **49**(2), 55–62.

Dietrich A. M., Thomas A., Zhao Y., Smiley E., Shanaiah N., Ahart M., Charbonnet K. A., DeYonker N. J., Alexander W. A. and Gallagher D. L. (2015). Partitioning, aqueous solubility, and dipole moment data for *cis*- and *trans*-(4-methylcyclohexyl)methanol, principal contaminants of the West Virginia chemical spill. *Environmental Science & Technology Letter*, **2**, 123–127.

Dravnieks A. (1985). Atlas of Odour Character Profiles. ASTM data series DS 61. ASTM, Philadelphia, USA.

Durand M. L. and Dietrich A. M. (2007). Contributions of silane crosslinked PEX pipe to chemical/solvent odours in drinking water. *Water Science and Technology*, **55**(5), 153–160.

Eccles R. (1994). Review: menthol and related cooling compounds. *Journal of Pharmacy and Pharmacology*, **46**, 618–630.

Engen T. (1960). Effect of practice and instruction on olfactory thresholds. *Perceptual and Motor Skills*, **10**, 195–198.

Environment Agency 2004 The Microbiology of Drinking Water (2004). Part 11, Taste, odour and related aesthetic problems. Environment Agency, United Kingdom. https://assets.publishing.service.gov.uk/government/uploads/system/uploads/attachment_data/file/755589/mdwpart112004_859972.pdf. (Accessed 08 January 2019).

Franzmann P. D., Heitz A., Zappia L. R., Wajon J. E. and Xanthis K. (2001). The formation of malodourous dimethyl oligosulphides in treated groundwater: the role of biofilms and potential precursors. *Water Research*, **35**(7), 1730–1738.

Freuze I., Brosillon S., Herman D., Laplanche A., Démocrate C. and Cavard J. (2004). Odourous products of the chlorination of phenylalanine in water: formation, evolution, and quantification. *Environmental Science & Technology*, **38**(15), 4134–4139.

Froese K. L., Wolanski A. and Hrudey S. E. (1999). Factors governing odourous aldehyde formation as disinfection by-products in drinking water. *Water Research*, **33**(6), 1355–1364.

Gallagher D. L., Phetxumphou K., Smiley E. and Dietrich A. M. (2015). Tale of two isomers: complexities of human odour perception for *cis*- and *trans*-4-methylcyclohexane methanol from the chemical spill in West Virginia. *Environmental Science & Technology*, **49**, 1319–1327.

Gerber N. N. (1968). Geosmin, from microorganisms, is *trans*-1,10-dimethyl-*trans*-9-decalol. *Tetrahedron Letter*, **9**(25), 2971–2974.

Gerber N. N. (1969). A volatile metabolite of actinomycetes, 2-methylisoborneol. *The Journal of Antibiotics*, **22**(10), 508–509.

Gerber N. N. (1971). Sesquiterpenoids from actinomycetes: cadin-4-ene-1-ol. *Phytochemistry*, **10**, 185–189.

Gerber N. N. (1979). Volatile substances from actinomycetes: their role in the odour pollution of water. *CRC Critical Reviews in Microbiology*, **7**, 191–214.

Gerber N. N. and LeChevalier H. A. (1965). Geosmin, an earthy smelling substance isolated from actino-mycetes. *Applied and Environmental Microbiology*, **13**(6), 935–938.

Giger W. and Schaffner C. (1981). Groundwater pollution by volatile organic chemicals. *Studies in Environmental Science*, **17**, 517–522.

Gittelman T. S. and Yohe T. L. (1989). Treatment of the Iodinated Haloform Medicinal Odours in Drinking Water. Proceedings of the AWWA Annual Conference, Los Angeles, CA, USA, pp. 105–123.

Glaze W. H., Weinberg H. S. and Cavanagh J. H. (1993). Evaluating the formation of brominated DBPs during ozonation. *Journal American Water Works Association*, **85**(1), 96–103.

Glindemann D., Dietrich A. M., Staerk H. J. and Kuschk P. (2006). The two odours of iron when touched or pickled: (Skin) carbonyl compounds and organophosphines. *Angewandte Chemie International Edition*, **45**, 7006–7009.

Guo L. (2007). Doing battle with the green monster of Taihu Lake. *Science*, **317**(5846), 1166.

Hack D. J. (1981). Common customer complaint: my hot water stinks. *Opflow*, **7**(9), 3.

Hayes K. P. and Burch M. D. (1989). Odourous compounds associated with algal blooms in South Australian waters. *Water Research*, **23**(1), 115–121.

Henley D. E. (1970). Odourous metabolite and other selected studies of Cyanophyta. *Doctoral dissertation*, North Texas State Univ., Denton, TX, USA.

Hettinger T. P., Myers W. E. and Frank M. E. (1990). Role of olfaction in perception of non-traditional 'taste' stimuli. *Chemical Senses*, **15**(6), 755–760.

Higgins M. J., Chen Y.-C., Yarosz D. P., Murthy S. N., Maas N. A., Glindemann D. and Novak J. T. (2006). Cycling of volatile organic sulfur compounds in anaerobically digested biosolids and its implications for odours. *Water Environment Research*, **78**, 243–252.

Hoehn R. C., Dietrich A. M., Farmer W. S., Orr M. P., Lee R. G., Aieta E. M., Wood D. W. I. I. I. and Gordon G. (1990). Household odours associated with the use of chlorine dioxide during drinking water treatment. *Journal American Water Works Association*, **82**(4), 166–172.

Hrudey S. E., Gac A. and Daignault S. A. (1988). Potent odour-causing chemicals arising from water disinfection. *Water, Science and Technology*, **20**(8/9), 55–61.

Hrudey S. E., Daignault S. A., Gac A. J., Poole D., Walker G. and Birkholz D. A. (1989). Swampy or Swimming Pool Aldehyde Odours Caused by Chlorination and

Chloramination. Proceedings of the AWWA Annual Conference, Los Angeles, CA, USA.

Hsieh W. H., Hung W. N., Wang G. S., Hsieh S. T. and Lin T. F. (2012). Effect of pH on the analysis of 2-MIB and geosmin in Water. *Water Air Soil Pollution*, **223**, 715–721.

Ibrahim E. A., Becker W. C., Capangpangan M. B. and Suffet I. H. (1990). Optimization of sequential unit operations for removal of organics from drinking water. In: Chemical Water and Wastewater Treatment, H. H. Hahn and R. Klute (eds), In Proceedings of the 4th International Gothenburg Symposium, Madrid, Spain. Springer-Verlag, Berlin, Germany, pp. 305–334, ISSN #3-540-53181-5.

Izaguirre G. (1992). A copper-tolerant *Phormidium* sp. from Lake Mathews, California, that produces 2-methylisoborneol and geosmin. *Water Science and Technology*, **25**(2), 217–223.

Izaguirre G., Hwang C. J., Krasner S. W. and McGuire M. J. (1982). Geosmin and 2-methylisobomeol from cyanobacteria in three water supply systems. *Applied and Environmental Microbiology*, **43**(3), 708–714.

Izaguirre G., Hwang C. J., Krasner S. W. and McGuire M. J. (1983). Production of 2-methylisoborneol by two benthic cyanophyta. *Water Science and Technology*, **15**(617), 211–220.

Jenkins D., Medsker L. L. and Thomas J. F. (1967). Odourous compounds in natural waters: some sulfur compounds associated with blue-green algae. *Environmental Science & Technology*, **1**(9), 731–735.

Jensen S. E., Anders C. L., Goatcher L. J., Perley T., Kenefick S. and Hrudey S. E. (1994). Actinomycetes as a factor in odour problems affecting drinking water from the North Saskatchewan River. *Water Research*, **28**(6), 1393–1401.

Joe W. H., Choi I. C., Baek Y. A., Choi Y. J., Park G. S. and Yu M. J. (2007). Advanced treatment for taste and odour control in drinking water: case study of a pilot scale plant in Seoul, Korea. *Water Science and Technology*, **55**(5), 111–116.

Juttner F. (1981). Detection of lipid degradation products in the water of a reservoir during a bloom of *Synura uvella. Applied and Environmental Microbiology*, **41**, 100–106.

Juttner F. (1983). Volatile odourous excretion products of algae and their occurrence in the natural aquatic environment. *Water Science and Technology*, **15**(6–7), 247–257.

Juttner F., Hoflacher B. and Wurster K. (1986). Seasonal analysis of volatile organic biogenic substances (VOBs) in freshwater phytoplankton populations dominated by *Dinobryon, Microcystis*, and *Aphani-zomenon. Journal of Phycology*, **22**(2), 169–175.

Karlsson S., Grimvall A. and Boren H. (1995). Drinking water off-flavour caused by the formation of short-chain fatty acids in slow sand filters. *Water Science and Technology*, **31**(11), 49–54.

Khiari D., Brenner L., Burlingame G. A. and Suffet I. H. (1992). Sensory gas chromatography for evaluation of taste and odour events in drinking water. *Water Science and Technology*, **25**(2), 97–104.

Khiari D., Suffet I. H. and Barrett S. E. (1995a). Extraction and identification of chemicals causing grassy odours in FPA reference standards. *Water Science and Technology*, **31**(11), 93–98.

Khiari D., Suffet I. H. and Barrett S. E. (1995b). The determination of compounds causing fishy/swampy odours in drinking water supplies. *Water Science and Technology*, **31**(11), 105–112.

Khiari D., Barrett S. E. and Suffet I. H. (1997). Determination of organic compounds causing decaying vegetation and septic odours in drinking water by sensory GC. *Journal American Water Works Association*, **89**(4), 150–161.

Khiari D., Brochet A., Gittelman T., Matia L., Barrett S. E., Suffet I. H. and Hund R. (1999a). Distribution-generated taste-and-odour phenomena. *Water Science and Technology*, **40**(6), 129–134.

Khiari D., Young C. C., Ye Q., Amar G., Atasi K., Huddleston J. and Suffet I. H. (1999b). Factors affecting the stability and behavior of grassy odours in drinking water. *Water Science and Technology*, **40**(6), 287–292.

Khiari D., Barrett S., Chinn R., Bruchet A., Piriou P., Matia L., Ventura F., Suffet I. H., Gittleman T. and Luitweiler P. (2002). Distribution System Generated Taste-and-Odour Phenomena. AWWA Research Foundation, No. 90897, AWWA, Denver, CO.

Kikuchi T., Mimura T., Harimaya K., Yano H., Arimoto T., Masada Y. and Inoue K. (1973a). Odourous metabolites of actinomycetes biwako-C and -D strain isolated from the bottom deposits of Lake Biwa. Identification of geosmin, methylisoborneol and furfural. *Chemical and Pharmaceutical Bulletin (Tokyo)*, **21**, 2339.

Kikuchi T., Mimura T., Harimaya K., Yano H., Arimoto T., Masada Y. and Inoue K. (1973b). Odourous metabolites of blue-green algae *Schizothrix mueller* Nageli collected in the southern basin of Lake Biwa. Identification of geosmin. *Chemical and Pharmaceutical Bulletin (Tokyo)*, **21**, 2342.

Kikuchi T., Mimura T., Moriwaki Y., Ando M. and Negoro K. (1986). Odourous components of the diatom, *Synedra rumpens* Kutz, isolated from the water in Lake Biwa. *Identification of n-hexanal. Yakugaku Zasshi*, **92**(12), 1567–1568.

Kikuchi T., Mimurn T., Moriwaki Y., Ando M. and Negoro K. I. (1974). Metabolites of a diatom, *Synedra rumpens* Kutz, isolated from the water in Lake Biwa. Identification of odourous compounds, *n*-hexanal, and *n*-heptanal, and analysis of fatty acids. *Chemical and Pharmaceutical Bulletin (Tokyo)*, **22**(4), 945–949.

Krasner S. W. and Barrett S. E. (1984). Aroma and Flavor Characteristics of Free Chlorine and Chloramines. Proceeding of AWWA Water Quality Technology Conference, Denver, CO, USA, pp. 381–398.

Krasner S. W., Barrett S. E., Dale M. S. and Hwang C. J. (1986). Free chlorine versus monochloramine for controlling off-tastes and -odours in drinking water. *Journal American Water Works Association*, **81**(2), 86–93.

Krasner S. W., Barrett S. E., Dale M. S. and Hwang C. J. (1989). Free chlorine versus monochloramine for controlling off-tastes and off-odours. *Journal American Water Works Association*, **81**(2), 86–93.

Krasner S. W., Glaze W. H., Weinberg H. S., Daniel P. A. and Najm I. N. (1993). Formation and control of bromate during ozonation of waters containing bromide. *Journal American Water Works Association*, **85**(1), 73–81.

Lawless H. T. (1999). Descriptive analysis of complex odours: reality, model of illusion? *Food Quality and Preference*, **10**, 325–332.

Lawless H. T. and Heymann H. (2010). Sensory Evaluation of Food: Principles and Practices. 2nd edn. Springer, New York, USA.

Lawless H. T., Shlake S., Smythe J., Lim J., Yang H., Chapman K. and Bolton B. (2004). Metallic taste and retronasal smell. *Chemical Senses*, **29**(1), 25–33.

Leavey S., Suffet I. H., Everitt M. and Piriou P. (2013). Threshold, Recognition, Objection and Rejection Concentrations to Set a Water Treatment Goal. Update of Proceedings of the American Water Works Association, Water Quality Technology Conference, Long Beach, CA, USA.

Lee C. F. (1967). Kinetics of reactions between chlorine and phenolic compounds. In: Principles and Applications of Water Chemistry, S. D. Faust and J. V. Hunter (eds), John Wiley and Sons, New York, NY, USA, pp. 54–74.

Leffingwell J. C. (2016). Chirality and odour perception: the 2-methylisoborneols. http://www.leffingwell.com/chirality/methylisoborneol.htm (accessed 30 March 2016).

Li X., Lin P., Wang J., Liu Y., Li Y., Zhang X. and Chen C. (2016). Treatment technologies and mechanisms for three odourants at trace level: IPMP, IBMP, and TCA. *Environmental Technology*, **37**(3), 308–315.

Lin T. F., Wong J. Y. and Kao H. P. (2002). Correlation of musty odour and 2-MIB in two drinking water treatment plants in South Taiwan. *Science of the Total Environment*, **289**, 225–235.

Lin T. F., Chang D. W., Lien S. K., Tseng Y. S., Chiu Y. T. and Wang Y. S. (2009). Effect of chlorination on the cell integrity of two noxious cyanobacteria and their releases of odourants. *Journal of Water Supply: Research and Technology – AQUA*, **58**(8), 539–551.

Lockhart E. E., Tucker C. L. and Merritt M. C. (1955). The effect of water impurities on the flavor of brewed coffee. *Food Research*, **20**, 598–605.

Lomans B. P., Pol A. and Camp H. J. M. O. (2002). Microbial cycling of volatile organic sulfur compounds in anoxic environments. *Water Science and Technology*, **45**, 55–60.

MacKenthun K. M. and Keup L. E. (1970). Biological problems encountered in water supplies. *Journal American Water Works Association*, **62**(8), 520–528.

Mackey E. D., Baribeau H., Crozes G. F., Suffet I. H. and Piriou P. (2004). Public thresholds for chlorinous flavors in U.S. tap water. *Water Science and Technology*, **49**(9), 335–340.

Mackey E. D., Suffet I. H. and Booth S. (2008). A Decision Tool for Earthy/Musty Taste and Odour Control. Water Works Association Research Foundation, Research Report #3032, Awwa Research Foundation, Denver, CO, USA.

Malleret L. and Bruchet A. (2002). A taste and odour episode caused by 2,4,6-tribromoanisole. *Journal American Water Works Association*, **94**(7), 84–95.

Malleret L., Bruchet A. and Hennion M. C. (2001). Picogram determination of 'earthy-musty' odourous compounds in water using modified closed loop stripping analysis and large volume injection GC/MS. *Analytical Chemistry*, **73**, 1485–1490.

Mallevialle J. and Suffet I. H. (1987). Identification and Treatment of Tastes and Odours in Drinking Water. AWWA Research Foundation and Lyonnaise Des Eaux Cooperative Research Report, American Water Works Assoc., Denver, CO, USA.

Mallevialle J., Lefebvre M., Rousseau C. and Saglier C. (1982). Pollution with petroleum products: qualitative and quantitative analysis of low concentrations. *Revue Française des Sciences de l'Eau*, **1**, 125–138.

Martin J. F., Fisher T. H. and Bennett L. W. (1988). Musty odour in chronically off-flavored channel catfish: isolation of 2-methylenebornane and 2-methyl-2-bornene. *Journal of Agricultural and Food Chemistry*, **36**, 1257–1260.

Martin J. F., Izaguirre G. and Waterstrat P. (1991). A planktonic *Oscillatoria* species from Mississippi cat-fish ponds that produces the off-flavor compound 2-methylisoborneol. *Water Research*, **25**(12), 1447–1451.

Matia L. (1995). Treatment of tastes in drinking water: causes and control. In: Advances in Taste-and-Odour: Treatment and Control, I. H. Suffet, J. Mallevialle and E. Kawczynski (eds), American Water Works Association Research Foundation, Denver, CO, USA.

Matsumoto A. and Tsuchiya Y. (1988). Earthy-musty odour producing cyanophytes isolated from five water areas in Tokyo. *Water Science and Technology*, **20**(8/9), 179–183.

McGuire M. J., Krasner S. W., Hwang C. J. and Izaguirre G. (1981). Closed-loop stripping analysis as a tool for solving taste and odour problems. *Journal American Water Works Association*, **73**(10), 530–537.

McGuire M. J., Hund R. and Burlingame G. (2005). A practical decision tree tool that water utilities can use to solve taste and odour problems. *Journal of Water Supply Research and Technology – AQUA*, **54**(5), 321–327.

McGuire M. J., Rosen J., Whelton A. J. and Suffet I. H. (2014a). An unwanted licorice odour in a West Virginia water supply. *Journal American Water Works Association*, **106**(6), 72–82.

McGuire M. J., Suffet I. H. and Rosen J. (2014b). Consumer panel estimates of odour thresholds for crude 4-methylcyclohexanemethanol. *Journal American Water Works Association*, **106**(10), E445–E458.

Medsker L. L., Jenkins D. and Thomas J. F. (1968). Odourous compounds in natural waters. An earthy-smelling compound associated with blue-green algae and actinomycetes. *Environmental Science & Technology*, **2**(6), 461–464.

Medsker L. L., Jenkins D., Thomas J. F. and Koch C. (1969). Odourous compounds in natural waters. 2-exo-hydroxy-2-methylbomate, the major odourous compound produced by actinomycetes. *Environmental Science & Technology*, **3**(5), 476–477.

Mills O. E., Chung S. Y. and Johnsen P. B. (1993). Dehydration products of 2-methylidoborneol are not responsible for off-flavor in the channel catfish. *Journal of Agricultural and Food Chemistry*, **41**, 1690–1692.

Mirlohi S., Dietrich A. and Duncan S. (2011). Age-associated variation in sensory perception of iron in drinking water and the potential for overexposure in the human population. *Environmental Science & Technology*, **45**(15), 6575–6583.

Miyazaki M., Yamashita T., Yusuke Suzuki Y., Saito Y., Soeta S., Taira H. and Suzuki A. (2006). A major urinary protein of the domestic cat regulates the production of felinine, a putative pheromone precursor. *Chemistry & Biology*, **13**, 1071–1079.

Montiel A. (1991). Study Investigating the Origin and Mechanism of Formation of Sapid Components Preceding Musty Taste in Our Water Supply. Proceedings of the AWWA Water Quality Technology Conference, Orlando, FL, USA.

Moskowitz H. R., Dravnieks A., Cain W. S. and Turk A. (1974). Standardized procedure for expressing odour intensity. *Chemical Senses and Flavor*, **1**, 235–237.

Narayan L. V. and Nunez W. J. (1974). Biological control: isolation and bacterial oxidation of the taste and odour compound geosmin. *Journal American Water Works Association*, **66**(9), 532–536.

National Institute for Occupational Safety and Health (2005). NIOSH Pocket Guide to Chemical Hazards. U.S. Dept. of Health and Human Services, Washington, DC, USA.

Negoro T., Ando M. and Ichikawa N. (1988). Blue-green algae in Lake Biwa which produce earthy-musty odours. *Water Science and Technology*, **20**(8/9), 117–123.

Noblet J., Schweitzer L., Ibrahim E., Stolzenbach K., Zhou L. and Suffet I. H. (1999). Evaluation of a taste and odour incident on the Ohio River. *Water Science and Technology*, **40**(6), 185–194.

Nystrom A., Grimvall A., Krantz-Rulcker C., Savenhed R. and Akerstrand K. (1992). Drinking water of off-flavour caused by 2,4,6-trichloranisole. *Water Science and Technology*, **25**(2), 241–249.

Oestman E., Schweitzer L., Tomboulian P., Corado A. and Suffet I. H. (2004). Effects of chlorine and chloramines on earthy and musty odours in drinking water. *Water Science and Technology*, **49**(9), 153–159.

Ömür-Özbek P. and Dietrich A. M. (2008). Developing hexanal as an odour reference standard for sensory analysis of drinking water. *Water Research*, **42**(10–11), 2598–2604.

Ömür-Özbek P. and Dietrich A. M. (2011). Retronasal perception and flavour thresholds of iron and copper in drinking water. *Journal of Water and Health*, **9**(1), 1–9.

Pangborn R. M., Berg H. W., Roessler E. B. and Webb A. D. (1964). Influence of methodology on olfactory response. *Perception Motor Skills*, **18**, 91–103.

Persson P. E. (1979a). Notes on musty odour vs. sensory properties of geosmin in water. *Aqua Fennica*, **9**, 53–56.

Persson P. E. (1979b). The source of muddy odour in Bream *(Abramis brama)* from the Porvoo Sea area (Gulf of Finland). *Journal of the Fisheries Research Board of Canada*, **36**(8), 883–890.

Persson P. E. (1992). A summary of problem areas in aquatic off-flavour research. *Water Science and Technology*, **25**(2), 335–339.

Peter A. and von Gunten U. (2007). Oxidation kinetics of selected taste and odour compounds during ozonation of drinking water. *Environmental Science & Technology*, **41**(2), 626–631.

Peter A., Koster O., Schildknecht A. and von Gunten U. (2009). Occurrence of dissolved and particle-bound taste and odour compounds in Swiss lake waters. *Water Research*, **43**, 2191–2200.

Pfaffmann C., Schlossberg H. and Cornsweet J. (1954). Variables affecting difference tests in food acceptance testing methodology. In: P. J. F. Pilgrim and M. S. Peterson (eds), National Academy of Science, Natl Res. Council, Washington DC, pp. 4–17.

Phattarapattamawong S., Echigo S. and Itoh S. (2011). Simultaneous control of bromate ion and chlorinous odour in drinking water using an advanced oxidation process (O_3/H_2O_2). *Ozone: Science & Engineering*, **33**, 136–142.

Piet G. J., Zoeteman B. C. J. and Kraayeveld A. J. A. (1972). Earthy smelling substances in surface waters of the Netherlands. *Water Treatment Examination Water*, **21**, 281–286.

Piriou P., Mackey E. D., Suffet I. H. and Bruchet A. (2004). Chlorinous flavor perception in drinking water. *Water Science and Technology*, **49**(9), 321–328.

Piriou P., Soulet C., Acero J. L., Bruchet A., Von Gunten U. and Suffet I. H. (2007). Understanding medicinal taste and odour formation in drinking waters. *Water Science and Technology*, **55**(5), 85–94.

Piriou P., Devesa R., De Lalande M. and Glucina K. (2009). European reassessment of MIB and geosmin perception in drinking water. *Journal of Water Supply: Research and Technology – AQUA*, **58**(8), 532–538.

Polak B. H. and Provasi J. (1992). Odour sensitivity to geosmin enantiomers. *Chemical Senses*, **17**(1), 23–26.

Powers J. J. and Shinholser K. (1988). Flavour thresholds for vanillin and predictions of higher or lower thresholds. *Journal of Sensory Studies*, **3**(1), 49–61.

Pretti G., Gittelman T. S., Staudte P. B. and Luitweiller P. (1993). Letting the nose lead the way. Malodourous compounds in water. *Analytical Chemistry*, **65**, 699A–702A.

Qin B., Zhu G., Gao G., Zhang Y., Li W., Paerl H. W. and Carmichael W. W. (2010). A drinking water crisis in lake taihu, china: linkage to climatic variability and lake management. *Environmental Management*, **45**, 105–112.

Rabin M. D. and Cain W. S. (1986). Determinants of measured olfactory sensitivity. *Perception & Psychophysics*, **39**, 281–286.

Rashash D. M. C., Hoehn R. C., Dietrich A. M. and Parker B. C. (1993). Isolation and Identification of Algal Odour Compounds by CLLE, GC/MS and FPA. Proceedings of the AWWA Water Quality Technology Conference, Miami, FL, USA, pp. 591–597.

Rashash D. M. C., Dietrich A. M. and Hoehn R. C. (1997). FPA of selected odourous compounds. *Journal American Water Works Association*, **89**(4), 131–141.

Richardson S. D. (2003). Disinfection by-products and other emerging contaminants in drinking water. *Trends in Analytical Chemistry*, **22**(10), 666–684.

Richardson S. D., Thruston A. D., Caughran T. V., Chen P. H., Collette T. W., Floyd T. L., Schenck K. M., Lykins B. W., Sun G. R. and Majetich G. (1999). Identification of new ozone disinfection byproducts in drinking water. *Environmental Science & Technology*, **33**(19), 3368–3377.

Rosen A. A., Safferman R. S., Mashni C. I. and Romano A. H. (1968). Identity of odourous substances produced by *Streptomyces griseoluteus*. *Applied Microbiology*, **16**, 178–179.

Rosen A. A., Mashni C. I. and Safferman R. S. (1970). Recent developments in the chemistry of odour in water: the cause of earthy/musty odour. *Water Treatment Examiantion*, **19**, 106–119.

Sagiura N., Yagi O. and Sudo R. (1983). Effect of various environmental factors on musty odor production by actinomyces (Streptomyces). *Chemical Abstracts*, **99**, 172570.

Safferman R. S., Rosen A. A., Mashni C. I. and Morriss M. E. (1967). Earthy-smelling substance from a blue-green alga. *Environmental Science & Technology*, **1**, 429–430.

Scholler C. E. G., Gurtler H., Pedersen R., Molin S. and Wilkins K. (2002). Volatile metabolites from actinomycetes. *Journal of Agricultural and Food Chemistry*, **50**(9), 2615–2621.

Schumann R. and Pendleton P. (1997). Dehydration products of 2-methylisoborneol. *Water Research*, **31**(5), 1243–1246.

Schweitzer L., Noblet J., Ruth E. and Suffet I. H. (1999a). The formation, stability, and odour characterization of 2-ethyl-4-methyl-1,3-dioxoane (2-EMD). *Water Science and Technology*, **40**(6), 293–298.

Schweitzer L., Noblet J., Ye Q., Ruth E. and Suffet I. H. (1999b). The environmental fate and mechanism of formation of 2-ethyl-5,5′-dimethyl-l,3-dioxane (2-EDD) – a malodourous contaminant in drinking water. *Water Science and Technolology*, **40**(6), 217–224.

Shen Y. F., Yoo L. J., Fitzsimmons S. R. and Yamamoto M. K. (1997). Threshold odor concentration of MTBE and other fuel oxygenates. *American Chemical Society, Division of Environmental Chemistry*, Preprints of Extended Abstracts, **37**(1), 407–409.

Shepherd G. (2006). Smell images and the flavour system in the human brain. *Nature*, **444**(16), 316–321.

Shih T., Rong Y., Harmon T. and Suffet I. H. (2004). Evaluation of the impacts of fuel hydrocarbons and oxygenates to groundwater resources. *Environmental Science & Technology*, **37**(1), 42–48.

Slater G. P. and Blok V. C. (1983). Volatile compounds of the cyanophyceae – a review. *Water Science and Technology*, **15**(6/7), 181–190.

Smith A. and Noble A. (1998). Effects of increased viscosity on the sourness and astringency of aluminum sulfate and citric acid. *Food Quality and Preference*, **9**(3), 139–144.

Stackelberg P. E., Furlong E. T., Meyer M. T., Zaugg S. D., Henderson A. K. and Reissman D. B. (2004). Persistence Of pharmaceutical compounds and other organic wastewater contaminants in a conventional drinking-water-treatment plant. *Science of the Total Environment*, **329**, 99–113.

Stocking A., Suffet I. H., McGuire M. J. and Kavanaugh M. (2001). Implications of a MTBE consumer threshold odour study for drinking water standard setting. *Journal American Water Works Association*, **93**(3), 95–105.

Stratford M., Plumridge A. and Archer D. B. (2007). Decarboxylation of sorbic acid by spoilage yeasts is associated with the PAD1 gene. *Applied and Environmental Microbiology*, **73**(20), 6534–6542.

Suffet I. H. (2007). A re-evaluation of the taste and odour of MTBE (methyl tertiary butyl ether) in drinking water. *Water Science and Technology*, **55**(5), 265–273.

Suffet I. H. and Rosenfeld P. (2007). The anatomy of odour wheels for odours of drinking water, wastewater, compost and the urban environment. *Water Science and Technology*, **55**(5), 335–344.

Suffet I. H., Anselme C. and Mallevialle J. (1986). Removal of Tastes and Odours by Ozonation. Proceedings of the AWWA Seminar 'Ozonation, Recent Advances and Research Needs', Denver, CO, USA.

Suffet I. H., Brady B., Bartels J. H. M., Burlingame G., Yohe T. L. and Mallevialle J. (1988). Development of the flavor profile method into a standard method for sensory analysis of water. *Water Science and Technology*, **20**(8/9), 1–9.

Suffet I. H., Ho J. and Mallevialle J. (1993). Off-flavors in raw and potable water. In: Food Taints and Off-Flavors, M. J. Saxby (ed.), Blackie and Sons Ltd. (Chapman and Hall), London, UK, pp. 89–116.

Suffet I. H., Corado A. and Khiari D. (1995a). Brominated organic compounds and tastes and odours in water supplies. *Water Supply: The Review Journal of the International Water Supply Association*, **13**(1), 85–99.

Suffet I. H., Ho J., Chou D., Khiari D. and Mallevialle J. (1995b). Taste-and-odour problems observed during drinking water treatment. In: Advances in Taste-and-Odour Treatment and Control, I. H. Suffet, J. Mallevialle and E. Kawczyski (eds), AWWA Research Foundation and Lyonnaise Des Eaux Cooperative Research Report, American Water Works Assoc., Denver, Colo., USA, p. 120, ISBN 0-89867-744-0.

Suffet I. H., Khiari D. and Mallevialle J. (1996). Taste and odour problems in drinking water. In: Food Taints and Off-Flavors, M. J. Saxby (ed.), 2nd edn, Blackie and Sons (Chapman and Hall), London, UK, pp. 107–138.

Suffet I. H., Khiari D. and Bruchet A. (1999). The drinking water taste and odour wheel for the millennium: beyond geosmin and 2-methylisoborneol. *Water Science and Technology*, **40**(6), 1–13.

Suffet I. H., Schweitzer L. and Khiari D. (2004). Olfactory and chemical analysis of taste and odour episodes in drinking water supplies. *Reviews in Environmental Science & Bio/Technology*, **3**, 3–13.

Suffet I. H., Burlingame G. and Mackey E. (2008). Controlling taste and odour events for the city of Philadelphia USA: a case study. *Water Science and Technology: Water Supply*, **8**(2), 135–141.

Suffet I. H., Decottignies V., Senante E. and Bruchet A. (2009). Sensory assessment and characterization of odour nuisance emissions during the composting of wastewater biosolids. *Water Environment Research*, **81**(7), 670–679.

Suffet I. H., Wiesenthal K., Crozes G., Chen T., Atasi K. Z. and Bruchet A. (2011). Removal and Production of Tastes and Odours by Ozonation. IOA IUVA World Congress and Exhibition, Paris, France, May 23–27.

Suffet I. H., Leavey S., Everitt M. and Piriou P. (2013). Development of a Methodology to Identify Odour Objection and Odour Rejection Concentrations for Taste and Odour Compounds in Drinking Water. Proceedings of the AWWA Annual Conference and Exposition, Denver, CO, USA.

Sugiura N., Yagi O. and Sudo R. (1983). Effects of various environmental factors on musty odour production by Actinomycetes (Steptomycetes). Suishitsu Odaku Kenko 6 77-88, via Chem. Abs. 99:172570.

Tabachek J. L. and Yurkowski M. (1976). Isolation and identification of blue-green algae producing muddy odour metabolites, geosmin and 2-methylisoborneol, in saline lakes in Manitoba. *Journal of the Fisheries Research Board of Canada*, **33**(1), 25–35.

E. T. Thiemer (ed.) (1982). *Fragrance Chemistry: The Science of the Sense of Smell.* Academic Press, New York, NY, USA.

Tomboulian P., Schweitzer L., Mullin K., Wilson J. and Khiari D. (2004). Materials used in drinking water distribution systems: contribution to taste-and-odour. *Water Science & Technology*, **49**(9), 219–226.

Tsuchiya Y., Matsumoto A. and Okamoto T. (1978). Volatile metabolites produced by actinomycetes isolated from Lake Tairo at Miyakejima. *Yakugaku Zasshi*, **98**(4), 545–550.

Tsuchiya Y., Matsumoto A. and Okamoto T. (1981). Identification of volatile metabolites produced by blue-green algae, *Oscillatoria splendida, O. amoena, O. germinata* and *Aphanizomenon* sp. *Yakugaku Zasshi*, **101**(9), 852–856.

Tucker C. L., Ömür-Özbek P. and Dietrichm A. M. (2007). Determination of the taste threshold of iron in water. NSF-REU proceedings of research. In: 2007 NSF REU Annual Report, T. Youno, V. K. Lohani and M. K. Licher (eds), Virginia Tech, Blacksburg, VA, USA, pp. 9–14.

UK Water Industry Research Ltd. (UKWIR) (1996). Formation and Occurrence of Bromophenol, Iodo-phenols, Bromoanisoles and Jodoanisoles in Drinking Water: An Investigation of Taste and Odour Potential. Report DW-05/L3, UKWIR, London, UK.

USEPA (2007). National Secondary Drinking Water Regulations. 40 Code of Federal Regulations, Part 143. United States Environmental Protection Agency, Washington,

DC. http://www.epa.gov/dwstandardsregulations and https://www.gov/dwstandardsregulations/secondary-drinking-water-standards-guidance-nuisance-chemicals

USEPA (2010). National Primary Drinking Water Regulations. 40 Code of Federal Regulations, Part 141. United States Environmental Protection Agency, Washington, DC. http://www.epa.gov/dwstandardsregulations

van Breemen L. W. C. A., Dits J. S. and Ketelaars H. A. M. (1992). Production and reduction of geosmin and 2-methylisoborneol during storage of river water in deep reservoirs. *Water Science and Technolology*, **25**(2), 233–240.

van Wezel A., Puijker L., Vink C., Versteegh A. and de Voogt P. (2009). Odour and flavour thresholds of gasoline additives (MTBE, ETBE and TAME) and their occurrence in Dutch drinking water collection areas. *Chemosphere*, **76**, 672–676.

Ventura F., Matia L., Romero J., Boleda M. R., Marti I. and Martin L. (1995). Taste and odour events in Barcelona's water supply. *Water Science and Technology*, **31**(11), 63–68.

Ventura F., Romero J. and Pares J. (1997). Determination of dicyclopentadiene and its derivatives as compounds in groundwater supplies. *Environmental Science & Technology*, **31**, 2368–2374.

Volk C., Dundore E., Schiermann J. and LeChevallier M. (1967). Practical evaluation of iron corrosion control in a drinking waterdistribution system. *Water Research*, **34**(6), 1967–1974.

Wajon J. E., Alexander R. and Kagi R. I. (1985a). Determination of trace levels of dimethyl polysulphides by capillary gas chromatography. *Journal of Chromatography*, **319**, 187–194.

Wajon J. E., Alexander R. and Kagi R. I. (1985b). Dimethyl trisulfide and objectional odours in potable water. *Chemosphere*, **14**(1), 85–89.

Wajon J. E., Kagi R. I. and Alexander R. (1985c). The Occurrence and Control of Swampy Odour in the Water Supply of Perth, Western Australia. Report submitted to Water Authority of Western Australia. School of Appl. Chem., Western Australian Inst. of Technol., Bentley, Australia.

Wajon J. E., Kavanagh B. V., Kagi R. I., Rosich R. S. and Alexander R. (1988). Controlling swampy odours in drinking water. *Journal American Water Works Association*, **80**(6), 77–83.

Watson S. B., Brownlee B., Satchwill T. and Hargesheimer E. E. (2000). Quantitative analysis of trace levels of Geosmin and MIB In source and drinking water using headspace SPME. *Water Research*, **34**(10), 2818–2828.

Welte B. and Montiel A. (1999). Study of the possible origins of chlorinous taste and odour episodes in a distribution network. *Water Science and Technology*, **40**(6), 257–264.

Whelton A. J. (2009). Advancing Potable Water Infrastructure through an Improved Understanding of Polymer Pipe Oxidation, Polymer–Contaminant Interactions, and Consumer Perception of Taste. Dissertation submitted to the Faculty of the Virginia Polytechnic Institute and State University, Blacksburg, Virginia, USA.

Whelton A. J. and Dietrich A. M. (2004). Relationship between intensity, concentration, and temperature for drinking water odourants. *Water Research*, **38**(6), 1604–1614.

Whelton A. J., Dietrich A. M., Burlingame G. A., Schechs M. and Duncan S. E. (2007). Minerals in drinking water: impacts on taste and importance to consumer health. *Water Science and Technology*, **55**(5), 283–291.

White G. C. (1980). Handbook of Chlorination. Van Nostrand Reinhold, New York, NY, USA.

Whitfield F. B. (1988). Chemistry of off-flavours in marine organisms. *Water Science and Technology*, **20**(8/9), 63–74.

Whitfield F. B., Last J. H., Shaw K. J. and Tindale C. R. (1988). 2,6-Dibromophenol: the cause of an iodoform-like off flavor in some Australian crustacea. *Journal of the Science of Food and Agriculture*, **46**(1), 29–42.

Wiesenthal K. E. and Suffet I. H. (2007). The effect of white or grey PVC pipe and its joint solvents (primer and cement) on odour problems in drinking water distribution systems. *Water Science and Technology*, **55**(5), 169–176.

Wiesenthal K. E., McGuire M. J. and Suffet I. H. (2007). Characteristics of salt taste and free chlorine or chloramine in drinking Water. *Water Science and Technology*, **55**(5), 293–300.

Wilkins K. (1996). Volatile metabolites from actinomycetes. *Chemosphere*, **32**(7), 1427–1434.

Williams P., Benton L., Warmerdam J. and Sheehans P. (2002). Comparative risk analysis of six volatile organic compounds in california drinking water. *Environmetal Science & Technology*, **36**(22), 4721–4728.

Wood S., Williams S. T. and White W. R. (1983). Microbes as a source of earthy flavours in potable waters – a review. *International Biodeterioration Bulletin*, **19**(3/4), 83–97.

World Health Organization (1985). Guidelines for Drinking Water Quality. WHO, Geneva, Switzerland, Vol. **2**.

World Health Organization (2003). Iron in Drinking-Water. Background Document for Development of WHO Guidelines for Drinking-Water Quality. WHO, Geneva, Switzerland.

World Health Organization (2011). Guidelines for Drinking Water Quality, 4th edn. WHO, Geneva, Switzerland.

Worley J. L., Dietrich A. M. and Hoehn R. C. (2003). Dechlorination techniques used to improve sensory odour testing of geosmin and 2-MIB. *Journal American Water Works Association*, **95**(3), 109–117.

Wu J. T. and Juttner F. (1988a). Effect of environmental factors on geosmin production by *Fischerella muscicola. Water Science & Technology*, **20**(8/9), 143–148.

Wu J. T. and Juttner F. (1988b). Differential partitioning of geosmin and 2-methylisoborneol between cellular constituents in Oscillatoria tenuis. *Archives of Microbiology*, **150**(6), 580–583.

Yang M., Yu J., Li Z., Guo Z., Burch M. and Lin T. F. (2008). Letter: Taihu Lake not to blame for Wuxi's woes. *Science*, **319**(5680), 158.

Yano H., Nakahara M. and Ito H. (1988). Water blooms of *Uroglena americana* and the identification of odourous compounds. *Water Science and Technology*, **20**(8/9), 75–80.

Young C. and Suffet I. H. (1999). Development of a standard method-analysis of compounds causing tastes and odours in drinking water. *Water Science and Technology*, **40**(6), 279–286.

Young C., Suffet I. H., Crozes G. and Bruchet A. (1999). Identification of a woody-hay odour-causing compound in a drinking water supply. *Water Science and Technology*, **40**(6), 273–278.

Young W. F., Horth H., Crane R., Ogden T. and Arnott M. (1996). Taste and odour threshold concentrations of potential potable water contaminants. *Water Research*, **30**(2), 331–340.

Yu J., Li Z., Cao N., Yang M., Ding J., Miao T. and Zhang J. (2007). Analyses on cause for odour and potential problems in water source during odour episode event in Wuxi. *Acta Scientiae Circumstantiae*, **27**, 1771–1777.

Yu J., An W., Cao N., Yang M., Gu J., Zhang D. and Ly N. (2014). Quantitative method to determine the regional drinking water odourant regulation goals based on odour sensitivity distribution: illustrated using 2-MIB. *Journal of Environmental Sciences*, **26**, 1389–1394.

Zacarias I., Yanez C. G., Araya M., Oraka C., Olivares M. and Uauy R. (2001). Determination of the taste threshold of copper in water. *Chemical Senses*, **26**, 85–89.

Zamyadi A., Henderson R., Stuetz R., Hofmann R., Ho L. and Newcombe G. (2015). Fate of geosmin and 2-methylisoborneol in full-scale water treatment plants. *Water Research*, **83**, 171–183.

Zhang G. R., Kiene L., Chan U. S. and Duguet J. P. (1992). Evolution of organoleptic water quality during treatment in Macau. *Environmental Technology*, **13**(10), 947–958.

Zhang X. J., Chen C., Ding J. Q., Hou A., Li Y., Niu Z. B., Su X. Y., Xu Y. J. and Laws E. A. (2010). The 2007 water crisis In Wuxi, China: analysis of the origin. *Journal of Hazardous Materials*, **182**(1–3), 130–135.

Chapter 3

Biological production of taste and odour compounds

Susan Watson[1] *and Friedrich Jüttner*[2]

[1]*Department of Biology, University of Waterloo, Waterloo, Ontario, Canada*
[2]*Limnological Station, Department of Plant and Microbial Biology, University of Zurich, Kilchberg, Switzerland*

3.1 INTRODUCTION

Most reported taste and odour (T&O) events are biological in origin (e.g. Jüttner, 1995; Watson, 2003, 2010), and these are the primary focus of this chapter. Tracing the sources of biological T&O is often challenging, and requires an understanding of both abiotic and biotic factors contributing to these events in order to investigate and interpret them effectively. Taste-odour may be caused by one or more of the many volatile organic compounds (VOCs) that are often present in source waters at any given time, although most odour events are caused by a relatively small number of volatiles (Watson, 2010; Table 3.1). VOC potency is often measured by odour threshold concentration (OTC; the minimum concentration at which VOC odour is detectable to humans), which span several orders of magnitude, with some of the more potent of these compounds detectable at less than nanomolar concentrations. With biological T&O events, the 'critical cell densities' of organisms required to produce OTC levels of these metabolites also range considerably, since species, and even strains of the same species, differ in *per capita* production (Rashash *et al.*, 1995; Watson, 2003, 2010).

Editor(s): Tsair-Fuh Lin, Susan Watson, Andrea M. Dietrich and I. H. (Mel) Suffet
doi: 10.2166/9781780406664_0063

Table 3.1 Major biological VOCs identified from field and culture material.

VOC	OTC (ug/L)	Bio-activity*	Odour	Prokaryotes			Algae (Eukaryotes)						
				Actino	Myxo	Cyan	Chl	Diat	Chry	Syn	Dino	Cryp	Rho
AMINES and other N-METABOLITES													
1-Aminopropane (propylamine)	0.061		Fishy[†] → Ammonical			•	•	•	•				
2-Aminopropane (isopropyl amine)	0.025		Ammonia						•				
Butanolamine			Fishy[†]				•						
Ethanolamine			Mildly ammoniacal						•				
Methylamine	0.035		Fishy[†] → Ammonical					•					
Trimethylamine	0.00003		Fishy[†] → ammonical						•				
Indole (2,3-benzopyrrole)	0.02		Floral → fecal			•							
Skatole (-methyl-2,3-benzopyrrole)	0.0008		Floral → fecal			•							
2,6-Dimethyl pyrazine			Cocoa, roasted nuts, coffee					•					
3-Methoxy-2-isopropyl pyrazine			Earthy/potato bin					•					
2,5-Dimethylpyrazine				⊙									
Pyrazine				•									
Methylpyrazine				⊙									
2,5-Dimethylpyrazine				⊙									
Vinyl pyrazine				⊙									
2-Methoxy-3-(1-methylpropylpyrazine				⊙									

Sulfur compounds										
1,2,4 Trithiolane									•	
4-Methylthio-1,2-dithiolane		G; P	Pungent				•			
5-Methylthio,-1,2,3,-trithiane		G; P	Pungent				•			
Dihydrogen sulfide	0.004	T	Rotten eggs			•				
Dimethylsulfide	<4.0		Intense, sulfurous			•				
Dimethyldisulfide	0.12[a]		Septic, garlic, putrid	•		•	•	•	•	
Dimethyltrisulfide	0.01		Septic, garlic, putrid, swampy	•	•	•	•	•	•	•
Dimethyltetrasulfide			Putrid	•	•	•	⊙		•	•
Dimethylpentasulfide			Septic, garlic, putrid, swampy	•			⊙			
Methane thiol	0.00033					•				
Isopropyl thiol		P	Strong ripe onion sulfurous			•				
Isopropyl disulfide	0.000006		Alliaceous			•				
Isopropyl methyl disulfide			Alliaceous			•				
Isopropyl trisulfide			Alliaceous			•				
S-Methyl ethanethioate			Sulfurous egg cabbage	•					•	•
S-Methyl propanethioate									•	
S-Methyl 2-methylpropanethioate									•	
S-Methyl 3-methylbutanethioate									•	•
Methyl propyl disulfide			Foul; cooked onion			•				

(Continued)

Table 3.1 Major biological VOCs identified from field and culture material (*Continued*).

VOC	OTC (ug/L)	Bio-activity*	Odour	Prokaryotes			Algae (Eukaryotes)						
				Actino	Myxo	Cyan	Chl	Diat	Chry	Syn	Dino	Cryp	Rho
Terpenes & terpenoids													
Avermitilol				⊙									
Avermitilone				⊙									
γ-Cadinene				⊙		•							
1,2-Dihydro-1,1,6-trimethylnapthalene		G; R/K											•
2,2,6-Trimethyl cyclohexanone	310		Tobacco, musty			•							
2-Methyl-2-bornene				•									
2-Methylisoborneol	0.002–0.02	M	Earthy, musty	•	•	•							
Borneol				⊙									
6-Methyl-5-hepten-2-ol			Fruity; ester-like			•							
6-Methyl-5-hepten-2-one	50	G	Fruity; ester-like	⊙	⊙	•	•	•	•				
6-Methylheptan-2-one				⊙									
				•									•
6,10,14-Trimethylpentadecan-2-one (hexahydrofarnesyl acetone)			Oily herbal jasmine, celery woody				•						
β-Cyclocitral	19.3	G; L	Tobacco, floral, smoky, mouldy			•	•		•	•			

Camphor	1290		Aromatic, musty, woody			•					
α–, δ–, ε–, γ-Cadinene				⊙							
β-Cedrene				•							
α-Copaene				•							
1,8-Cieneol	12		Camphor, spicy, cool	⊙			•				
β-Farnesene				⊙							
Geosmin	0.006–0.01	M	Earthy,/musty	•	•	•					
Geranyl acetone	60	G	Flowery	⊙	⊙					•	
Geraniol (trans-3,7-dimethyl-2,6-octadien-1-ol)	0.05–0.75	G	Sweet, floral, fruity	⊙		•	•				
(–) Germacrene D			Grassy	⊙		•	•				
Germacradien-4-ol; Germacradien-11-ol				⊙	⊙						
α-Ionone	0.007		Violets				•				
β-Ionone	0.007	G	Violets			•	•		•	•	•
Epoxy-β-Ionone			Violets				•				
Limonene	4		Citrus	⊙	⊙		•	•			
Linalool	6		Grassy, floral	•			•				
α–, β–, γ– Muurolene				•							
Nerol (2-cis-3,7-dimethyl-2,6-octadien-1-ol)		G	Sweet, rose-like; bitter			•					
β-Myrcene				⊙		•					
α–, β–Pinene (+)				⊙							
Squalene	65		Faint, agreeable			•					
Styrene	65		Sweet, balsamic					⊙	•		

(*Continued*)

Table 3.1 Major biological VOCs identified from field and culture material (*Continued*).

VOC	OTC (ug/L)	Bio-activity*	Odour	Prokaryotes			Algae (Eukaryotes)						
				Actino	Myxo	Cyan	Chl	Diat	Chry	Syn	Dino	Cryp	Rho
OXYLIPINS and other Hydrocarbons													
3-Methyl butanal			Rancid, putrid				•		•				
2-Butanone	0.0054		Ethereal	⊙		•			•				⊙
3-Methylbut-2-en-1-ol	0.002–0.02		Alcoholic			•	•	•					
cis-2-Octene									•				
2,4-Decadienal	0.07–0.02	R, F, EC, G, T	Fishy[†], rancid					⊙	•	•	•	•	
2,4,7-Decatrienal	1.5	R, F, EC, G, T	Fishy[†], swampy					•	•	•			
Ethyl acetate	5.0		Pineapple, ethereal							•			
Dictyopterene C' (6-butyl-cyclohepta-1,4-diene)		P	Seaweed					•					
Dictyopterene A (trans-1-(trans-1-hexenyl)-2-vinylcyclopropane)		Ph	Seaweed					•					
Ectocarpene (6-(Z1-butenyl)-1,4-cycloheptadiene)		Ph	Tomato leaf like					•					
Ethyl propionate	0.001		Sweet, fruity, fragrant, ethereal			•							
1-Heptadecene							•						
Heptadec-cis-5-ene						•							
2,4-Heptadienal	5	T	Fishy[†]; cod-liver oil			•		•	•	•			
N-Heptane			Sweet-ethereal					⊙					

1-Heptanol		Fresh, light green, nutty		•					
2-Heptanol		Fruity-green, herbaceous					•		
2-Heptanone		Banana, slightly spicy	⊙				•		⊙
N-Hexanal	4.5	Grassy, fatty		•	•	•		•	
1-Hexadecane		Gasoline-like			•				
Hexanol-1	200	Grassy		•	•				
cis-3-Hexen-1-ol	70	Grass cuttings			•				
Isoamyl acetate (3-methyl-1-butyl acetate)								•	
Isobutyl acetate (2-methylpropyl acetate						⊙		•	
Isobutyl alcohol								•	
Isopropyl acetate								•	
Isopropyl alcohol							•	•	
Methyl acetate		Glue, nail polish remover						•	
Methyl butanol						•			
Methyl propionate		Pungent		•			•	•	
Methyl isobutyrate							•		
Methyl butyrate				•		•		•	
Methyl octadecanoate					•				
Methyl octadec-9-enoate					•				
2-Hexanone			⊙						
3-Hexanone			⊙						

(Continued)

Table 3.1 Major biological VOCs identified from field and culture material (*Continued*).

				Prokaryotes			Algae (Eukaryotes)						
VOC	**OTC (ug/L)**	**Bio-activity***	**Odour**	**Actino**	**Myxo**	**Cyan**	**Chl**	**Diat**	**Chry**	**Syn**	**Dino**	**Cryp**	**Rho**
4-Methyl-2-hexanone				⊙									
5-Methyl-4-hexen-3-one				⊙									
3-Methyl -1- butanol (isoamyl alcohol)			Smoky, mouldy; fusel oil			•			•				
3-Methylbutan-2-one									•				
3-Methylbut-2-enal			Rancid; putrid						•				
2-Methylbut-2-en-1-ol									•				
Methyl octadec-9,12-dienoate							•						
2-Methylpent-2-enal	290		Green apples				•						
2-Methylpropan-1-ol			Penetrating, winey-sweet, alcoholic							•			
N-Octadecane			Sweet, fruity, fusel-like			•			•				
N-Nonadecane			Faint leather note					•		•			
2,4-Nonadienal		T								•			
2,6-Nonadienal	0.08	T	Grassy; cucumber → rancid				•			•			
3-Nonenone				⊙									
1,3-Octadiene			Earthy, mushroom; humus			•		•					
Octan-1-ol			Sweet, fresh, orange-rose			•		⊙	•	•			

Octan-3-one			Strong, fruity, lavender			•				•	⊙
Octadiene	5600		Geranium					•			
1,3,5-Octatriene (fucoserratene)		Ph	Geranium			•		•			
2,4,6-Octatriene						•		•			
2-Octenal			Green-leafy, orange, honey			•			•	•	
Oct-1-ene			Gasoline-like					⊙	•		
Octan-3-ol	0.0012		Herbaceous			•					
Oct-2-en-1-ol	0.04[b]		Green			•			•		
Oct-1-en-3-ol			Sweet, earthy, herbaceous, mushroom						•		
Oct-1-en-3-one			Mushroom			•		⊙		⊙	
3-Octanone			Strong, fruity, lavender	⊙							
7-Methyloctan-3-one					•				•		
Pentanal			Slightly fruity, nutty → acrid, pungent					•			
Methyl pentanoate			Green-fruity, apple-pineapple							•	
1-Pentanol			Harsh, faint fusel oil			•					
2-Pentenal						•			•		
Pent-1-en-3-ol						•					
Pent-1-en-3-one			Pungent				•	•		•	

(Continued)

Table 3.1 Major biological VOCs identified from field and culture material (*Continued*).

VOC	OTC (ug/L)	Bio-activity*	Odour	Prokaryotes			Algae (Eukaryotes)						
				Actino	Myxo	Cyan	Chl	Diat	Chry	Syn	Dino	Cryp	Rho
3-Methyl-2- pentanone					⊙								
Propenal			Acrid, burnt fat					⊙		•			
Undecan-2-one			Sweet. Slightly peachy						•				
Fermentation products													
4-Methylpent-3-en-2-one						•	•						
2-Propanone (acetone)			Acetone		•	•						•	
Propyl acetate (propyl ethanoate)			Pears					⊙		•			
Methyl butyrate				•					•				
Methyl-3-methyl-2-butenoate				•					•				
2-Pentylfuran			Buttery, green bean							•			
2-Acetylfuran				⊙	⊙								
3-Methylfuran				⊙									
2,3-Dimethyl-tetrahydrofuran				•									
Bis-3,4-hydroxymethylfurfural				•									
Furfuryl isobutyrate				⊙									
Dimethyl 2,4-furan-dicarboxylate				•									

Methyl-2-furan-carboxylate			•		
2,3-Dimethyl-tetrahydrofurane			⊙		
Benzaldehyde	0.5	Almond			•
Benzyl acetate		Jasmine		•	
Benzyl alcohol		Sharp burning taste, faint aromatic		•	
2-Phenylphenol				•	
Phenethyl alcohol (benzyl carbinol)		Rose		•	
2-Phenylphenol				•	
p-Cresol		Medicinal		•	

Notes: OTC = odour threshold concentrations (in water); Actino = Actinomycetes; Myxo = Myxobacteria; Cyano = Cyanobacteria; Chl = Chlorophyta (green algae), Diat = Diatoms; Chry = Chrysophyceae; Syn = Synurophyceae, Dino = Dinophyta, Cryp = Cryptophyta, Rho = Rhodophyta. Closed circles (•) = known production; open circles (⊙) = inferred production. Arrows indicate change in odour characteristics between low and high concentrations.

[a] http://www.aroxa.com/water

[b] http://www.leffingwell.com

*Bioactivity: mutagen (M), Kairomone (K), Pheromone (Ph); grazer Repellent (R); protectant (P: temperature, salinity, oxidants), cell Lysin (L), Toxin (T); Abnormal/reduced: Growth (G), Egg Cleavage (EC), Fecundity (F), Larval development (L); (see also Table 3-2).

[†] 'Fishy' odour descriptor can be ambiguous; amines typically have an odour resembling 'old' herring or fish-markets while polyunsaturated aldehydes (PUAs) are redolent of aged or heated oily fish such as mackerel.

Planktonic algal communities are a frequent source of T&O and often the first to be investigated, although all organisms may indirectly contribute to T&O during their death and bacterial degradation (Ma *et al.*, 2013a; Watson, 2010). Plankton communities are often highly diverse, and the predominant species in a given assemblage may not necessarily be the main sources of odour. The general tendency to focus on plankton overlooks the often significant odour potential of less visible sources such as deep-living populations, or species in benthic mats and littoral biofilms (Izaguirre, 2007; Ridal *et al.*, 2007). This is an essential consideration for rivers, shallow lakes and impoundments, where the more stable VOCs can be transported from remote (e.g. upstream) or hidden sources over considerable distances by water movement to shorelines and intakes (Figure 3.2; e.g. Bruder *et al.*, 2014; Burlingame *et al.*, 1992). Non-algal taxa (e.g. fungi, bacteria, and zooplankton) also produce VOCs as cell metabolites and/or as a result of other activities (e.g. degradation of organic material, grazing activity).

A number of factors may affect the levels and types of VOCs identified in the lab and the odours perceived in the finished water, which can impede successful source tracking. T&O events often remain untraced due to their complexity and ineffective sampling and diagnostic/analytical procedures. Some of the key factors that complicate source-tracking include: (a) non-biological sources of T&O; (b) VOC stability and transport away from sites of origin; (c) incomplete knowledge of confirmed odour producers; (4) species misidentification; and (5) variation in VOC output by cells at different growth stages or under different environmental conditions. Weather, basin morphometry, thermal regime and hydrology influence the dominant biological sources and the eventual fate of these compounds (e.g. Bruder *et al.*, 2014; Newcombe *et al.*, 2009; Watson *et al.*, 2007a). The timing and severity of T&O events experienced by water treatment plants (WTPs) can also be affected by the position of raw water intakes relative to surface or deep-living VOC producers (Figure 3.3). In addition, the VOC concentrations and chemistry in samples collected from both source waters and WTPs are affected by the site location(s) and sample depth (s), the timing and frequency of sampling, the species present and their stage of growth and/or decay, the length and conditions of sample storage, and the analytical methods and treatment processes used (Chapters 4 to 8). Prolonged or inappropriate storage of unprocessed raw samples between sampling and analyses can result in changes in VOC concentrations through lysis and degradation of cyanobacterial and algal cells, headspace loss, photo-oxidation, or chemical/biological activity. Some VOCs have different perceived odour characteristics at different concentrations, which can lead to the conclusion that several compounds and/or species are involved over the course of a T&O event. For example, *trans*-2-*cis*-6-nonadienal has a 'sweet, green cucumber' odour at low concentrations and a 'waxy, oily, rancid' smell at higher levels

(Burlingame *et al.*, 1992). VOC mixes may act synergistically to produce a 'bouquet', or mask the odours of individual odour components in a sample (Burlingame *et al.*, 2007; Guo *et al.*, 2018; Höckelmann *et al.*, 2004; Marsili & Laskonis, 2014; Miyazawa *et al.*, 2008; Wise *et al.*, 2007). Photoisomerization, and/or the activation of catalytic cell enzymes (e.g. lipoxygenases) during cell rupture (from natural or treatment-based mechanisms) may release, alter or induce VOC production in the final sample (e.g. Jüttner, 1995; Yano *et al.*, 1988). As such, some of the processes used to treat drinking water may actually intensify T&O problems in the finished water, for example, the application of algicides to control blooms, which can release intracellular VOCs (and toxins) and induce VOC production from catalytic enzymes. Filter bleed and biotic activity associated with filter beds and distribution systems may also introduce organoleptic compounds into the finished water (Jüttner & Watson, 2007; Karlsson *et al.*, 1995; Ridal *et al.*, 2001). Last, but not least, anthropogenically-derived organoleptic contaminants from disinfection by-products, solvents, columns, etc., or pollutants (e.g. halogenated and aromatic compounds and alkanes such as undecane, nonane, decane, benzothiozole; Jüttner, 1983, 1992), and VOCs produced as analytical artefacts (e.g. Watson & Jüttner, 2017) need to be distinguished from those of biogenic origin.

An extensive list of VOCs has been attributed to biological sources (see e.g. Watson, 2003, 2010 and Table 3.1), although only a few of these compounds are frequent T&O sources. Some VOCs have been attributed to the dominant species present in source-waters during a T&O event but not verified (see below), or identified from non-axenic cultures where they may have been produced by bacterial or fungal activity (e.g. Jüttner, 1995). Current techniques used to identify the biological source(s) of T&O are often inconclusive. Microscopy is commonly used to monitor source waters for cyanobacterial and algal producers but not applicable to cases of actinomycete (or other bacterial/fungal) impairment and, at best, yields tentative diagnostics that need to be verified using other methods (see below). This approach is more successful with *a priori* knowledge of likely producers based on historical data from the source-water (if recorded) or other studies. Expertise in microscopic identification is both scarce and inconsistent. Conventional Historically, species identification has been based largely on morphological traits, but these can vary with environmental conditions, growth phase (and preservative used). Modern revised systematics now adopt a 'polyphasic approach' which integrates morphology with other characteristics using molecular and biochemical diagnostics (e.g. Johansen & Casamatta, 2005; Komárek & Mareš, 2012), but the considerable lag in the adoption of these revised systematics by most workers can lead to confusion in taxonomic identities amongst earlier and later reports (e.g. Komárek, 2002). For example, *Anabaena* has now been reclassified into the aerotype ('gas-vesicle')-forming planktonic subgenus *Dolichospermum* and the benthic morphotypes *Anabaena*;

Oscillatoria as the aerotype-forming *Planktothrix*, *Planktothricoides* and benthic *Oscillatoria* etc. (Komárek, 2003; Komárek & Komárková, 2004; Komárek & Zapomelová, 2007, 2008).

Traditional approaches to identify the biological sources of T&O have often been based on the method outlined by Persson (1983), notably: (i) co-occurrence of a given species and VOC in a source-water; and (ii) demonstrated production of this VOC by cultured isolates of that species from the same field samples. However, few agencies have the resources to carry out this procedure, which has a number of limitations. It cannot be applied for rapid diagnostics and requires different protocols for different organisms such as cyanobacteria or actinomycetes. There can be considerable variance in production among different isolates of a morphospecies – even within the same sample – while culture conditions (e.g. growth media, light etc.) may not be conducive to T&O production or robust growth. Factors such as nutrients, light, temperature, and bacterial interactions moderate both the production and loss of biogenic VOCs in source waters (see e.g. Watson, 2010, and references therein). This suggests that the genetic capacity for production of different odour compounds may be widespread but varies in the degree to which it is expressed (e.g. as may be the case with geosmin and 2-MIB; see below), thus a potential producer cannot necessarily be verified by presence/absence of detectable VOCs in cultures strains. Given these limitations, there has been an increasing interest in the development of more sensitive, powerful, and rapid diagnostic tools such as genetics, that can be applied to T&O and other biologically produced impairments (e.g. toxins) by the water industry and other agencies. Nevertheless, in most cases these techniques rely on some preliminary diagnostics to identify candidate VOCs and biota, as outlined below.

It is useful to consider biogenic VOCs according to their mode of production and when they are produced and excreted during the growth cycle, which provides insight into the source biota, timing and longevity of T&O events (e.g. Watson, 2010; Table 3.2). Biogenic T&O can be produced by: (I) active production of secondary metabolites (as intra- and extracellular fractions over growth (terpenoids, thiols, esters, short chain hydrocarbons); (II) induced production triggered by activation of catabolic enzymes at cell damage and/or death by external agents (e.g. oxylipins, nor-carotenoids); (III) non-specific VOC release from (all) decaying biomass by heterotrophic bacteria, fungi and other microorganisms (e.g. methyl sulfides, fermentation products).

There is some overlap among these categories, for example as seen with fermentation products and nor-carotenoids. Both phototrophs (cyanobacteria, eukaryotic algae, liverworts and some land plants) and heterotrophs (notably actinomycetes and other bacteria; fungi) produce VOCs by the above mechanisms. The general characteristics of each of these modes of production are discussed below with reference to specific compounds that are most problematic as T&O sources.

Table 3.2 General modes of biological VOC production, source biota, triggers/controls, biochemistry, known bioactivity and environment(s).

Mode	I	II	III
Biosynthesis	During active growth	Only at cell damage or death	Only at cell degradation
Producers	Bacteria (myxobacteria, actinomycetes), Cyanobacteria, Rhodophyta, Chlorophyta, moulds, bryophytes	Cyanobacteria, Chrysophyta (diatoms, Chrysophyceae, Synurophyceae), Cryptophyta, Dinophyta Chlorophyta, Prymnesiophyta, Pheophyta	Most dead organic material derived both from T&O-producing and non-producing taxa and other (e.g. terrestrial)
Production	Continuous (cell-bound or excreted)	Products of catalytic enzyme activity; activated by damage	Catalytic enzyme products
Triggers	Internal (metabolic) or external	Exogenous (grazers, treatment, environmental stress)	Heterotrophic microbial activity
VOCs	Biogenic sulfides; terpenoids	Oxylipins, nor carotenoids	Hydrocarbons, amines, terpenes sulfides etc.
Bioactivity	Protectants (temperature, salinity, ROS[‡]), growth inhibitors, attractants, deterrents, carcinogens	Pheromones, grazer deterrents, teratogens, toxins	Toxins
Environ	Benthic and planktonic; biofilms; terrestrial (soil microbiota) secondary treatment basins; biologically active water treatment filter beds	Benthic and planktonic; biofilms	Shorelines, beaches, benthos, rivers, stagnant waters, swamps, wastewater effluent.

[‡]Reactive oxygen species (after Watson, 2010; 2012 and cited references therein).

3.2 VOCs PRODUCED AS SECONDARY METABOLITES DURING GROWTH (TERPENOIDS, THIOLS, ESTERS, SHORT CHAIN HYDROCARBONS)

These odour-causing compounds are usually produced during active cell metabolism, although in many cases their production is heightened with the loss of cell membrane integrity, with the release of intracellular VOCs or the up-regulation of enzyme-catalysed VOC production. The biosynthesis and internal storage and release of different compounds by individual cells vary significantly among species and with environmental conditions that regulate growth, induce stress and/or modify extracellular yield. These moderating factors are, for the most part, poorly understood, but recent advances in analytical and molecular methods have provided important insight into this variance, which may eventually allow more proactive control of T&O events. The most problematic T&O compounds in this group are terpenoids, nor-carotenoids and thiols, as discussed below.

3.2.1 Terpenoids

Terpenoids (isoprenoids) are a structurally diverse class of compounds composed of modified isoprene units. They include pigments and important mono- and sesquiterpenoids produced by algae, cyanobacteria and other microorganisms.

3.2.1.1 Geosmin and 2-MIB

3.2.1.1.1 Occurrence

Geosmin (trans-1,10-dimethyl-trans-9-decalol) and 2-MIB (2-methylisoborneol) are widespread and primary causes of off flavours in water, fish and other foods and the subject of an ever-increasing number of studies, particularly in the last decade (Figure 3.1). These two important terpenoids are produced by select species of heterotrophic bacteria (notably Actinobacteria and Myxobacteria), Cyanobacteria and fungi, bryophytes and a small number of plant and protozoan taxa (e.g. Izaguirre, 2007; Jüttner & Watson, 2007; Silvey & Roach, 1953; Watson, 2010; Zaitlin *et al.*, 2003a; Zaitlin & Watson, 2006; Table 3.1). Geosmin and 2-MIB are non-toxic to human and most aquatic biota at the concentrations found in most open waters (e.g. Burgos *et al.*, 2014a, b; Watson, 2012), and non-essential for growth, but may function at the foodweb levels in a number of semiochemical interactions (e.g. Nakajima *et al.*, 1996; Ogura *et al.*, 2000; Tosi & Sola, 1993; Watson *et al.*, 2007b).

3.2.1.1.2 Biological sources

Cyanobacteria are generally considered to be the main sources of geosmin and 2-MIB in surface waters, particularly in eutrophic systems where cyanobacterial blooms are common, and the risk of severe odour outbreaks is high (Davies *et al.*, 2004; Jüttner & Watson, 2007; Olsen *et al.*, 2016; Parinet *et al.*, 2013; Smith *et al.*, 2002; Watson

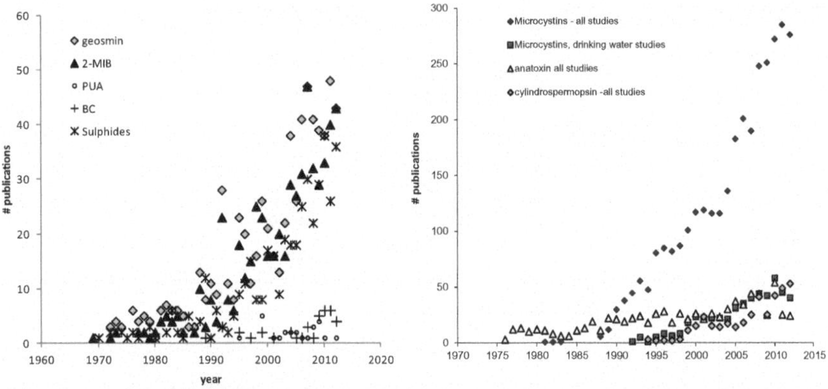

Figure 3.1 Number of publications in the refereed literature since 1960 focussed on (right) cyanobacterial toxins and (left) major biogenic T&O compounds (geosmin, 2-MIB, polyunsaturated aldehydes (PUA), β-cyclocitral (BC) and biogenic sulfides based on a literature search from major bibliographical databases (PubMed©, Scopus©).

et al., 2015). However, high nutrient levels are not always a prerequisite for geosmin/2-MIB related T&O events, and both compounds can be problematic in a variety of lakes, rivers and reservoirs (e.g. Dzialowski *et al.*, 2009; Sugiura *et al.*, 2004; Watson *et al.*, 2007a, 2008; Yu *et al.*, 2009). Among the cyanobacteria surveyed to date, geosmin production has been almost exclusively associated with planktonic and benthic filamentous and pseudo-filamentous taxa, which include nitrogenfixers and non-fixers (e.g. *Anabaena, Aphanizomenon, Dolichospermum, Fischerella, Geitlerinema, Planktothrix, Lyngbya, Oscillatoria, Phormidium*). More recently, however, Godo *et al.* (2011, 2017) reported production by the colonial species *Coelosphaerium* sp. (Synechoccoccales) isolated from Lake Shinji, Japan. 2-MIB has been reported from both filamentous and (a few) coccoid taxa, but to date, has not been detected in N-fixing cyanobacteria. A few known cyanobacteria (e.g. *Oscillatoria tenuis*; syn *Phormidium neotenue*) produce both compounds (Wu & Jüttner, 1988).

The importance of other microbial producers such as fungi and actinomycetes in geosmin and 2-MIB events is still debated and often specific to the type of source water (e.g. Zaitlin *et al.* 2003b; Asquith *et al.*, 2013). Fungi and moulds such as *Aspergillus, Basidiobolus, Chaetonium* and *Penicillium* can produce a range of volatiles including geosmin and 2-MIB, and are commonly found in drinking water distribution systems, but the significance of their contribution to T&O is not well characterised (Hageskal *et al.*, 2006; Kelley *et al.*, 2003). Actinomycetes (notably species of *Streptomyces*) also produce these and other VOCs, and there is evidence that in some cases these organisms may contribute to source-water odour, for example during runoff events, in aquaculture facilities or where there

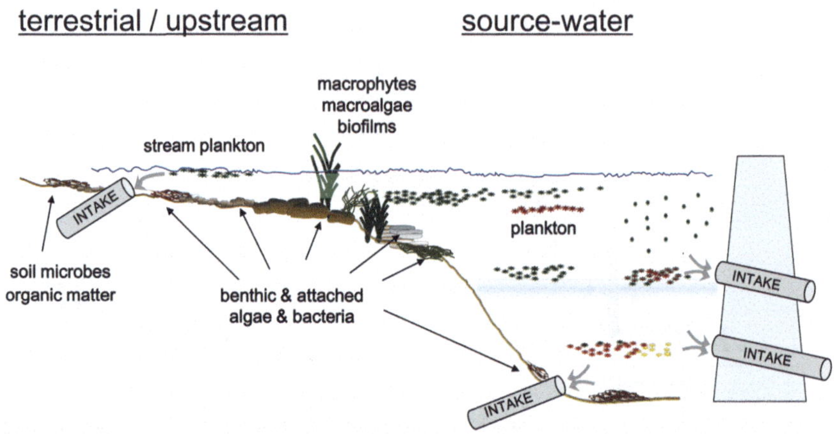

Figure 3.2 Schematic of different potential biological sources of T&O in source-water (plankton, benthos, attached algae etc.) relative to common locations of raw water intakes.

is significant water-sediment or shoreline interaction (Jensen *et al.*, 1994; Klausen *et al.*, 2004, 2005; Jorgensen, 2005; Park *et al.*, 2016; Wood *et al.*, 1985). Geosmin/2-MIB producing strains of *Streptomycetes* (*S. coelicolor; S. avermitilis* are facultative or obligate aerobes, which may limit their growth and/or T&O production in some of the environments where they are found (e.g. sediments; Van Keulen *et al.*, 2007; Zaitlin & Watson 2006). Traditional methods indicate that in most source waters, culturable, odour-producing strains are not abundant in open water, but prevalent in sediments (e.g. Zaitlin *et al.*, 2003a,b; Zuo *et al.*, 2010). Such studies have reported significant geosmin and/or 2-MIB production by actinomycetes from lake sediment samples incubated under aerobic conditions in sediments or with various substrates or media (Sugiura *et al.*, 1987, 1994; Sugiura & Nakano, 2000; Zuo *et al.*, 2009), although production was, in general, minimal under the low temperatures typical of many sediment surfaces in deeper stratified lakes. However, the extent to which these bacteria are in fact growing and metabolically active in sediments or largely present as dormant spores remains unresolved, particularly in eutrophic systems where sediment-water interface is often anoxic (e.g. Asquith *et al.*, 2013).

3.2.1.1.3 Production

Early mechanistic and labelling work with cyanobacteria suggested geosmin and 2-MIB as overflow or by-products of pigment synthesis, but reported inconsistent relationships with pigment content, which were modified by temperature and growth phase (e.g. Bafford *et al.*, 1993; Utkilen & Froshaug, 1992; Zhang *et al.*, 2009). Later work indicated a much more fundamental link between 2-MIB and cell energetics (Zimba *et al.*, 1999). In a recent study Gust *et al.* (2003) suggested

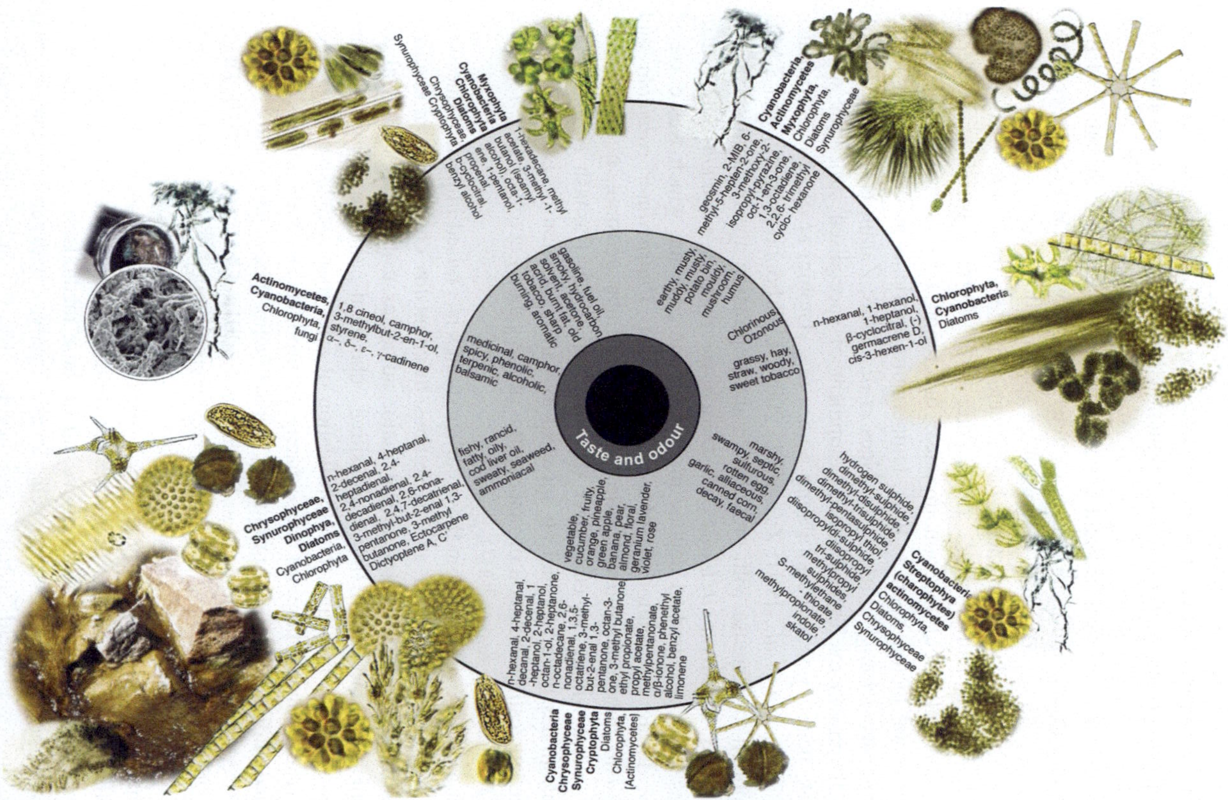

Figure 3.3 Biological Taste-and-Odour Wheel. Candidate source biota known to be associated with characteristic odours and VOCs in perimeter text, along with representative images (*Note*: non-biological 'Taste only' categories in original T&O wheel (Chapter 2) not included).

an important feedback between these morphological traits and the genes involved in geosmin production (see below). Both the nature and importance of these apparent links require further elucidation.

Culture- and field-based studies have observed significant variance in intra- and extracellular geosmin and 2-MIB production by different taxa, with important implications for treatment and control. Among cyanobacteria, both isoprenoids are associated with thylakoid and cytoplasmic membrane proteins, but 2-MIB is less closely bound and more readily excreted than geosmin (Wu & Jüttner 1988). Differences in extra- and intracellular production over growth in both actinomycetes and cyanobacteria have been reported from both field and lab populations (e.g. Durrer *et al.*, 1999; Peter *et al.*, 2009; Rashash *et al.*, 1996) while total cell yield in actinomycetes, fungi and cyanobacteria varies with environmental factors such as light, temperature, nutrients and stressors such as heavy metals (Dionigi *et al.*, 1990; Naes *et al.*, 1988; Zaitlin & Watson, 2006), and with growth cycles and the development of specialised structures such as akinetes, heterocysts and sporulating bodies (e.g. Dionigi *et al.*, 1992; Komatsu *et al.*, 2008; Wu *et al.*, 1991; Zuo *et al.*, 2009). Cell-bound geosmin and 2-MIB are reported to decrease with growth rate and population age for some taxa, e.g. studies with *Anabaena circinalis* (syn. *Dolichospermum sigmoideum*) and *A. laxa* cultures showed a gradual decline in intracellular geosmin, which was excreted mainly during the stationary phase. In contrast, *Phormidium calcicola* and *Fischerella muscicola* showed little change in cellular geosmin, most of which was retained within the cells over time (Jüttner, 1995; Rashash *et al.*, 1996). Several studies have reported an apparent loss in production by isolates held in culture, suggesting mutation or inactivation of the gene(s) involved (e.g. Izaguirre 2007; Jüttner & Watson, 2007).

The accrued evidence from these earlier studies has raised important questions about the key endogenous and exogenous mechanisms controlling production, providing the basis for some recent major advances using new analytical and molecular diagnostics. Highly sensitive chemical analyses, isotopic and fluorescence labelling, and the identification of a far more extensive list of biota (notably cyanobacteria) with the capacity to produce geosmin and 2-MIB has provided information on producers, biosynthetic rates and pathways and field-based modelling studies (e.g. Bruder *et al.*, 2014; Chen *et al.*, 2010; Davies *et al.*, 2004; Dzialowski *et al.*, 2009; Peter *et al.*, 2009; Qi *et al.*, 2012; Watson *et al.*, 2007a). Efforts to use fluorescence *in situ* hybridization (FISH) and immunofluorescence techniques are limited by the non-specificity of these methods, which detect both producers and non-producers (Klausen *et al.*, 2005; Nielsen *et al.*, 2006). As noted above, all of these methods are largely unable to elucidate production at the cell level, resolve between producers and non-producers in field samples and account for the substantial variance in production among and within species, highlighting the need to combine these traditional approaches with newly advanced molecular methods such as

Denaturing Gradient Gel Electrophoresis (DGGE), cloning and recombinant DNA technology, quantitative and real time Polymerase Chain Reaction (PCR), metagenomics and bioinformatic analysis (for example, Auffret *et al.*, 2011; Kutovaya & Watson, 2014; Otten *et al.*, 2016; Suurnäkki *et al.*, 2015; Tsao *et al.*, 2014; Wang *et al.*, 2011, 2016; Watson *et al.*, 2016).

3.2.1.1.4 Biosynthetic pathways and genetic control

The basic building blocks for all terpenoids are the two C_5 isoprenes, isopentenyl diphosphate (IPP) and dimethylallyl diphosphate (DMAPP). Changes in central cell carbon processing and IPP/DMAPP supply, and the requirement for other terpenoid products (e.g. pigments) will affect the yield of terpenoid secondary metabolites such as geosmin and 2-MIB, which show strong but seemingly disparate links with cell pigment content as noted above (e.g. Jüttner & Watson, 2007; Giglio *et al.*, 2011). Isoprenoids are derived from products of primary carbon metabolism (notably acetyl-CoA and pyruvate) by two major independent biosynthetic routes, the MVA (mevalonate) and MEP (methylerythritol phosphate) pathways (e.g. Jüttner & Watson, 2007). Over the past decade there have been significant advances in the elucidation of the genes and metabolic pathways regulating the production of these two terpenoids in both cyanobacteria and actinomycetes (e.g. Watson *et al.*, 2016, and references therein). This knowledge may provide some insight into the variance in the capacity to produce these compounds, questions which require research into the role of these VOCs in intra- and extracellular signalling (i.e. why they are produced) coupled with molecular approaches to elucidate differences in genomic sequences and factors which induce or up-regulate expression.

3.2.1.2 Geosmin

Geosmin is an irregular sesquiterpenoid (i.e. derived from 3 isoprene units), and was first identified from actinomycetes in the early 1960s. More recently, the structure, biosynthesis and genetic regulation of geosmin production were also largely elucidated from these taxa by a series of studies with *Streptomyces* (*S. tendae, S coelicolor* A3 and *S. avermitilis*) which combined ^{2}H-labelled precursors, genetics and other methods (e.g. Cane *et al.*, 2006; Cane & Watt, 2003; Dickschat *et al.*, 2007; Dionigi *et al.*, 1990; Gust *et al.*, 2003; Jiang *et al.*, 2006). These authors have shown that geosmin is produced from the universal sesquiterpene precursor farnesyl diphosphate (FPP) in a two-step Mg^{2+}-dependent reaction, yielding geosmin, (4S,7R)-germacra-1(10)-diene-11-ol(3), and the bicyclic hydrocarbons germacrene D(4), and octalin-5. Research with *S. coelicolor* showed that this two-step cyclization is catalyzed by a single bi-functional enzyme germacradienol/geosmin synthase (geoA), encoded by the single SCO6073 (*cyc2*) (Cane & Watt, 2003; Gust *et al.*, 2003; He & Cane, 2004). Enzyme and gene sequences (e.g. identified in *Streptomyces avermitilis* (*geoA*), *S. peucetius* (*Spterp13*)) are highly conserved in myxobacteria and other *Streptomyces* strains (e.g. Auffret *et al.*, 2011; Cane *et al.*, 2006; Ghimire *et al.*, 2008), and subsequent studies with cyanobacteria clearly demonstrated the presence of homologous geosmin synthases and genes in strains of the species *Nostoc punctiforme,* and the three genera *Phormidium, Anabaena,* and *Geitlerinema* (Giglio *et al.*, 2008). This gene has two catalytically independent N- and C-terminal domains (with germacradienol/germacrene D and geosmin synthase activity, respectively). Knockout experiments pointed to mutation as a possible explanation for loss of production by strains and reported variation in the relative yield of geosmin to germacradienol (Giglio *et al.*, 2008). Intriguingly, work by Gust *et al.* (2003) with *Streptomyces* suggested a link between sporulation and the genes involved in geosmin production, consistent with earlier studies showing higher production in sporulating vs. non-sporulating cultures (Dionigi *et al.*, 1992), but this link has not, to our knowledge, been followed up.

This new knowledge has led to the development and application of molecular probes to identify the presence and expression of geosmin synthase genes in lab and field populations and identify the biological sources, with an increasingly advanced array of techniques. Molecular studies have already provided significant insight into the genetic and environmental regulation of geosmin production and possible evolutionary relationships among producers, and identified unresolved issues where work is still required (see reviews by Asquith *et al.*, 2013; Watson *et al.*, 2016). In an early study, DGGE was applied successfully in conjunction with 16SrRNA to resolve among species of *Anabaena* and other taxa as major geosmin producers in Lake Ontario, demonstrating the potential value of this method to resolve among homologous taxa (Gill *et al.*, 2006). Ludwig *et al.* (2007) identified putative *geoA1* and *geoA2* gene sequences and expression in *Phormidium* sp. isolated from a reservoir subject to geosmin events, and applied this to test for the prevalence of homologous sequences in environmental samples. Giglio and co-workers developed and applied a PCR-based probe (combined with GC-MS validation) to screen a series of cyanobacterial isolates and field samples for geosmin synthase genes; these authors subsequently applied qPCR to study the expression of this gene relative to changes in intra- and extracellular production by *Anabaena circinalis* AWQC318 under different growth regimes (Giglio *et al.*, 2008, 2011). A molecular probe based on taxon-specific PCR and RT-PCR assays was developed and applied to cyanobacteria and actinomycete cultures and open water, benthic mat and sediment samples from the Great Lakes to identify both the presence and expression of geosmin synthase genes and the associated taxa. These authors established a phylogenetic tree showing homologies among different cyanobacterial and actinomycete groupings, providing insight into within and among species mutations (Kutovaya & Watson, 2014). The combined evidence from these studies indicates that geosmin synthase is constitutively expressed, but may not yield geosmin as an end product if there are mutational changes in the C-terminal domain.

3.2.1.2.1 2-Methylisoborneol

2-MIB is an irregular monoterpenoid (i.e. a methylation product derived from two isoprene units). The biochemistry, enzymes and genes involved in synthesis were also first elucidated from actinomycetes, although studies with other taxa have since demonstrated the universality of this pathway. As with geosmin, 2-MIB synthesis also involves a two-step process, but differs in that this process is regulated by a multi-gene operon. An early study by Bentley and Meganathan (1981) reported that 2-MIB produced by *Streptomyces* is derived from the universal monoterpene precursor geranyl pyrophosphate (GPP). Later work with the myxobacterium *Nannocystis exedens* showed that GPP undergoes S-adenosylmethionine (SAM)-dependent methylation to 2-methyl-GPP, with subsequent cyclization to 2-MIB (Dickschat *et al.*, 2007). Working with *Streptomyces coelicolor*, Wang and Cane (2008) showed that this process is catalyzed by two enzymes: geranyl diphosphate 2-methyltransferase (GPPMT) first catalyses the SAM-dependent methylation of GPP, which is then cyclised by 2-MIB synthase (MIBS). These authors identified two genes in the operon controlling the two-stage MIB biosynthesis, a methyltransferase gene *sco7701* and a monoterpene cyclase gene *sco7700*. Komatsu and co-workers (2008) independently confirmed this operon in six other actinomycete strains using both PCR and bioinformatic analysis, and by deleting part of this operon showed a loss of 2-MIB synthesis, pointing to mutation as a potential cause of loss in 2-MIB production by long-held culture strains.

Using whole genome sequencing, Giglio and co-workers (2011) compared these bacterial 2-MIB producers with three cyanobacteria (*Pseudanabaena limnetica* (from Castaic Lake and Lake Biwa) and *Oscillatoria limosa*) and identified orthologous MIBS and GPPMT enzymes in cyanobacteria and a third nucleotide-binding protein gene (*cnb*) in the operon of uncertain function. Significantly, they identified a sequence reversal in the cyanobacterial MIB operon compared to the majority of actinomycetes, and using a phylogenetic analysis of MIB synthases concluded that sequence rearrangement has occurred several times. A parallel study further elucidated the structure and phylogenetic evolution of this operon and proposed a functional role for the *cnb* nucleotide-binding protein gene in the inducible expression of the 2-MIB operon (Wang *et al.*, 2011). These authors proposed at least two evolutionary branches in genes encoding 2-MIB synthesis arising first in actinomycetes, most of which cluster into a lineage that is distinct from cyanobacteria. Working with the cyanobacteria *Pseudanabaena* sp.dqh15 and *Planktothricoides raciborskii* CHAB 3331, Wang *et al.* (2011) also identified a methyl transferase gene (*mtf*) and a 2-MIB cyclase gene (*mic*) but showed two *cnb* genes flanking the operon, which, they proposed, play a regulatory role. BLAST and CCD database searches aligned the nucleotide-binding proteins with the Crp-Fnr family, an important family of transcription regulators found in certain groups of bacteria which allow rapid adaptation to transitory imbalances in activities such as photosynthesis, N_2-fixation, N-C regulation and respiration (for a review, see Korner *et al.*, 2003). Wang and coauthors proposed that the *cnb* genes may regulate 2-MIB biosynthesis (and/or MVA or MEP pathways), possibly in response to short term changes in the GPP pool. They observed rapid and reversible changes in 2-MIB gene expression by *Pseudanabaena* within three hours of exposure to different light regimes, consistent with observations from earlier studies (Bowmer *et al.*, 1992; Zhang *et al.*, 2009; Zimba *et al.*, 1999).

This has important implications for 2-MIB production and control in source waters, which is often associated with benthic or low-light tolerant taxa. For example, exposure to increased light, for example via aeration or water level changes may modify cell production. To date, few studies have applied molecular methods to identify and quantify the genetic capacity for 2-MIB production in the field, and this will likely change as this technology is adopted by drinking water and other sectors. The value of this approach, coupled with analytical validation, has been recently demonstrated by its application to recirculating aquaculture systems (RAS) by Auffret and co-workers (2011), who used genetic sequences from databases and eleven *Streptomyces* strains to develop and apply qPCR probes for quantification of both geosmin and 2-MIB -producing *Streptomyces* in sand and biofilters in two RAS.

3.2.2 Biogenic sulfides

Biogenic sulfides are increasingly reported as a major source of odours in drinking water supplies, notably those derived from eutrophic waters (e.g. Figures 3.1 and 3.3; Table 3.1 and Section 3.4). Sulfur compounds are fundamental to many abiotic and biotic processes which occur across wide temporal and spatial scales in aquatic systems (e.g. Hay & Kubanek, 2002; Simó, 2001; Watson & Jüttner, 2017). The global sulfur cycle is a primary example of the importance of this compound in the biosphere, affecting climate change and acid rain, and occurs through biological reduction of sulfate to sulfides and emission of important volatile organic sulfur compounds (VOSCs) such as dimethyl sulfide (DMS), coupled with atmospheric abiotic (photochemical) DMS oxidation to sulfate and sulfur oxyacids (Bentley & Chasteen, 2004). Production and emission of VOSCs – notably those produced in marine systems such as DMS – have been extensively characterized (e.g. Groene, 1995). At the cellular level, increasing evidence is demonstrating the importance of sulfides and thiols as antioxidants, cryogens and grazer repellents/toxins (Kirst *et al.*, 1991; Wolfe *et al.*, 1997; Watson & Jüttner, 2017). Many VOSCs are natural flavourants (e.g. onion and cabbage families, Alliaceae, Brassicaceae) and used extensively in the food industry, for example DMS is used to enhance corn-related flavors. On the other hand, volatile sulfides are also primary sources of malodours; DMS and other methyl sulfides are major constituents of municipal wastewater odours (Cheng *et al.*, 2005; Suffet & Rosenfeld, 2007).

In most cases, source-water VOSCs are attributed to heterotrophic bacterial activity (see Section 3.4; Bechard & Rayburn, 1979; Richards *et al.*, 1994), a process which is becoming a major issue for drinking water supplies impacted by domestic and industrial wastewater and agricultural inputs (Brownlee *et al.*, 1995; Cheng *et al.*, 2005; Filipy *et al.*, 2006; Headley, 1987; Kangas *et al.*, 1986; Schiffman *et al.*, 2001). VOSCs generated by the massive decay of algal blooms, notably those dominated by *Microcystis*, can cause major drinking water impairments, particularly in hypereutrophic systems as seen, for example, in the 'Wuxi event' in China where the anoxic degradation of dense macrophyte and bloom material affected drinking water supplied to some 2 million consumers (Ma *et al.*, 2013a, b; Zhang *et al.*, 2010). Some cyanobacteria and algae actively synthesize volatile sulfides during cell growth. For example, littoral and benthic mats of 'stinkweed' – or stoneworts (green algae belonging to the Streptophyta; formerly grouped with the charophytes) *Chara aspera*, *C. globularis* and *Nitellopsis obtusa*) – produce potent thiols (4-methylthio-1,2-dithiolane and particularly 5-methylthio-1,2,3-trithiane; Anthoni *et al.*, 1980; Berger & Schagerl, 2004). Strains of the bloom-forming cyanobacteria *Microcystis (M. aeruginosa, M. flos-aquae)* synthesize potent thiols (isopropyl thiol, isopropyl disulfide, isopropyl methyl sulfide, and isopropyl methyl disulfide; Hofbauer & Jüttner, 1988). These compounds can produce strong airborne odours, but are rarely problematic in treated drinking water supplies as they are removed by chlorination (Tsuchiya *et al.*, 1992).

The overwhelming evidence to date indicates that autotrophic VOSC production is not a major source of T&O in most drinking water supplies. Improved analytical techniques, largely adapted from the food and wine industries, have provided more sensitive measures of these alkyl thiols and have been applied recently to study the chemical ecology of their production (Watson & Jüttner, 2017).

3.2.3 Amines and other nitrogenous VOCs

Nitrogen-containing VOCs are not well recovered by analytical techniques routinely used for T&O and are rarely measured or reported from drinking water supplies. Amines in particular can have highly potent fishy or putrid odours under alkaline conditions. They can be produced through bacterial degradation of amino acids from plant and animal material in source waters, or as by-products of drinking water treatment (Suffet & Rosenfeld, 2007), and their occurrence in raw and treated waters is generally attributed to these processes. Nevertheless, amines and other N-compounds are also produced by algae during active growth. In particular, significant levels (50–300 μg/L) of odourous amines such as diethylamine, trimethylamine, 1-aminopropane, 2-aminopropane and ethanolamine have been reported from cultures of cyanobacteria, green algae and chrysophytes (Herrman & Jüttner, 1977; Pohnert & Von Elert, 2000), and in at least one of these studies, it was demonstrated that these amines were produced by the algae, not their bacterial contaminants. In one reported case, amines were reported from a *Chlamydomonas* bloom (Ruhr River, Germany) and cultured isolates from this bloom (Herrman & Jüttner, 1977).

A diversity of other volatile N-compounds (heterocyclic compounds, aromatics, alkoxyamines and amides) have been reported from (non-axenic) strains of the cyanobacterium *Aphanizomenon flos-aquae* (Dembitsky *et al.*, 2000). These VOCs have similarly not been investigated as sources of T&O. *A. flos-aquae* and related species are particularly problematic in numerous small and large eutrophic and mesotrophic lakes (e.g. Lin *et al.*, 1979; Peterson *et al.*, 1995). While *A. flos-aquae* does not produce geosmin or 2-MIB (although geosmin production by *Aphanizomenon gracile* was reported by Durrer *et al.*, 1999), this species can be a source of alkaloid neurotoxins and commonly associated with 'mouldy, grassy, vile' source water odours.

3.3 VOC INDUCED BY LOSS OF CELL INTEGRITY AND/OR DEATH AND THE ACTIVATION OF CATABOLIC ENZYMES

3.3.1 Oxylipins and polyunsaturated fatty acid (PUFA) derivatives

3.3.1.1 Chemistry

Oxylipins and PUFA derivatives are an important source of T&O. Most fatty acids (FAs) are not present in free form in the aquatic environment but as structural cell components, and are catabolically oxidized by enzymes and non-enzymatic processes at cell rupture (cf. Jüttner 2001). FAs are fundamental to cell structure and physiology, for example, as key components of cell and thylakoid membranes, and these molecules and their derivatives play important roles in many intra- or intercellular semiochemical interactions (i.e. intra- or intercellular chemical signaling, eliciting behavioural or physiological responses; Watson *et al.*, 2010). FA oxidation is typically induced by loss of cell membrane integrity, often via disruptive agents (e.g. invertebrate predators, disease agents, temperature, photolytic or osmotic stress, freezing, dehydration, chemical treatment). As detailed below, these processes activate lipases, lipoxygenases (LOX) and other catalytic enzymes to yield a group of important derivatives termed oxylipins. Depending on the precursor FAs and double-bond cleavage pattern, products can include dienes, dienals and trienes, together with related alcohols, alkenes, aldehydes, ketones and branched hydrocarbons (Jüttner, 1995). Cleavage yields *trans, cis* alkene isomers which are subsequently photo- or thermally isomerized to a mixture of *trans* and *cis* isomers and the more stable, generally less odorous (and less bioactive) all-*trans* molecules (e.g. Yano *et al.*, 1988). Geometric isomers of the same compound may also have different odour descriptors (e.g. Burlingame *et al.*, 1992; Meijboom & Stroink, 1972), an important consideration for sensory analysis.

Some oxylipins in particular are known throughout the food and drinking water industries for their strong fishy, rancid, oily or cucumber odours, caused largely by unsaturated and polyunsaturated aldehydes (PUAs) and other related alkenes derived from unsaturated FA cleavage (Watson, 2003; Watson *et al.*, 2010). Oxylipins such as 2,4,7-decatrienal, 2,4-decadienal, 2,6-nonadienal, 2,4-heptadienal can be a significant T&O sources in range of source waters spanning oligotrophic through eutrophic/ dystrophic conditions (e.g. Chorus *et al.*, 1990; Jüttner *et al.*, 1986; Watson *et al.*, 2001a, b; Zhao *et al.*, 2013). These compounds can also be responsible for 'grassy/fruity/floral' odours and are predominantly formed via homolytic cleavage (Gardner, 1989; Jüttner, 1995). Other oxylipins are also implicated in some cases of T&O: 4-heptenal, pentanal and heptanal are also linked to fishy odours, *trans*-2-nonenal and *trans,cis*-2,6-nonadienal with 'cucumber' odours and *cis*-3-Hexenal and 3-hexenol produce grassy odours (Cotsaris *et al.*, 1995; Jüttner *et al.*, 1986; Khairi *et al.*, 1995).

3.3.1.2 Occurrence and biota

Odorous oxylipins are produced by select members of the cyanobacteria and most eukaryotic algal groups, particularly those with a high cell content of unsaturated FAs (e.g. diatoms and flagellates belonging to the Chrysophyta, Dinophyta and Cryptophyta; e.g. Lang, 2007; Figure 3.3). Important producers include eukaryotic algae such as species of *Fragilaria, Synedra, Aulacoseira, Melosira, Stephanodiscus, Tabellaria, Uroglena*, *Dinobryon*, *Mallomonas*, *Hydrurus*, *Synura* and *Poterioochromonas* (see Table 3.1; more extensive species listings can be found in Watson, 2003, 2010). Each of these taxa can produce varying amounts of several oxylipins, the yield and chemistry of which varies with species, growth and environment (Watson *et al.*, 2001a, b; Watson & Satchwill, 2003; Watson, 2003).

Evidence of heterolytic cleavage has been detected among species of the scaled colonial chrysophytes *Synura*, which produce the unconjugated polyunsaturated aldehyde (PUA) *trans,cis*-2,6-nonadienal, together with non-conjugated unsaturated derivatives such as 2,4,7-decatrienal and 2,4-decadienal (Watson *et al.*, 2001a, b; Wee *et al.*, 1994). *Trans,cis*-2,6-nonadienal is a primary cause of 'cucumber' T&O events in drinking water supplies (Burlingame *et al.*, 1992), and identified as a species-specific organoleptic 'fingerprint' of the scaled colonial chrysophyte *Synura petersenii* by Wee *et al.* (1994) who surveyed a number of *Synura* species. However, field evidence and our recent work with other *Synura* species shows production of this VOC by *S. uvella* (Jüttner, 1981b), possibly reflecting inter- or intraspecific variation in enzyme cleavage specificity or taxonomic ambiguity. Traditionally, scaled chrysophytes are identified by scale and colony morphology, a method which requires high resolution microscopy or TEM/SEM. It is well known that these and other morphological features vary both among species and with environmental conditions (e.g. Kynčlová *et al.*,

2010; Pichrtová & Němcová, 2011), which can result in misidentification of taxa in field samples and cultures. Recently, advances have been made to evaluate this issue using molecular diagnostics. Kynčlová *et al.* (2010) compared differences in ITS (internal transcribed spacer) rDNA sequences and morphological diagnostics among strains of *Synura petersenii* and other *Synura* and important scaled chrysophyte species (*Mallomonas, Chrysosphaerella*), and used these data to re-evaluate their phylogenetic groupings. Importantly, this work has identified significant variance among different strains of the single species complex *S. petersenii* in SSU (small subunit) rDNA sequences which showed a robust alignment with scale morphology, providing the basis for a revision of this species complex into six pseudo-cryptic species (Škaloud *et al.*, 2013).

Species of *Synura, Mallomonas* and *Chrysosphaerella* are implicated elsewhere in oxylipin-related T&O events, and can form noxious blooms which may or may not be correlated with nutrients and eutrophication (e.g. Watson *et al.*, 2001a, b; Lee *et al.*, 2013). Combined with the above-discussed taxonomic ambiguity, this makes management at the source-water level a challenging task, particularly since many algal and cyanobacterial taxa have yet to be evaluated for oxyplin production. For example, a recent study reported cucumber odours from the lipid-rich unicellular heterokont *Nannochloropsis oculata* and colonial green alga *Botryococcus braunii*. The latter also produced 2,4,6-nonatrienal, but neither species yielded detectable amounts of *trans, cis*-2,6-nonadienal among the many other volatiles they produced (Van Durme *et al.*, 2013). As seen in the section below, a multiplex approach (i.e. combining morphological, biochemical and molecular metrics) is clearly the path forward to elucidate both the species and the LOX and other catalytic enzymes that underlie these different patterns in VOC production.

3.3.1.3 Biosynthesis

The chemistry and yield of PUFA derivatives depend on the precursor FA supply and molecular configurations, which in turn varies among and within different taxonomic groups. Freshwater cyanobacteria and algae fall into two major groups: cyanobacteria and chlorophytes have a relatively low cell lipid content predominated by saturated and monounsaturated C_{16} to C_{18} FAs, while chrysophytes, diatoms, cryptophytes, prymnesiophytes and dinoflagellates have a higher PUFA cell content, typically C_{18} up to C_{22}, and PUFA derivatives are among the predominant volatiles produced by these groups. Eicosapentaenoic acid (EPA; 20:5ω3) is a common precursor of 2,4,7-decatrienal, 2,4-heptadienal and other oxylipins (d'Ippolito *et al.*, 2004; Pohnert, 2000), while hexadecatrienoic acid C16:3 n-4 and hexadecatetraenoic (HTA) C16:4 n-1 are precursors of octadienal and octatrienal, respectively, (D'Ippolito *et al.*, 2002).

Variance in PUFA content and oxylipin chemistry and yield also occurs within these different taxonomic groups (e.g. Lang *et al.*, 2011; Piepho *et al.*, 2012;

Wacker *et al.*, 2016). For example, chlorophytes and most cyanobacteria contain a predominance of α-linolenic acid (18:3ω3), but species of the cyanobacterium *Spirulina* and most other taxonomic groups produce only γ-linolenic (18:3ω6; Ahlgren *et al.*, 1992) which has allellopathic properties (see below). Cryptomonads, dinoflagellates and some chrysophytes have a high content of EPA, but only a few of these are known to produce derivatives of this fatty acid (*trans*, *cis* 1,3,5-octatriene and 2,4,7-decatrienal) (Jüttner, 1995; Rashash *et al.*, 1995; Watson *et al.*, 1999). At the individual cell level, both FA composition and PUFA derivative chemistry and yield are affected by growth phase, carbon sources (inorganic vs. organic), and environmental factors such as light intensity and spectrum, temperature, pH and nutrients, heavy metals, chlorine and pesticides (e.g. Jüttner, 1995; Piepho *et al.*, 2012; Rashash *et al.*, 1995; Wacker *et al.*, 2016; Watson & Satchwill, 2003).

Several studies have elucidated lipoxygenase-driven oxylipin pathways in diatoms, (reviewed, for example, in Watson *et al.*, 2010 and references cited therein). Tissue disruption induces lipase activity, which releases substrate PUFAs from phospholipid and glycerolipid membrane lipid pools. These PUFAs are further activated via lipoxygenase-mediated transformation and the positional specific insertion of O_2 to generate reactive hydroperoxides. LOX are non-heme iron-containing dioxygenases produced by both prokaryotic and eukaryotic organisms and represent a ubiquitous and extremely important group of region- and stereo-specific enzymes. In diatoms, for example, 9-LOX and 13-LOX catalysis yields 9- and 13-hydroperoxy FAs, which are further oxidized by different catabolic enzymes via homolytic or heterolytic hydroperoxide cleavage to generate PUAs and (i) hydrocarbons, (ii) alcohols and (ii) halogenated hydrocarbons (ibid) (Hombeck & Boland, 1998; Hombeck *et al.*, 1999; Pohnert & Boland, 1996).

Extensive research has elucidated the chemistry and bioactivity of oxylipins and associated LOX for terrestrial plants, where their role in wound-activated defence has been well characterised, and also among marine microalgae and macroalgae where early studies using genome sequencing suggested that in fact LOX enzymes may not be distributed ubiquitously (Armbrust *et al.*, 2004). For example, recent studies have combined advanced analytical tools with bioassays and genetics (e.g. semi-quantitative RT-PCR) to elucidate the substrates, enzyme characteristics and expression and oxylipin products of wound-activated responses in marine seaweeds such as *Gracilaria chilensis* (Weinberger *et al.*, 2011). However comparatively little work has focussed on the LOX family and biochemistry and genetic regulation of oxylipin production among freshwater taxa (Lang & Flessner, 2007; Watson *et al.*, 2010), although with the increased industrial use of these taxa as food and biofuels sources, research in this area can provide the basis for further work in drinking water T&O applications. Lang (2007) conducted an extensive survey of over 2,300 freshwater microalgal strains, using advanced biochemical and molecular tools to characterise PUFAs and oxylipins. Using genomic databases to identify putative LOX sequences in two cyanobacteria, Lang further elucidated and cloned two 13-LOXs (NpLOX1 and NpLOX2) from *Nostoc punctiforme* PCC 73102, and the candidate enzyme protein NspFP from the cyanobacterium *Nostoc* sp. which produced only 9-LOX derived FA hydroperoxides. Importantly, the identification of key differences in LOX specificity and products among two closely related taxa is of prime interest and relevance to observed differences in oxylipin products among taxa. This important study also established an extensive database of microalgal FAs, and provided the basis for further research to elucidate stereo- and regiospecific LOX-based reactions using FA profiling, substrate labelling and 16SrDNA and 18SrDNA sequence analyses.

The transitional and labile nature of lipid peroxidation products has presented challenges to sensitive diagnostics and measurement. Advances in analytical methods have improved sensitivity and reduced storage and analytical artefacts, for example using O-(2,3,4,5,6-pentafluorobenzyl) hydroxylamine hydrochloride (PFBHA·HCl) derivatization to stabilize labile PUAs in field samples, allowing in situ measures of production (D'Ippolito *et al.*, 2002; Wichard & Pohnert, 2006). More recently, reverse phase liquid chromatography and tandem quadrupole-TOF mass spectrometry (qTOF LC-MS/MS) has provided a means to investigate LOX positional specificity (Cutignano *et al.*, 2011) and this technique was recently combined with molecular and phylogenetic analyses in a study of the important marine diatom *Pseudo-nitzschia* by Lamari *et al.* (2013). Using this combined approach, these authors showed significant inter- and intra-specific diversity in LOX-based EPA metabolism even among genetically related taxa, which corresponded to specific oxylipin markers, highlighting the potential use of oxylipin profiles in species identification along with molecular and morphological traits. This approach provides a powerful new multiplex diagnostic tool to investigate endogenous and exogenous mechanisms and controls of oxylipin production and bioactivity across other species, with important potential application to T&O studies.

3.3.1.4 Biological activity

Oxylipins are widely generated across the plant and animal kingdom and are known for their bioactivity. In aquatic systems they can function as teratogens, grazer deterrents and toxins at many foodweb levels, and their semiochemistry has been an intensive area of research with marine taxa, notably diatoms (e.g. Fink, 2007; Jüttner, 2001, 2005; Pohnert & Boland, 2002; Watson, 2012; Watson *et al.*, 2010; Weinberger *et al.*, 2011). Although this offers a highly promising and productive approach, comparatively little work has been done in freshwaters, particularly in relationship to T&O. A few studies have demonstrated the importance of some PUAs in freshwater grazer defence interactions (e.g. 2,4-decadienal, 2,4,7-decatrienal) (Jüttner, 2001, 2005). Other work has shown that some benthic diatom taxa produce oxylipins that are structurally identical to marine seaweed pheromones (brown algae), but as yet there is no direct evidence that these compounds have a similar mode of reproductive-related activity among the known freshwater producers, which include both (asexual) prokaryotic and eukaryotic taxa. For example, fucoserratene (*trans*, *cis*-1,3,5-octatriene) is produced by both the planktonic diatom *Asterionella formosa* (for which sexual reproduction has yet to be observed; Mann 1989) and the benthic cyanobacteria *Anabaena oscillarioid*es (Jüttner & Müller, 1979; Möhren & Jüttner, 1983). Nevertheless, intriguing patterns of production e.g. of aliyclic C_{11} oxylipins (hormosirene, ectocarpene and dictyopterenes) by benthic diatoms (e.g. *Gomphonema parvulum*) but not planktonic species merits further investigation, since as noted below, insight into the functional role(s) of any T&O compound will provide a far better understanding of the timing and mechanisms regulating production (Jüttner & Dürst, 1997; Wendel & Jüttner, 1996).

3.3.2 Carotenoid derivatives

3.3.2.1 Chemistry and odour properties

Carotenoid derivatives (nor-carotenoids) are widely distributed across all phyla. They have a variety of odours, many of which are pleasant (e.g. floral, fruity; see Table 3.1 and Figure 3.3), and are used extensively in the fragrance and flavour industries. Nor carotenoids are important signalling molecules which act as chromophores, grazer deterrents, growth inhibitors, toxins etc. (see below). They can often contribute to the overall 'bouquet' of VOCs generated by biofilms and plankton (e.g. Chen *et al.*, 2010; Höckelmann *et al.*, 2004; Jiang *et al.*, 2006) and to source water T&O, but with the exception of β-cyclocitral (below) are rarely identified as primary odour sources. Among the most commonly reported nor-carotenoids from both field samples and cultures are 6-methylhept-5-en-2-one, nerol, dihydroactinidiolide, β-Ionone and α- and β-cyclocitral (e.g. Chorus *et al.*, 1990; Henatsch & Jüttner, 1983, 1985; Jüttner 1995; Rashash *et al.*, 1996). Their OTCs vary considerably, ranging over six orders of magnitude (e.g. Table 3.1; Watson, 2010).

3.3.2.2 Occurrence and biological sources

Nor-carotenoids are derived from carotenes (e.g. α- and β-carotene) and xanthophylls (e.g. zeaxanthin), pigments which are widely distributed among cyanobacteria, algae, fungi, actinomyces and other microbiota, where they participate in important functions such as light capture and dissipation (quenching), membrane fluidity and protection from reactive oxygen species (ROS; e.g. oxygen ions and peroxides), UV and other oxidants (e.g. Scherzinger & Al-Babili, 2008). Carotenoid levels in cells (and the production of nor-carotenoids) change in response to a number of factors, and can vary independently to chlorophyll-a (chl*a*), a commonly measured parameter in many monitoring programmes. For example, exposure to extreme high light can result in a decrease in chl*a* and either a decrease or increase in photo- and reactive oxygen species (ROS) protective carotenoids; carotenoid/chl*a* ratios may vary among and within major taxa, plankton and benthos, and are affected by temperature, water transparency, colour (e.g. DOM) and other factors (e.g. Bonilla *et al.*, 2009; Lohscheider *et al.*, 2011; Mandal *et al.*, 2011; Zeeshan & Prasad, 2009). Thus, the abundance of carotenoids in surface waters may change quite rapidly in response to taxonomic and environmental shifts (e.g. Soma *et al.*, 1993), as may nor-carotenoids such as β-cyclocitral, β-Ionone, and 6-methylhept-5-en-2-one (e.g. Jüttner *et al.*, 1986; Peter *et al.*, 2009).

3.3.2.3 Biosynthesis

Nor-carotenoids are produced by carotenoid cleavage dioxygenases (CCDs) which catalyse the breakdown of cell carotenoids. During this reaction, oxygen is introduced into a double-bond cleavage site along the carotenoid backbone, producing ketone or aldehyde nor-carotenoid products (e.g. Marasco *et al.*, 2006). CCDs have a high degree of specificity to carotenoid substrates and the position of the double bond cleavage (Cui *et al.*, 2012). This characteristic may have some value for tracing select T&O compounds and problematic biota in source waters. For example Hofbauer and Jüttner (1988) studied the link between blooms of *Microcystis* and high source-water levels of β-cyclocitral ('tobacco' odour). They isolated a carotene oxidase from the cyanobacterium *Microcystis* with high specificity for β-carotene and zeaxanthin at the 7,8 and 7′,8′ cleavage sites, producing β-cyclocitral and hydroxy-β-cyclocitral, respectively and a second derivative, crocetindial. They also observed that enzyme activity is induced by loss of cell integrity, with no detectable release of these nor-carotenoids by intact cells and no activity towards other carotenoids (echinenone and myxoxanthophyll). β-carotene is a major pigment in cyanobacteria and many algal classes but the 7,8 (7′,8′) carotene oxygenase is, to date, only identified from *Microcystis* spp. Subsequent examination of a range of *Microcystis* spp. in lab and field populations has observed β-cyclocitral as the major VOC produced by these taxa (e.g. Jüttner *et al.*, 2010), far more than any other taxa examined to date, suggesting orthologous CCDs and associated genes. This discovery has particular significance in view of the increasing prevalence these cyanobacteria in toxic blooms in major, highly eutrophic source-waters such as Lake Taihu and Lake Erie (Steffen *et al.*, 2014; Tao *et al.*, 2012). Furthermore, β-cyclocitral is now known to play an important role in grazer defence (Jüttner *et al.*, 2010). Recent work has also identified trace amounts of β-cyclocitral from several eukaryotic algae (*Botryococcus braunii*, *Chlorella*, *Nannochloropsis oculata*, *Rhodomonas* sp; Van Durme *et al.*, 2013), but have not identified the enzymes involved. Other nor-carotenoids, which are more ubiquitous, may be useful to track certain species where there is prior knowledge of the source water and dominant algal/cyanobacterial taxa. For example, Peter *et al.* (2009) observed a correlation between β-Ionone and *Planktothrix rubescens* in Swiss lakes, although high levels of this compound have been reported from other waterbodies where this cyanobacterium is absent and the source untraced (Yu *et al.*, 2009).

3.3.2.4 Factors inducing production

As noted earlier, carotenoid cleavage is important in many cell activities, for example, photoadaptive regulation of cell pigments. Carotenoids are also vulnerable to non-specific oxidation by LOX, peroxidases, ROS, UV, and other stressors which may generate carotenoid oxygenase derivatives. During healthy cell growth, these processes typically do not generate significant extracellular levels of nor-carotenoids in source water (e.g. Peter *et al.*, 2009). However, cell lysis (brought on by grazers, pollutants and other environmental stressors, senescence etc.) induces heightened CCD and other oxidative enzyme activity and can generate more significant nor-carotenoids levels. Recent work with cyanobacteria suggests that CCDs may also function to regulate intracellular nor-carotenoids both in cell membranes and cytosol (Scherzinger & Al-Babili, 2008) possibly acting to minimise adverse effects of these highly bioactive derivatives on the host cell. As noted above, however, production of many nor carotenoids is heightened at cell rupture with the induction of catalytic enzymes and also induced by many analytical extraction processes and reported 'cell bound' fractions in fact represent maximum potential yield at lysis.

3.3.2.5 Genetic regulation

From the drinking water perspective, the question remains: what is the functional role of these carotenoid derivatives and to what extent can they diagnose specific nuisance taxa? Many of these compounds have been reported from both eukaryotic and prokaryotic microorganisms (algae, bacteria, fungi), but the mechanisms that lead to their release are not well characterised. Some of the answers to these questions may lie in differences in CCDs. Molecular and protein diagnostics have provided considerable advancement in the field and have been widely applied to plant and animal taxa, and recently several studies have focussed on algae and cyanobacteria. As with geosmin and 2-MIB, this approach provides the ability to characterise homologies and differences in enzymes and their genetic coding, their distribution, phylogeny, and structure, substrate/cleavage specificity and predict the nature and function of the nor carotenoid derivatives.

There are several different classes of CCDs, depending on their substrate specificity and position or *cis-trans* configuration of the double bond cleavage site (Bouvier *et al.*, 2005). Eukaryotic CCD homologs have been found in all sequenced cyanobacterial genomes, with several homologues with different substrate specificity in individual species. In a recent study, Cui *et al.* (2012) characterised the evolutionary phylogeny and structure of *ccd* genes encoding for the carotene dioxygenases produced by 37 cyanobacteria, providing new insight into differences among closely related taxa. In a more focussed study of the benthic cyanobacterium *Nostoc*, Marasco *et al.* (2006) identified and sequenced three CCDs encoded by separate genes which showed different specificities. They determined that two of these CCDs acted on full chain carotenoids at different sites while the third acted only on nor-carotenoid products, also with site specificity. This explains low levels through growth and also suggests a functional mechanism that may help minimise the biological activity of these derivatives on cell.

3.4 NON-SPECIFIC VOC RELEASE FROM (ALL) DECAYING BIOMASS, HETEROTROPHIC BACTERIA, FUNGI AND OTHER MICROORGANISMS

Environmental conditions may also affect the predominant metabolic pathways and by-products. Fermentation products synthesized by both cyanobacteria and eukaryotic algae during cell growth (Jüttner, 1995) can impart significant organoleptic properties to water. These compounds include alcohols, esters and alkyl sulfides, many of which are derived from amino acids.

3.4.1 Biogenic sulfides-II

Alkyl sulfides are especially odorous products of fermentation (Jüttner, 1995; Karlsson *et al.*, 1995; Wnorowski & Scott, 1992), responsible, for example, for the putrid odour produced during bacterial degradation of algae and macrophyte biomass. Again, environmental factors such oxygen levels and light influence the synthesis of fermentation products, and in open water most of these compounds may originate from more anoxic zones such as sediment surfaces. Methyl sulfides produced by bacterial activity are primary contributors to sulfurous/septic and putrid odours (Figure 3.3) produced by rotting algae and plant material from bloom decay, standing water, sediments, shorelines, organic-laden runoff, groundwater and distribution systems (e.g. Cotsaris *et al.*, 1995; Franzmann *et al.*, 2001; Jones & Korth, 1995). Blooms of the prymnesiophyte *Chrysochromulina breviturrita* have been described to produce 'rotten cabbage' or 'garbage dump' odours characteristic of methyl sulfides (Nicholls, 1982). Recently, sulfides have become a major issue in drinking water derived from heavily impacted source waters where significant levels of alkyl sulfides are produced during the degradation of dense cyanobacterial blooms and other organic material (e.g. Ma *et al.*, 2013a, b).

3.4.2 Microbial fermentation products

Decaying algal/cyanobacterial or plant material along shorelines, sediments, filter beds etc. are rich sources of volatile fermentation products, produced by heterotrophic organisms such as bacteria and fungi. This process may also release or transform significant amounts of cell-bound VOCs such as geosmin. Volatile fermentation products are reported from mixed or non-axenic field and lab samples, and have been generally attributed to bacteria. Nevertheless, careful work has shown that alcohols and other odorous fermentation products are produced from nor-carotenoids and short chain aliphatic aldehydes (e.g. heptanal, hexanal and pentanal) by both axenic and monoxenic cyanobacteria (Jüttner, 1981a; Jüttner & Hahne, 1981). Among eukaryotic algae, *Chlamydomonas peterfii* was shown to actively convert organic supplements such as peptone to isovaleric acid, benzyl acetate and phenethyl alcohol (Rashash *et al.*, 1996). Cultures of chrysophytes (*Synura, Ochromonas* and *Poterioochromonas*) and other eukaryotic taxa (*Botryococcus braunii, Chlorella, Nannochloropsis oculata, Rhodomonas* sp.) are reported to produce fermentation products such as isoamyl alcohol, thio- and O-methyl esters and sulfides (e.g. Collins & Kalnins, 1965; Jüttner & Hahne, 1981; Van Durme *et al.*, 2013). One chrysophyte (*Ochromonas danica*) is reported to actively excrete fermentation products such as O- and S-methyl esters via a light-dependent process, and to re-absorb them in the dark (Jüttner *et al.*, 1982). Cotsaris and co-workers (1995) also found trace amounts of alcohol derivatives in cultures of non-axenic chlorophytes and chrysophytes although these may have been bacterial in origin. In field samples, breakdown products of amino acids (4-cresol, indole, skatole) have been identified in natural cyanobacterial samples and rotting mats of green algae (*Cladophora*; Brownlee *et al.*, 1984).

3.5 TASTE AND ODOUR: RELATIONSHIP WITH CYANOBACTERIAL TOXINS

Taste and odour is often used by consumers as a measure of drinking water safety, although there are few cases where odour is actually diagnostic of health hazards. Nevertheless, there is increasing concern with the global increase in toxic blooms, which in inland waters are largely caused by species of cyanobacteria. A large number of cyanobacterial toxins that have now been identified which are difficult to detect and predict; a major concern with water providers. In recent years, the number of studies published on cyanobacterial toxins has far exceeded those related to T&O (Figure 3.1), although in fact very few human fatalities have been directly attributed to cyanotoxins, while aesthetics (T&O, appearance) continues to be the primary criteria used by consumers to judge drinking water acceptability, quality and safety.

Several studies have reported correlations between cyanobacterial VOCs such as geosmin, 2-MIB, β-cyclocitral and toxins (e.g. Graham *et al.*, 2010).

Nevertheless, while both the risk of toxins and T&O increases with cyanobacterial abundance (e.g. Downing *et al.*, 2001; Watson *et al.*, 2008), the presence of these toxins cannot be determined by simple visual inspection or using surrogate measures such as T&O. In essence, not all odour-causing blooms are toxic, not all toxic blooms are odourous, and both toxins and T&O can occur in the absence of visible blooms. While many blooms lead to increased source water T&O (directly or indirectly), not all blooms are produced by cyanobacteria. As with VOCs, toxin production varies considerably among cyanobacterial species while strains of the same species can include producers and non-producers. Toxins are colourless and tasteless, and there are no known direct physiological links between odour and toxin production. Thus, in many cases, outbreaks of T&O are not concurrent with (or caused by) toxic biota and thus are often poorly correlated, as shown, for example, in recent studies from North America and China (Li *et al.*, 2010; Watson *et al.*, 2008).

3.6 CHANGING FACE OF BIOLOGICAL ODOUR PRODUCTION: LOOKING FORWARD

It is now known that many volatile compounds are produced as important semiochemicals, which modify cell functioning (e.g. as antioxidants, pheromones, autoregulation) and food web interactions (grazer deterrents, inhibitors, toxins, attractants). Much of the information on the chemical ecology of these compounds is derived from marine studies and applied research (e.g. food industry), where the bioactivity of many VOCs has been extensively investigated. Several reviews of this topic focussing on the relevance of this functional approach to drinking water T&O have also been published (e.g. Jüttner, 2005; Watson, 2003, 2012; Watson *et al.*, 2010). As noted throughout this chapter, future advancement in the control of biogenic T&O requires an understanding of the function of the VOCs, and the cellular pathways and genes that control their production. This is fundamental to our ability to predict and influence shifts in species and their VOCs and optimise source water control and treatment in the face of global eutrophication, climate change and other factors. Significant advancement in T&O research and monitoring can be developed from other disciplines, and needs to adopt a multifaceted approach. This may combine a well planned sampling effort, initial screening with sensory analyses and a modified version of the T&O wheel, further diagnostics using molecular probes to detect the presence and up-regulation of genes encoding VOC production (e.g. Tsao *et al.*, 2014), biosensors, 'omics' (e.g. metagenomics, proteomics) and sensitive chemical analyses. Figure 3.3 shows a new 'Biological' version of the T&O wheel (see Chapter 2), which includes an additional layer linking odour types and associated VOCs with likely biological sources, shown in the outer circle and adjacent representative images.

Other applications have adapted behavioural and toxicity assays. For example, Watson *et al.* (2007b) used an inline toximeter installed at a WTP (Zurich, Switzerland) to detect several major T&O compounds, while Sorichetti *et al.* (2014) adapted the rainbow trout gill (RTgill-W1) cytotoxicity assay (RCA) to detect VOCs such as geosmin, 2-MIB, β-cyclocitral and two PUAs in source waters and cultures. These tools can be combined with other monitoring devices to provide early warning indicators e.g. fluorescence profiling, image recognition etc., and represent new and potentially powerful methods to address T&O in source and drinking waters.

Taste and odour management requires an understanding of both production and losses, including VOC biodegradation, yet research into biological T&O has largely focussed on production. In the treatment industry, considerable work has investigated the degradation potential of the flora associated with biologically active filters (BAFs) using both traditional and more recently, molecular methods (e.g. Guttman & van Rijn, 2012; Hoefel *et al.*, 2009; Zhou *et al.*, 2011). Other studies have investigated these processes in source waters using field samples enriched with nutrients and/or BAF-derived microflora, while more recently Tang *et al.* (2013) made the important next step to advance this understanding to natural waters, and measured biodegradation rates of cyanobacterial metabolites (toxins; geosmin and 2-MIB) from unamended field samples. These authors measured degradation rates that varied between 5–10 days, with significant seasonal variance in capacity and rates, likely attributable to changes in both the microbiota, temperature and water quality. Combinations of such approaches with control measures such as water level regulation (Su *et al.*, 2017), biological control at the source (e.g. Jonns *et al.*, 2017), modelling (Otten *et al.*, 2016; Parinet *et al.*, 2010; Qi *et al.*, 2012) advanced genomics and other techniques will provide a more holistic approach to our understanding and ability to predict, model and manage sources and fates of biological T&O in drinking water supplies.

REFERENCES

Ahlgren G., Gustafsson I. B. and Boberg M. (1992). Fatty acid content and chemical composition of freshwater microalgae. *Journal of Phycology*, **28**, 37–50.

Anthoni U., Christophersen C., Madsen J. O., Wium-Andersen S. and Jacobsen N. (1980). Biologically active sulfur compounds from the green alga *Chara globularis*. *Phytochemistry*, **19**, 694–695.

Armbrust E., Berges J., Bowler C., Green B., Martinez D., Putnam N., Zhou S., Allen A., Apt K., Bechner M., Brzezinski M., Chaal B., Chiovitti A., Davis A. K., Demarest M., Detter J., Glavina T., Goodstein D., Hadi M., Hellsten U., Hildebrand M., Jenkins B., Jurka J., Kapitonov V., Kröger N., Lau W., Lane T., Larimer F., Lippmeier J., Lucas S., Medina M., Montsant A., Obornik M., Parker M., Palenik B., Pazour G., Richardson P., Rynearson T., Saito M., Schwartz D., Thamatrakoln K., Valentin K., Vardi A., Wilkerson F. and Rokhsar D. (2004). The genome of the diatom *Thalassiosira pseudonana*: ecology, evolution, and metabolism. *Science*, **306**(5693), 79–86.

Asquith E. A., Evans C. A., Geary P. M., Dunstan R. H. and Cole B. (2013). The role of Actinobacteria in taste and odour episodes involving geosmin and 2-methylisoborneol in aquatic environments. *Journal of Water Supply: Research and Technology – AQUA*, **62**(7), 452–467.

Auffret M. A., Pilote A., Proulx É, Proulx D., Vandenberg G. and Villemur R. (2011). Establishment of a real-time PCR method for quantification of geosmin-producing *Streptomyces* spp. in recirculating aquaculture systems. *Water Research*, **45**, 6753–6762.

Bafford R. A., Seagull R. W., Chung S.-Y. and Millie D. F. (1993). Intracellular localization of the taste/odor metabolite 2-methylisoborneol in *Oscillatoria limosa* (Cyanophyta). *Journal of Phycology*, **29**, 91–95.

Bechard M. and Rayburn W. (1979). Volatile organic sulfides from freshwater algae. *Journal of Phycology*, **15**, 379–383.

Bentley R. and Chasteen T.-G. (2004). Environmental VOSCs – formation and degradation of dimethyl sulfide, methanethiol and related materials. *Chemosphere*, **55**, 291–317.

Bentley R. and Meganathan R. (1981). Geosmin and methylisoborneol biosynthesis in streptomycetes. Evidence for an isoprenoid pathway and its absence in non-differentiating isolates. *FEBS Letters*, **125**, 220–222.

Berger J. and Schagerl M. (2004). Allelopathic activity of Characeae. *Biologia Bratislava*, **59**, 9–15.

Bonilla S., Rautio M. and Vincent W. (2009). Phytoplankton and phytobenthos pigment strategies: implications for algal survival in the changing Arctic. *Polar Biology*, **32**, 1293–1303.

Bouvier F., Isner J.-C., Dogbo O. and Camara B. (2005). Oxidative tailoring of carotenoids: a prospect towards novel functions in plants. *Trends in Plant Science*, **10**, 187–194.

Bowmer K. H., Padovan A., Oliver R. L., Korth W. and Ganf G. G. (1992). Physiology of geosmin production by *Anabaena circinalis* isolated from the Murrumbidgee River, Australia. *Water Science and Technology*, **25**(2), 259–267.

Brownlee B. G., Kenefick S. L., MacInnis G. A. and Hrudy S. E. (1995). Characterization of odorous compounds from bleached kraft pulp mill effluent. *Water Science and Technology*, **31**, 35–40.

Bruder S., Babbar-Sebens M., Tedesco L. P. and Soyeux E. (2014). Use of fuzzy logic models for prediction of taste and odor compounds in algal bloom-affected inland water bodies. *Environmental Monitoring and Assessment,* **186**(3), 1525–1545.

Burgos L., Lehmann M., de Andrade H. H. R., de Abreu B. R. R., de Souza A. P., Juliano V. B. and Dihl R. R. (2014a). In vivo and in vitro genotoxicity assessment of 2-methylisoborneol, causal agent of earthy–musty taste and odor in water. *Ecotoxicology and Environmental Safety*, **100**(Supplement C), 282–286.

Burgos L., Lehmann M., Simon D., de Andrade H. H. R., de Abreu B. R. R., Nabinger D. D., Grivicich I., Juliano V. B. and Dihl R. R. (2014b). Agents of earthy-musty taste and odor in water: evaluation of cytotoxicity, genotoxicity and toxicogenomics. *Science of the Total Environment*, **490**(Supplement C), 679–685.

Burlingame G., Muldowney J. and Maddrey R. E. (1992). Cucumber flavor in Philadelphia's drinking water. *Journal American Water Works Association*, **84**, 92–97.

Cane D. E. and Watt R. M. (2003). Expression and mechanistic analysis of a germacradienol synthase from *Streptomyces coelicolor* implicated in geosmin biosynthesis. *Proceedings of the National Academy of Science USA*, **100**(4), 1547–1551.

Cane D. E., He X., Kobayashi S., Omura S. and Ikeda H. (2006). Geosmin biosynthesis in *Streptomyces avermitilis*. Molecular cloning, expression, and mechanistic study of the germacradienol/geosmin synthase. *Journal of Antibiotics (Tokyo)*, **59**(8), 471–479.

Chen J., Xie P., Ma Z., Niu Y., Tao M., Deng X. and Wang Q. (2010). A systematic study on spatial and seasonal patterns of eight taste and odor compounds with relation to various biotic and abiotic parameters in Gonghu Bay of Lake Taihu, China. *Science of the Total Environment,* **409**(2), 314–325.

Cheng X., Peterkin E. and Burlingame G. A. 2005. A study on volatile organic sulfide causes of odors at Philadelphia's northeast water pollution control plant. *Water Research*, **39**, 3781–3790.

Chorus I., Klein G. and Rotard W. (1990). Volatile substances associated with algal blooms of different trophic state and in bank filtrate. *Internationale Vereinigung fur Theoretische und Angewandte Limnologie,* **24**, 270–273.

Collins R. and Kalnins K. 1965. Volatile constituents produced by the alga *Synura petersenii.* II: Alcohols, esters and acids. *International Journal of Air and Water Pollution*, **9**, 501–504.

Cotsaris E., Bruchet A., Mallevialle J. and Bursill D. (1995). The identification of odorous metabolites produced from algal monocultures. *Water Science and Technology*, **31**, 251–258.

Cui H., Wang Y. and Qin S. A. (2012). Genome-wide analysis of carotenoid cleavage dioxygenases in unicellular and filamentous cyanobacteria. *Comparative and Functional Genomics*, Article #164690.

Cutignano A., Lamari N., d'Ippolito G., Manzo E., Cimino G. and Fontana A. (2011). Lipoxygenase products in marine diatoms: a concise analytical method to explore the functional potential of oxylipins. *Journal of Phycology*, **47**(2), 233–243.

D'Ippolito G., Romano G., Iadicicco O., Miralto A., Ianora A., Cimino G. and Fontana A. (2002). New birth-control aldehydes from the marine diatom *Skeletonema costatum*: characterization and biogenesis. *Tetrahedron Letters*, **43**(35), 6133–6136.

D'Ippolito G., Tucci S., Cutignano A., Romano G., Cimino G., Miralto A. and Fontana A. (2004). The role of complex lipids in the synthesis of bioactive aldehydes of the marine diatom *Skeletonema costatum. Biochimica Biophysica Acta,* **1686**, 104–107.

Davies J. M., Roxborough M. and Mazumder A. (2004). Origins and implications of drinking water odours in lakes and reservoirs of British Columbia, Canada. *Water Research*, **38**, 1900–1910.

Dembitsky V. M., Shkrob I. and Lev O. (2000). Occurrence of volatile nitrogen-containing compounds in nitrogen-fixing cyanobacterium *Aphanizomenon flos-aquae. Journal of Chemical Ecology*, **26**, 1359–1366.

Dickschat J. S., Nawrath T., Thiel V., Kunze B., Müller R. and Schulz S. (2007). Biosynthesis of the off-flavor 2-methylisoborneol by the myxobacterium. *Nannocystis exedens. Angewandte Chemie – International Edition*, **46**(43), 8287–8290.

Dionigi C. P., Greene, Millie D. and Johnsen P. (1990). Mixed function oxidase inhibitors affect production of the off-flavor microbial metabolite geosmin. *Pesticide Biochemistry and Physiology*, **38**, 76–80.

Dionigi C. P., Mille D. F., Spanier A. M. and Johnsen P. B. (1992). Spore and geosmin production by *Streptomyces tendae* on several media. *Journal of Agricultural and Food Chemistry*, **40**, 122–125.

Downing J. A., Watson S. B. and McCauley E. (2001). Predicting cyanobacteria dominance in lakes. *Canadian Journal of Fisheries and Aquatic Sciences*, **58**, 1905–1908.

Durrer M., Zimmermann U. and Jüttner F. (1999). Dissolved and particle-bound geosmin in a mesotrophic lake (Lake Zurich): spatial and seasonal distribution and the effect of grazers. *Water Research*, **33**, 3628–3636.

Dzialowski A. R., Smith V. H., Huggins D. G., deNoyelles F., Lim N.-C., Baker D. S. and Beury J. H. (2009). Development of predictive models for geosmin-related taste and odor in Kansas, USA, drinking water reservoirs. *Water Research*, **43**(11), 2829–2840.

Filipy J., Rumburg B., Mount G., Westberg H. and Lamb B. (2006). Identification and quantification of volatile organic compounds from a dairy. *Atmospheric Environment*, **40**, 1480–1494.

Fink P. (2007). Ecological functions of volatile organic compounds in aquatic systems. *Marine and Freshwater Behaviour and Physiology*, **40**, 155–168.

Franzmann P., Heitz A., Zappia F., Wajon J. and Xanthis K. (2001). The formation of malodorous dimethyl oligosulfides in treated groundwater. *Water Research*, **35**(7), 1730–1738.

Gardner H. W. (1989). Oxygen radical chemistry of polyunsaturated fatty acids. *Free Radical Biology and Medicine*, **7**(1), 65–86.

Ghimire G. P., Oh T. J., Lee H. C., Kim B. G. and Sohng J. K. (2008). Cloning and functional characterization of the germacradienol synthase (spterp13) from *Streptomyces peucetius* ATCC 27952. *Journal of Microbiology and Biotechnology*, **18**, 1216–1220.

Giglio S., Jiang J., Saint C. P., Cane D. E. and Monis P. T. (2008). Isolation and characterization of the gene associated with geosmin production in cyanobacteria. *Environmental Science & Technology*, **42**, 8027–8032.

Giglio S., Saint C. P. and Monis P. T. (2011). Expression of the geosmin synthase gene in the cyanobacterium *Anabaena circinalis* AWQC318. *Journal of Phycology*, **47**(6), 1338–1343.

Gill A., Muller K. M. and Watson S. B. (2006). Identification of geosmin-producing Cyanobacteria in Lake Ontario using DGGE. *Journal of Phycology*, **24**(42), 8.

Godo T., Ohtani S., Saki Y. and Ishitobi Y. (2011). Detection of geosmin from *Coelosphaerium kuetzingianum* separated by a step density gradient medium from suspended materials in water in Lake Shinji. *Japanese Limnology*, **12**, 253–260.

Godo T., Saki Y., Nojiri Y., Tsujitani M., Sugahara S., Hayashi S., Kamiya H., Ohtani S. and Seike Y. (2017). Geosmin-producing species of *Coelosphaerium* (Synechococcales, Cyanobacteria) in Lake Shinji. *Japanese Scientific Reports*, **7**, 41928, doi: 10.1038/srep41928.

Graham J. L., Loftin K. A., Meyer M. T. and Ziegler A. C. (2010). Cyanotoxin mixtures and taste-and-odor compounds in cyanobacterial blooms from the midwestern United States. *Environmental Science & Technology*, **44**, 7361–7368.

Groene T. (1995). Biogenic production and consumption of dimethylsulfide (DMS) and dimethylsulfoniopropionate (DMSP) in the marine epipelagic zone: a review. *Journal of Marine Systems*, **6**, 191–209.

Guo Q., Yu J., Su M., Wang C., Yang M., Cao N., Zhao Y. and Xia P. (2018). Synergistic effect of musty odorants on septic odor: verification in Huangpu River source water. *Science of the Total Environment*, **653**, doi: 10.1016/j.scitotenv. 2018.11.062.

Gust B., Challis G. L., Fowler K., Kieser T. and Chater K. F. (2003). PCR-targeted *Streptomyces* gene replacement identifies a protein domain needed for biosynthesis of the sesquiterpene soil odor geosmin. *Proceedings of the National Academy of Science USA*, **100**, 1541–1546.

Guttman L. and van Rijn J. (2012). Isolation of bacteria capable of growth with 2-methylisoborneol and geosmin as the sole carbon and energy sources. *Applied and Environmental Microbiology*, **78**(2), 363–370.

Hageskal G., Knutsen A. K., Gaustad P., de Hoog G. S. and Skaar I. (2006). Diversity and significance of mold species in Norwegian drinking water. *Applied and Environmental Microbiology*, **72**, 7586–7593.

Hay M. E. and Kubanek J. (2002). Community and ecosystem level consequences of chemical cues in the plankton. *Journal of Chemical Ecology*, **28**(10), 2001–2016.

He X. and Cane D. E. (2004). Mechanism and stereochemistry of the germacradienol/germacrene D synthase of *Streptomyces coelicolor* A3(2). *Journal of the American Chemical Society*, **126**, 2678–2679.

Headley J. V. (1987). GC/MS Identification of organosulfur compounds in environmental samples. *Biomedical Environmental Mass Spectrometry*, **14**, 275–280.

Henatsch J. and Jüttner F. (1983) Volatile odorous excretion products of different strains of *Synechococcus* (Cyanobacteria). *Water Science and Technology,* **15**(6–7), 259–266.

Henatsch J. and Jüttner F. (1985). Evidence of beta carotene 7,8,7′,8′-oxygenase beta cyclocitral crocetindial generating in *Microcystis*. *Archives of Microbiology*, **141**(4), 37–343.

Herrman V. and Jüttner F. (1977). Excretion products of algae. Identification of biogenesis amines by gas-liquid chromatography and mass spectrometry of their trifluoroacetamides. *Analytical Biochemistry*, **78**, 365–373.

Höckelmann C., Moens T. and Jüttner F. (2004). Odor compounds from cyanobacterial biofilms acting as attractants and repellents for free-living nematodes. *Limnology and Oceanography*, **49**, 1809–1819.

Hoefel D., Ho L., Monis P. T., Newcombe G. and Saint C. P. (2009). Biodegradation of geosmin by a novel Gram-negative bacterium; isolation, phylogenetic characterisation and degradation rate determination. *Water Research*, **43**, 2927–2935.

Hofbauer B. and Jüttner F. (1988). Occurrence of isopropylthio compounds in the aquatic ecosystem Lake Neusiedl Austria as a chemical marker for *Microcystis flos-aquae*. *FEMS Microbiology Ecology*, **53**, 113–122.

Hombeck M. and Boland W. (1998). Biosynthesis of the algal pheromone fucoserratene by the freshwater diatom *Asterionella formosa* (Bacillariophyceae). *Tetrahedron*, **54**, 11033–11042.

Hombeck M., Pohnert G. and Boland W. (1999). Biosynthesis of dictyopterene A: stereoselectivity of a lipoxygenase/hydroperoxide lyase from *Gomphonema parvulum* (Bacillariophyceae). *Chemical Communications*, **3**, 243–244.

Izaguirre G. (2007). Geosmin- and MIB-producing cyanobacteria found in the United States. In: Early Warning and Management of Surface Water Taste-and-Odor Events, W. D. Taylor, R. F. Losee, M. Torobin, G. Izaguirre, D. Sass, D. Khiari and K. Atasi (eds), American Water Works Association Research Foundation, Denver, CO, USA, pp. 59–119.

Jensen S. E., Anders C. L., Goatcher L. J., Perley T., Kenefick S. and Hrudey S. E. (1994). Actinomycetes as a factor in odour problems affecting drinking water from the North Saskatchewan River. *Water Research*, **28**(6), 1393–1401.

Jiang J., He X. and Cane D. E. (2006). Geosmin Biosynthesis. *Streptomyces coelicolor* Germacradienol/Germacrene D synthase converts farnesyl diphosphate to geosmin. *Journal of the American Chemical Society*, **128**, 8128–8129.

Johansen J. R. and Casamatta D. A. (2005). Recognizing cyanobacterial diversity through adoption of a new species paradigm. *Algological Studies*, **117**(1), 71–93.

Jones G. and Korth W. (1995). In situ production of volatile odour compounds by river and reservoir phytoplankton populations in Australia. *Water Science and Technology*, **31** (11), 145–151.

Jonns J. A., Brooks P. R., Exley P., Poole S. and Kurtböke Dİ. (2017). Streptophage-mediated control of off-flavour taint producing streptomycetes isolated from barramundi ponds. *Synthetic and Systems Biotechnology*, **2**(2), 105–112.

Jorgensen N. (2005). Abundance of actinobacteria and production of geosmin and 2-methylisoborneol in Danish streams and fish ponds. *FEMS Microbiology Ecology*, **52**(2), 265–278.

Jüttner F. (1981a). Biologically active compounds released during algal blooms. *Internationale Vereinigung fur Theoretische und Angewandte Limnologie*, **21**, 227–230.

Jüttner F. (1981b). Detection of lipid degradation products in the water of a reservoir during a bloom of *Synura uvella*. *Applied and Environmental Microbiology*, **41**(1), 100–106.

Jüttner F. (1983). Volatile odourous excretion products of algae and their occurrence in the natural aquatic environment. *Water Science and Technology*, **15**, 247–257.

Jüttner F. (1992). Flavour compounds in weakly polluted rivers as a means to differentiate pollution sources. *Water Science and Technology*, **25**(2), 155–164.

Jüttner F. (1995). Physiology and biochemistry of odourous compounds from freshwater cyanobacteria and algae. *Water Science and Technology*, **31**, 69–78.

Jüttner F. (2001). Liberation of 5,8,11,14,17-eicosapentenoic acid and other polyunsaturated fatty acids from lipids as a grazer defense reaction in epilithic diatom biofilms. *Journal of Phycology*, **37**, 744–755.

Jüttner F. (2005). Evidence that polyunsaturated aldehydes of diatoms are repellents for pelagic crustacean grazers. *Aquatic Ecology*, **39**, 271–282.

Jüttner F. and Dürst U. (1997). High lipoxygenase activities in epilithic biofilms of diatoms. *Archiv für Hydrobiologie*, **138**, 451–463.

Jüttner F. and Hahne B. (1981). Volatile excretion products of *Poteriochromonas malhamensis*. Identification and formation. *Zeitschrift für Pflanzenphysiologie*, **103**, 403–412.

Jüttner F. and Müller H. (1979). Excretion of octadiene and octatrienes by a freshwater diatom. *Naturwissenschaften*, **66**, 363–364.

Jüttner F. and Watson S. B. (2007). Biochemical and ecological control of off-flavours of aquatic cyanobacteria and algae. *Applied and Environmental Microbiology*, **73**(14), 4395–4406.

Jüttner F., Wiedemann E. and Wurster K. (1982). Excretion of S- and O- methyl esters and other volatile compounds by *Ochromonas danica*. *Phytochemistry*, **21**(9), 2185–2188.

Jüttner F., Hoflacher B. and Wurster K. (1986). Seasonal analysis of volatile organic biogenic substances in freshwater phytoplankton populations dominated by *Dinobryon, Microcystis* and *Aphanizomenon*. *Journal of Phycology*, **22**, 169–175.

Jüttner F., Watson S. B., Von Elert E. and Köster O. (2010). β-cyclocitral, a grazer defence signal unique to the cyanobacterium *Microcystis*. *Journal of Chemical Ecology*, **36** (12), 1387–1397.

Kangas J., Nevalainen A., Manninen A. and Savolainen H. (1986). Ammonia, hydrogen sulfide and methyl mercaptides in finnish municipal sewage plants and pumping stations. *Science of the Total Environment*, **57**, 49–55.

Karlsson S. A., Grimvall A. and Borén H. (1995). Drinking water off-flavour caused by the formation of short-chain fatty acids in slow sand filters. *Water Science and Technology*, **31**, 49–54.

Kelley J., Kinsey G., Paterson R., Brayford D., Pitchers R. and Rossmore H. (2003). Identification and Control of Fungi in Distribution Systems. AWWA Research Foundation and American Water Works Association, Denver, CO, USA.

Khairi D., Suffet I. and Barrett S. E. (1995). The determination of compounds causing fishy/swampy odours in drinking water supplies. *Water Science and Technology*, **31**, 105–112.

Kirst G. O., Thiel C., Wolff H., Nothnagel J., Wanzek M. and Ulmke R. (1991). Dimethylsulfonio-propionate (DMSP) in ice-algae and its possible biological role. *Marine Chemistry*, **35**, 381–388.

Klausen C., Jørgensen N., Burford M. and O'Donahue M. (2004). Actinomycetes may also produce taste and odour. *Water*, **31**(5), 58–62.

Klausen C., Nicolaisen M. H., Strobel B. W., Warnecke F. and Nielsen J. L. (2005). Abundance of actinobacteria and production of geosmin and 2-methylisoborneol in Danish streams and fish ponds. *FEMS Microbiology Ecology*, **52**, 265–278.

Komárek J. (2002). Problems in cyanobacterial taxonomy; implication for most common toxin producing species. In: Workshop the Blooms of Toxic Algae in Fresh Waters: Health Emergency and Control Measures, S. Melchiorre, E. Viaggiu and M. Bruno (eds), National Institute of Health, Rome, Italy, pp. 6–43.

Komárek J. (2003). Planktic oscillatorialean cyanoprokaryotes (short review according to combined phenotype and molecular aspects). *Hydrobiologia*, **502**, 367–382.

Komárek J. and Komárková J. (2004). Taxonomic review of the cyanoprokaryotic genera *Planktothrix* and *Planktothricoides*. *Czech Phycology*, **4**, 1–18.

Komárek J. and Mareš J. (2012). An update to modern taxonomy (2011) of freshwater planktic heterocystous cyanobacteria. *Hydrobiologia*, **698**, 327–351.

Komárek J. and Zapomělová E. (2007). Planktic morphospecies of the cyanobacterial genus *Anabaena* = subg. *Dolichospermum* – 1. part: coiled types. *Fottea*, **7**(1), 1–31.

Komárek J. and Zapomělová E. (2008). Planktic morphospecies of the cyanobacterial genus *Anabaena* = subg. *Dolichospermum* – 2. part: straight types. *Fottea*, **8**(1), 1–14.

Komatsu M., Tsuda M., Omura S., Oikawa H. and Ikeda H. (2008). Identification and functional analysis of genes controlling biosynthesis of 2-methylisoborneol. *Proceedings of the National Academy of Science USA*, **105**(21), 7422–7427.

Korner H., Sofia H. and Kumft W. (2003). Phylogeny of the bacterial superfamily of Crp-Fnr transcription regulators: exploiting the metabolic spectrum by controlling alternative gene programs. *FEMS Microbiology Reviews*, **27**, 559–592.

Kutovaya O. A. and Watson S. B. (2014). Development and application of a molecular assay to detect and monitor geosmin-producing cyanobacteria and actinomycetes in the Great Lakes. *Journal of Great Lakes Research*, **40**, 404–414.

Kynčlová A., Škaloud P. and Škaloudová M. (2010). Unveiling hidden diversity in the *Synura petersenii* species complex (Synurophyceae, Heterokontophyta). *Nova Hedwigia*, **136**, 283–298.

Lamari N., Ruggiero M. V., d'Ippolito G., Kooistra W., Fontana A. and Montresor M. (2013). Specificity of lipoxygenase pathways supports species delineation in the marine diatom genus *Pseudo-nitzschia. PLoS One*, **8**(8), e73281.

Lang I. (2007). New Fatty Acids, Oxylipins and Volatiles in Microalgae. PhD thesis, Georg August University, Göttingen, Germany. http://ediss.uni-goettingen.de/handle/11858/00-1735-0000-0006-ACE4-3

Lang I. and Feussner I. (2007). Oxylipin formation in *Nostoc punctiforme* (PCC73102). *Phytochemistry*, **68**, 1120–1127.

Lang I., Hodac L., Friedl T. and Feussner I. (2011). Fatty acid profiles and their distribution patterns in microalgae: a comprehensive analysis of more than 2000 strains from the SAG culture collection. *BMC Plant Biology*, **11**, 1–16.

Lee K. L., Kim J. H., Yoon H.-S. and Kim H. S. (2013). Growth characteristics of two bloom-forming synurophytes (*Mallomonas elongata* and *Synura petersenii*) at different nitrate and phosphate concentrations. *Nordic Journal of Botany*, **30**(1), 104–108.

Li Z., Yu J., Yang M., Zhang J., Burch M. D. and Han W. (2010). Cyanobacterial population and harmful metabolites dynamics during a bloom in Yanghe Reservoir, North China. *Harmful Algae*, **9**, 481–488.

Lin S. D., Evans R. L. and Beuscher D. B. (1979). Phytoplankton in the Fox chain of lakes. *Journal of the American Water Resources Association*, **15**, 974–983.

Lohscheider J.-N., Strittmatter M., Kupper H. and Adamska I. (2011). Vertical distribution of epibenthic freshwater cyanobacterial *Synechococcus* spp. strains depends on their ability for photoprotection. *PLoS One*, **6**(5), e20134.

Ludwig F., Medger A., Bornick H., Opitz M., Lang K., Gottfert M. and Roske I. (2007). Identification and expression analyses of putative sesquiterpene synthase genes in *Phormidium sp.* and prevalence of geoA-like genes in a drinking water reservoir. *Applied and Environmental Microbiology*, **73**, 6988–6993.

Ma Z., Niu Y., Xie P., Chen J., Tao M. and Deng X. (2013a). Off-flavour compounds from decaying cyanobacterial blooms of Lake Taihu. *Journal of Environmental Sciences*, **25** (3), 495–501.

Ma Z. M., Xie P., Chen J., Niu Y., Tao M., Qi M., Zhang W. and Deng X. W. (2013b). *Microcystis* blooms influencing volatile organic compounds concentrations in Lake Taihu. *Fresenius Environmental Bulletin*, **22**, 95–102.

Mandal S., Rath J. and Adhikary S. P. (2011). Adaptation strategies of the sheathed cyanobacterium *Lyngbya majuscula* to ultraviolet-B. *Photochemistry and Photobiology*, **102**, 115–122.

Mann D. G. (1989). Why didn't Lund see sex in *Asterionella*? A discussion of the diatom life cycle in nature. In: Algae and the Aquatic Environment, F. E. Round (ed.), Biopress, Bristol, UK, pp. 385–412.

Marasco E., Vay K. and Schmidt-Dannert C. (2006). Identification of carotenoid cleavage dioxygenases from *Nostoc* sp. PCC 7120 with different cleavage activities. *Journal of Biological Chemistry*, **281**, 31583–31593.

Marsili R. T. and Laskonis C. R. (2014). Odorant synergy effects as the cause of fishy malodors in algal marine oils. *Journal of Agricultural and Food Chemistry,* **62**, 9676–9682, doi: 10.1021/jf502252q.

Meijboom P. and Stroink J. (1972). 2-*trans*,4-*cis*,7-cis-decatrienal, the fishy off-flavor occurring in strongly autoxidized oils containing linolenic acid or (3,6,9, etc.) fatty acids. *Journal of the American Oil Chemists Society*, **49**, 555–558.

Miyazawa T., Gallagher M., Preti G. and Wise P. M. (2008). Synergistic mixture interactions in detection of perithreshold odors by humans. *Chemical Senses*, **33**, 363–369.

Möhren S. and Jüttner F. (1983). Odorous compounds of different strains of *Anabaena* and *Nostoc* (Cyanobacteria). *Water Science and Technology*, **15**, 221–223.

Naes H., Utkilen H. and Post A. (1988). Factors influencing geosmin production by the cyanobacterium *Oscillatoria brevis*. *Water Science and Technology*, **20**, 125–131.

Nakajima M. T., Ogura Y., Kusama N., Iwabuchi T., Imawaka A., Araki T., Sasaki T. and Hirose E. (1996). Inhibitory effects of odor substances, geosmin and 2-methylisoborneol, on early development of sea urchins. *Water Research*, **30**, 2508–2516.

Newcombe G., House J., Ho L., Baker P. and Burch M. (2009). Management Strategies for Cyanobacteria (Blue-Green Algae) and their Toxins: A Guide for Water Utilities. CRC for Water Quality and Treatment/WQRA Research Report 74. Water Quality Research Australia. http://www.wqra.com.au/WQRA_publications.htm

Nicholls K. H. (1982). Lakewide odours in Ontario and New-Hampshire caused by *Chrysochromulina breviturrita* Nich. (Prymnesiophyceae). *Hydrobiologia*, **96**, 91–95.

Nielsen J. L., Klausen C., Nielsen P. H., Burford M. and Jørgensen N. (2006). Detection of activity among uncultured Actinobacteria in a drinking water reservoir. *FEMS Microbiology Ecology*, **55**, 432–438.

Ogura T., Sunairi M. and Nakajima M. (2000). 2-methylisoborneol and geosmin, the main sources of soil odor, inhibit the germination of Brassicaceae seeds. *Soil Science and Plant Nutrition*, **46**, 217–227.

Olsen B. K., Chislock M. F. and Wilson A. E. (2016). Eutrophication mediates a common off-flavor compound, 2-methylisoborneol, in a drinking water reservoir. *Water Research*, **92**, 228–234.

Otten T. G., Graham J. L., Harris T. D. and Dreher T. W. (2016). Elucidation of taste-and-odor producing bacteria and toxigenic cyanobacteria by shotgun metagenomics in a Midwestern drinking water supply reservoir. *Applied and Environmental Microbiology*, **82**, 5410–5420.

Parinet J., Rodriguez M. J. and Sérodes J.-B. (2010). Influence of water quality on the presence of off-flavour compounds (geosmin and 2-methylisoborneol). *Water Research*, **44**(20), 5847–5856.

Parinet J., Rodriguez M. and Sérodes J. B. (2013). Modelling geosmin concentrations in three sources of raw water in Quebec, Canada. *Environmental Monitoring and Assessment*, **185**(1), 95–111.

Park T. J., Yu M. N., Kim H. S., Cho H. S., Hwang M. Y., Yang H. J., Lee J. C., Lee J. K. and Kim S. J. (2016). Characteristics of actinomycetes producing geosmin in Paldang Lake, Korea. *Desalination and Water Treatment*, **57**(2), 888–899.

Persson P.-E. (1983). Off-flavours in aquatic ecosytems – an introduction. *Water Science and Technology*, **15**, 1–11.

Peter A., Koster O. and von Gunten U. (2009). Occurrence of dissolved and particle-bound taste and odor compounds in Swiss lake waters. *Water Research*, **43**(8), 2191–2200.

Peterson H. G., Hrudey S. E., Cantin I. A., Perley T. R. and Kenefick S. L. (1995). Physiological toxicity, cell membrane damage and the release of dissolved organic carbon and geosmin by *Aphanizomenon flos-aquae* after exposure to water treatment chemicals. *Water Research*, **29**, 1515–1523.

Pichrtová M. and Němcová Y. (2011). Effect of temperature on size and shape of silica scales in *Synura petersenii* and *Mallomonas tonsurata* (Stramenopiles). *Hydrobiologia*, **673**, 1–11.

Piepho M., Arts M. and Wacker A. (2012). Species-specific variation in fatty acid concentrations of four phytoplankton species: does phosphorus supply influence the effect of light intensity or temperature? *Journal of Phycology*, **48**, 64–73.

Pohnert G. and Boland W. (1996). Biosynthesis of the algal pheromone hormosirene by the freshwater diatom *Gomphonema parvulum* (Bacillariophyceae). *Tetrahedron*, **52**(30), 10073–10082.

Pohnert G. (2000). Wound-activated chemical defense in unicellular planktonic algae. *Angewandte Chemie International*, **39**, 4352–4354.

Pohnert G. and Boland W. (2002). The oxylipin chemistry of attraction and defense in brown algae and diatoms. *Natural Product Reports*, **19**, 108–122.

Pohnert G. and Von Elert E. (2000). No ecological relevance of trimethylamine in fish – *Daphnia* interactions. *Limnology and Oceanography*, **45**, 1153–1156.

Qi M., Chen J., Sun X., Deng X., Niu Y. and Xie P. (2012). Development of models for predicting the predominant taste and odor compounds in Taihu Lake, China. *PLoS One*, **7**(12), e51976.

Rashash D., Dietrich A. R., Hoehn D. and Parker B. (1995). The influence of growth conditions on odor-compound production by two chrysophytes and two cyanobacteria. *Water Science and Technology*, **31**(11), 165–172.

Rashash D. M., Hoehn R. C., Dietrich A., Grizzard T. J. and Parker B. C. (1996). *Identification and Control of Odorous Algal Metabolites*. American Water Works Association Research Foundation, Denver, CO, USA, 172 p.

Richards S. R., Rudd J. W. M. and Kelly C. A. (1994). Organic volatile sulfur in lakes ranging in sulfate and dissolved salt concentration over five orders of magnitude. *Limnology and Oceanography*, **39**, 562–572.

Ridal J., Brownlee B., McKenna G. and Levac N. (2001). Removal of taste and odour compounds by conventional granular activated carbon filtration. *Water Pollution Research Journal of Canada*, **36**(1), 43–54.

Ridal J., Watson S. B. and Hicky B. (2007). Macrophytes: key substrates for taste and odour producers in the St. Lawrence River. *Water Science and Technology*, **55**(5), 15–21.

Scherzinger D. and Al-Babili S. (2008). In vitro characterization of a carotenoid cleavage dioxygenase from *Nostoc* sp. PCC 7120 reveals a novel cleavage pattern, cytosolic localization and induction by highlight. *Molecular Microbiology*, **69**(1), 231–244.

Schiffman S. S., Bennett J. L. and Raymer J. H. (2001). Quantification of odours and odorants from swine operations in North Carolina. *Agricultural and Forest Meteorology*, **108**, 213–240.

Silvey J. K. and Roach A. W. (1953). Actinomycetes in the Oklahoma City water supply. *Journal of the American Waterworks Association*, **45**, 409–416.

Simó R. (2001). Production of atmospheric sulfur by oceanic plankton: biogeochemical, ecological and evolutionary links. *Trends in Ecology and Evolution*, **16**(6), 287–294.

Škaloud P., Kristiansen J. and Škaloudová M. (2013). Developments in the taxonomy of silica-scaled chrysophytes – from morphological and ultrastructural to molecular approaches. *Nordic Journal of Botany*, **31**, 385–402.

Smith V. H., Sieber-Denlinger J., deNoyelles Jr. F., Campbell S., Pan S., Randtke S. J., Blain G. T. and Strasser V. A. (2002). Managing Taste and Odor Problems in a Eutrophic Drinking Water Reservoir. *Lake and Reservoir Management*, **18**(4), 319–323.

Soma Y., Imaizumi T., Yagi K. and Kasuga S. (1993). Estimation of algal succession in lake water using HPLC analysis of pigments. *Canadian Journal of Fisheries and Aquatic Sciences*, **50**, 1142–1146.

Sorichetti R. J., McLaughlin J. T., Creed I. F. and Trick C. G. (2014). Suitability of a cytotoxicity assay for detection of potentially harmful compounds produced by freshwater bloom-forming algae. *Harmful Algae*, **31**, 177–187.

Steffen M., Belisle S., Watson S. B., Boyer G. L. and Wilhelm S. W. (2014). Status, causes and controls of cyanobacterial blooms in Lake Erie. *Journal of Great Lakes Research*, **40** (2), 215–225.

Su M., Jia D., Yu J., Vogt R. D., Wang J., An W. and Yang M. (2017). Reducing production of taste and odor by deep-living cyanobacteria in drinking water reservoirs by regulation of water level. *Science of The Total Environment*, **574**(Supplement C), 1477–1483.

Suffet I. H. and Rosenfeld P. (2007). The anatomy of odour wheels for odours of drinking water, wastewater, compost and the urban environment. *Water Science and Technology*, **55**(5), 335–344.

Sugiura N. and Nakano K. (2000). Causative microorganisms for musty odor occurrence in the eutrophic Lake Kasumigaura. *Hydrobiologia*, **434**, 145–150.

Sugiura N., Yagi O. and Sudi R. (1987). Effect of carbohydrate in the sediment on the musty odor production by actinomycetes. *Environmental Technology Letters*, **8**(2), 95–103.

Sugiura N., Inamori Y., Hosaka Y., Sudo R. and Takahashi G. (1994). Algae enhancing musty odor production by actinomycetes in Lake Kasumigaura. *Hydrobiologia*, **288**(1), 57–64.

Sugiura N., Utsumi M., Wei B., Iwami N., Okano K., Kawauchi Y. and Maekawa T. A. (2004). Assessment for the complicated occurrence of nuisance odours from phytoplankton and environmental factors in a eutrophic lake. *Lakes and Reservoirs*, **9**, 195–201.

Suurnäkki S., Gomez-Saez G. V., Rantala-Ylinen A., Jokela J., Fewer D. P. and Sivonen K. (2015). Identification of geosmin and 2-methylisoborneol in cyanobacteria and molecular detection methods for the producers of these compounds. *Water Research*, **68**, 56–66.

Tang T., Hoefel D., Mosisch T. and Ho T. (2013). Assessing the fate and biodegradation of cyanobacterial metabolites in Australian waters. *Water Practice and Technology*, **7**(4), doi: 10.2166/wpt.2012.064.

Tao M., Xie P., Chen J., Qin B., Zhang D., Niu Y., Zhang M., Wang Q. and Wu L. (2012). Use of a generalized additive model to investigate key abiotic factors affecting microcystin cellular quotas in heavy bloom areas of Lake Taihu. *PLoS One*, **7**(2), e32020.

Tosi L. and Sola C. (1993). Role of geosmin, a typical inland water odour, in guiding glass eel *Anguilla anguilla* (L.) migration. *Ethology*, **95**, 177–185.

Tsao H.-W., Michinaka A., Yen H.-K., Giglio S., Hobson P., Monis P. and Lin T.-F. (2014). Monitoring of geosmin producing *Anabaena circinalis* using quantitative PCR. *Water Research*, **49**, 416–425.

Tsuchiya Y., Watanabe M. F. and Watanabe M. (1992). Volatile organic sulfur compounds associated with blue-green algae from inland waters of Japan. *Water Science and Technology*, **25**, 128–130.

Utkilen H. C. and Froshaug M. (1992). Geosmin production and excretion in a planktonic and benthic *Oscillatoria*. *Water Science and Technology*, **25**(2), 199–206.

Van Durme J., Goiris K., De Winne A., De Cooman L. and Muylaert K. (2013). Evaluation of the volatile composition and sensory properties of five species of microalgae. *Journal of Agricultural and Food Chemistry*, **61**(46), 10881–10890.

Van Keulen G., Alderson J., White J. and Sawers R. G. (2007). The obligate aerobic actinomycete *Streptomyces coelicolor* A3(2) survives extended periods of anaerobic stress. *Environmental Microbiology*, **9**, 3143–3149.

Wacker A., Piepho M., Harwood J. L., Guschina I. A. and Arts M. T. (2016). Light-induced changes in fatty acid profiles of specific lipid classes in several freshwater phytoplankton species. *Frontiers in Plant Science*, **7**, 264.

Wang C. M. and Cane D. (2008). Biochemistry and molecular genetics of the biosynthesis of the earthy odorant methylisoborneol in *Streptomyces coelicolor. Journal of the American Chemical Society*, **130**, 8908–8909, doi: 10.1021/jf403112k.

Wang Z., Xu Y., Shao J., Wang J. and Li R. (2011). Genes associated with 2-methylisoborneol biosynthesis in cyanobacteria: isolation, characterization, and expression in response to light. *PLoS One*, **6**(4), e18665.

Wang Z., Song G., Shao J., Tan W., Li Y. and Li R. (2016). Establishment and field applications of real-time PCR methods for the quantification of potential MIB-producing cyanobacteria in aquatic systems. *Journal of Applied Phycology*, **28**(1), 325–333.

Watson S. B. (2003). Cyanobacterial and eukaryotic algal odour compounds: signals or by-products? A review of their biological activity. *Phycologia*, **42**(4), 332–350.

Watson S. B. (2010). Algal Taste and Odour. Chapter 15 in Algae: Source to Treatment. Manual of Water Supply Practices – M57. American Water Works Association, Denver, CO, ISBN 978-1-58321-787-0.

Watson S. B. (2012). Ecotoxicity of taste and odour compounds. In: Encyclopaedia of Aquatic Ecotoxicology, J. F. Férard and C. Blaise (eds), Springer, New York, USA.

Watson S. B. and Jüttner F. (2017). Malodorous volatile organic sulfur compounds: sources, sinks and significance in inland waters. *Critical Reviews in Microbiology*, **43**(2), 210–237.

Watson S. B. and Satchwill T. (2003). Chrysophyte odour production: resource-mediated changes at the cell and population levels. *Phycologia*, **42**(4), 393–405.

Watson S. B., Brownlee B., Satchwill T. and McCauley E. (1999). The use of solid phase microextraction (SPME) to monitor for major organoleptic compounds produced by chrysophytes in surface waters. *Water Science and Technology*, **40**(6), 251–256.

Watson S. B., Satchwill T. and McCauley E. (2001a). Drinking water taste and odour: a chrysophyte perspective. *Nova Hedwigia*, **122**, 119–146.

Watson S. B., Satchwill T., Dixon E. and McCauley E. (2001b). Under-ice blooms and source-water odour in a nutrient-poor reservoir: biological, ecological and applied perspectives. *Freshwater Biology*, **46**(11), 1553–1567.

Watson S. B., Charlton M., Rao Y. R., Howell T., Ridal J., Brownlee B., Marvin C. and Millard S. (2007a). Off flavours in large waterbodies: physics, chemistry and biology in synchrony. *Water Science and Technology*, **55**(5), 1–8.

Watson S. B., Jüttner F. and Köster O. (2007b). *Daphnia* behavioural responses to taste and odour compounds: ecological significance and application as an inline treatment plant monitoring tool. *Water Science and Technology*, **55**(5), 23–31.

Watson S. B., Ridal J. and Boyer G. L. (2008). Taste and odour and cyanobacterial toxins: impairment, prediction, and management in the Great Lakes. *Canadian Journal of Fisheries and Aquatic Sciences*, **65**(8), 1779–1796.

Watson S. B., Caldwell G. and Pohnert G. (2010). Fatty acids and oxylipins as semiochemicals. In: Lipids in Aquatic Ecosystems, M. Arts, M. Brett and M. Kainz (eds), Springer-Verlag, New York.

Watson S. B., Whitton B. A., Higgins S. N., Paerl H. W., Brooks B. W. and Wehr J. D. (2015). Harmful algal blooms. In: Freshwater Algae of North America, J. D. Wehr, R. Sheath and P. Kociolek (eds), 2nd edn, Elsevier, the Netherlands, pp. 873–905.

Watson S. B., Monis P., Baker P. and Giglio S. (2016). Biochemistry and genetics of taste and odor producing cyanobacteria. *Harmful Algae*, **54**, 112–127.

Wee J. L., Harris S. A., Smith J. P., Dionigi C. P. and Millie D. F. (1994). Production of the taste/odour compound, *trans*-2,*cis*-6-nonadienal within the Synurophyceae. *Journal of Applied Phycology*, **6**, 365–369.

Weinberger F., Lio U., Delage L., Kloareg B., Potin P., Beltran J., Flores V., Faugeron S., Correas J. and Pohnert G. (2011). Up-Regulation of lipoxygenase, phospholipase, and oxylipin-production in the induced chemical defense of the red alga *Gracilaria chilensis* against epiphytes. *Journal of Chemical Ecology*, **37**, 677–686.

Wendel T. and Jüttner F. (1996). Lipoxygenase-mediated formation of hydrocarbons and unsaturated aldehydes in freshwater diatoms. *Phytochemistry*, **41**, 1445–1449.

Wichard T. and Pohnert G. (2006). Formation of halogenated medium chain hydrocarbons by a lipoxygenase/hydroperoxide halolyase-mediated transformation in planktonic microalgae. *Journal of the American Chemical Society*, **128**, 7114–7115.

Wise P. M., Miyazawa T., Gallagher M. and Preti G. (2007). Human odor detection of homologous carboxylic acids and their binary mixtures. *Chemical Senses*, **32**, 475–482.

Wnorowski A. U. and Scott W. E. (1992). Incidence of off-flavors in South African surface waters. *Water Science and Technology*, **25**, 225–232.

Wolfe G. V., Steinke M. and Kirst G. O. (1997). Grazing-activated chemical defence in a unicellular marine alga. *Nature London*, **387**, 894–897.

Wood S., Williams S. T. and White W. R. (1985). Potential sites of geosmin production by streptomycetes in and around reservoirs. *Journal of Applied Microbiology*, **58**, 319–326.

Wu J. T. and Jüttner F. (1988). Differential partitioning of geosmin and 2-methylisoborneol between cellular constituents in *Oscillatoria tenuis*. *Archives of Microbiology*, **150**, 580–583.

Wu J. T., Ma P. I. and Chou T. L. (1991). Variation of geosmin content in *Anabaena* cells and its relation to nitrogen utilization. *Archives of Microbiology*, **157**, 66–69.

Yano H., Nakahara M. and Ito H. (1988). Water blooms of *Uroglena americana* and the identification of odorous compounds. *Water Science and Technology*, **20**, 75–80.

Yu J. W., Zhao Y. M., Yang M., Lin T.-F., Guo Z. H., Gu J. N., Li S. and Han W. (2009). Occurrence of odour-causing compounds in different source waters of China. *Journal of Water Supply: Research and Technology – AQUA*, **58**(8), 587–594.

Zaitlin B. and Watson S. B. (2006). Actinomycetes in relation to taste and odour in drinking water: myths, tenets and truths. *Water Research*, **40**(9), 1741–1753.

Zaitlin B., Watson S. B., Ridal J., Satchwill T. and Parkinson D. (2003a). Actinomycetes in Lake Ontario: habitats and geosmin and MIB production. *Journal of the American Water Works Association*, **95**(2), 113–118.

Zaitlin B., Watson S. B., Dixon J. and Steel D. (2003b). Actinomycetes in the Elbow River basin, Alberta, Canada. *Water Quality Research Journal of Canada*, **38**(1), 115–125.

Zeeshan M. and Prasad S. M. (2009). Differential response of growth, photosynthesis, antioxidant enzymes and lipid peroxidation to UV-B radiation in three cyanobacteria. *South African Journal of Botany*, **75**, 466–474.

Zhang T., Li L., Song L. and Chen W. (2009). Effects of temperature and light on the growth and geosmin production of *Lyngbya kuetzingii* (Cyanophyta). *Journal of Applied Phycology*, **21**, 279–285.

Zhang X.-J., Chen C., Ding J.-Q., Hou A., Li Y., Niu Z.-B., Su X.-Y., Xu Y.-J. and Laws E. A. (2010). The 2007 water crisis in Wuxi, China: analysis of the origin. *Journal of Hazardous Materials*, **182**(1–3), 130–135.

Zhao Y., Yu J., Su M., An W. and Yang M. (2013). A fishy odor episode in a north China reservoir: occurrence, origin, and possible odor causing compounds. *Journal of Environmental Sciences*, **25**, 2361–2366.

Zhou B., Yuan R., Shi C., Yu L., Gu J. and Zhang C. (2011). Biodegradation of geosmin in drinking water by novel bacteria isolated from biologically active carbon. *Journal of Environmental Sciences*, **23**, 816–823.

Zimba P. V., Dionigi C. P. and Millie D. M. (1999). Evaluating the relationship between photopigment synthesis and 2-methylisoborneol accumulation in cyanobacteria. *Journal of Phycology*, **35**(6), 1422–1429.

Zuo Y., Li L., Wu Z. and Song L. (2009). Isolation, identification and odour-producing abilities of geosmin/2-MIB in actinomycetes from sediments in Lake Lotus, China. *Journal of Water Supply: Research and Technology – AQUA*, **58**(8), 552–561.

Zuo Y., Li L., Zhang T., Zheng L., Dai G., Liu L. and Song L. (2010). Contribution of *Streptomyces* in sediment to earthy odor in the overlying water in Xionghe Reservoir, China. *Water Research*, **44**(20), 6085–6094.

Chapter 4

Chemical analytical techniques for taste and odour compounds

Auguste Bruchet

CIRSEE, 38 Rue du President Wilson, 78230 Le Pecq, France

4.1 INTRODUCTION

The initial isolation (Gerber & Lechevalier, 1965) and identification of geosmin, one of the most common problem-odour compounds in natural waters (as noted in the introduction and Chapter 2), was a very tedious analytical process. The analysis used preparative gas chromatography of concentrated extracts from actinomycetes cultures followed by the use of rather insensitive analytical methods such as infrared spectroscopy and nuclear magnetic resonance spectroscopy (Gerber, 1968). The first detection of geosmin in real waters quickly followed (Rosen *et al.*, 1970). Forty years later, many utility laboratories are able to routinely detect geosmin and the equally important 2-methylisoborneol (2-MIB) at low ppt (parts per trillion; ng/L) levels in water samples using a fully automated solid-phase-microextraction method (SPME) coupled with gas-chromatography mass spectrometry (GC/MS).

During these five decades, progress in our understanding of the origins of taste-and-odour (T&O) episodes has been closely linked with advances in the development of new sample preparation techniques combined with the ever increasing sensitivity of separation-detection-identification methods. For example, the invention of the closed-loop-stripping (CLS) extraction technique (Grob & Zürcher, 1976) together with the wider availability of low resolution quadrupole

 Taste and Odour in Source and Drinking Water: Causes, Controls, and Consequences
Editor(s): Tsair-Fuh Lin, Susan Watson, Andrea M. Dietrich and I. H. (Mel) Suffet
doi: 10.2166/9781780406664_0113

GC/MS instruments at the beginning of the 1980s, allowed the routine determination of geosmin and 2-MIB with detection limits close to their odour-detection thresholds of 1–10 ng/L (see Table 2.3 in Chapter 2). Using the same CLS analysis (CLSA)-GC/MS approach, many new families of water odourants were discovered during the 1980–1990 decade; in particular, swampy odours in drinking-water distribution systems could be attributed to dimethyl trisulfide, a compound with an odour threshold of 10 ng/L (Wajon *et al.*, 1985). Fruity odours in ozonated waters were linked with the formation of C7–C14 aliphatic aldehydes (Anselme *et al.*, 1988). The compound responsible for the dominant cucumber odour of a surface bloom of *Synura petersenii* was identified for the first time as *trans*-2,*cis*-6-nonadienal (Hayes & Burch, 1989) and subsequently during an important cucumber-odour episode in Philadelphia drinking water (Burlingame *et al.*, 1992). Despite rather low recoveries, CLSA-GC/MS, or its open loop modification, also allowed the discovery of iodinated trihalomethanes (THMs) in disinfected drinking waters as the cause of typical medicinal odours (Bruchet *et al.*, 1989; Hansson *et al.*, 1987).

Due to their sub ng/L odour detection thresholds, halogenated anisoles often remain undetected during CLSA-GC/MS analyses. The first detection of 2,3,6-trichloroanisole (TCA) after a slow sand filter pilot plant spiked with 2,3,6-trichlorophenol was accomplished using a tedious XAD-4 resin extraction procedure followed by HPLC-UV analysis of the resin eluate and confirmation by GC/MS (Montiel *et al.*, 1987). Again, extraction by open-loop stripping followed by preparative GC guided by sensory-GC, then GC with electron-capture detection (ECD) for quantitation and GC/MS for confirmation were necessary to identify 2,4,6-TCA at 0.1–3 ng/L in various Swedish waters (Nyström *et al.*, 1992). Coupling a CLSA extraction with a large volume injection (LVI) of the extract finally allowed detection of 2,4,6-TCA and other earthy-musty odourants at levels below 50 pg/L, that is below its odour-detection threshold (Malleret *et al.*, 2001). The same approach allowed the first identification of 2,4,6-tribromoanisole, another musty odourant, in drinking water (Malleret & Bruchet, 2002).

At the turn of the new millennium, CLSA extraction was challenged by the use of more modern, more automated extraction techniques such as SPME or stir-bar-sorptive-extraction (SBSE) for many T&O compounds such as iodinated THMs (Cancho *et al.*, 1999), earthy-musty and other odourants (Benanou *et al.*, 2004; Watson *et al.*, 2000). On the other hand, the simultaneous-distillation-extraction technique that helped solve an important episode caused by plastic additives from polyethylene pipes during the 1980s (Anselme *et al.*, 1985) had fallen into disuse by the end of the 1990s.

In contrast, exemplary techniques that have been used throughout the history of T&O research, and continue to be used today include sensory-GC and liquid–liquid extraction (LLE). Since its invention by Fuller *et al.* (1964), sensory-GC with its application in detecting drinking water odours by Khiari *et al.* (1992) was often decisive in pinpointing odourous compounds in complex chromatograms and eventually led to the identification of real odour causative agents such as geosmin in wastewaters (Agus *et al.*, 2011), dioxanes and dioxolanes (Preti *et al.*, 1993) or compounds causing grassy odours (Khiari *et al.*, 1999). Since the determination of the odour characteristics of various chlorinated phenols (Burttschell *et al.*, 1959), LLE has been the most common method for extracting this group of medicinal odour causative agents at low ppt levels. It also remains the extraction method of choice when dealing with various chemical odours encountered during chemical spills.

The purpose of this chapter is to show the principles and areas of applications in the T&O field of analytical methods in use today. Whenever possible, these methods will be illustrated with detailed practical examples of application. Perspectives on new methods will also be discussed.

4.2 TARGET CHEMICALS AND DETECTION LIMITS

Table 4.1 lists some of the most common odourants in waters, with their odour descriptors and odour threshold concentrations (OTC) or taste threshold concentrations (FTC) and octanol/water partition coefficients ($\log_{10} K_{ow}$). The latter parameter is an indication of the polarity/hydrophobicity of a chemical. The higher this parameter, the easier the extraction of the compound will be. $\log K_{ow}$ values were as reported by Young and Suffet (1999), and Pirbazari *et al.* (1992) or were calculated theoretically by means of a commercial computer program that breaks down the molecule and assigns a value to each fragment according to a database. The final value is the sum of the partial values with a final correction after considering the structure of the molecule. A few values were retrieved from the web. Compounds with a $\log K_{ow}$ above 3 are considered as hydrophobic and will be easy to extract from water while compounds with a $\log K_{ow}$ below 2 have a high water solubility and will be more difficult to extract.

Due to variations in published odour thresholds, references for the values given in Table 4.1 have been indicated. Whenever a given reference indicated several values for the same compound, the lowest one was considered. Thresholds included in Table 4.1 are odour-threshold levels, except for the halogenated phenols that are perceived at lower concentrations by taste. As a general rule, the target LOD (limit of detection) for an odourant should be set at one tenth or one fifth of the OTC to account for differences in sensitivity in the general population. For compounds with a much higher OTCs, the target LOD can be set at a higher fraction of the odour threshold. In some extreme cases such as 2,4,6-TCA and 2,4,6-tribromoanisole that have the lowest known OTCs amongst water odourants, reaching a LOD equal to or slightly below the OTCs is the only realistic goal at the present time.

Table 4.1 Sensory and physicochemical characteristics of common odourous compounds as shown on the Taste and Odour Wheel, Chapter 2.

Compound	Descriptor	OTC or FTC (ng/L)	log K_{ow}	Target Limit of Detection (LOD, ng/L)
Geosmin	Earthy	4[1]	3.7	1
2-MIB	Musty	9[1]	3.13	1
2-Methoxy-3-isopropylpyrazine	Earthy-musty	2[1]	2.41	1
2-Methoxy-3-isobutylpyrazine	Earthy-musty	2[1]	2.72	1
2,3-6-Trichloroanisole (TCA)	Musty	7[1]	–	1
2,4,6-Trichloroanisole	Musty	0.02–0.08[2]	3.91	0.02
2,4,6-Tribromoanisole	Musty	0.03[12]		0.02
Iodoform	Medicinal	20[1]–32[2]	3.03	10
Bromodiiodomethane	Medicinal	130[2]	3.52	50
2,6-Dichlorophenol	Medicinal	14[3]	2.75	5
2,6-Dibromophenol	Medicinal	6[3]	3.36	5
2-Bromophenol	Medicinal	14[3]	2.35	5

Hexanal	Lettuce-Fruity	300[1]	1.78	100
Heptanal	Rancid-Fishy-Fruity	3,000[1]	2.26	100
Octanal	Rancid-Fruity	700[1]	2.78	100
Nonanal	Orange	1,000[1]	3.27	100
Decanal	Orange-Fruity	100[1]	3.76	20
2-Ethyl-4-methyl-1,3-dioxolane (2-EMD)	Tutti-frutti/Olive oil	5–10[4]	1.76	5
2-Ethyl-5,5-dimethyl-1,3-dioxane (2-EDD)	Tutti-frutti/Olive oil	10[5]	1.96	5
trans,trans-2,4-Heptadienal	Fishy	2,500[6]	2.11	1,000
trans,2-*cis*-Nonadienal	Cucumber	2.2[7]	3.19	1
cis-3-Hexen-1-ol	Grassy	50,000[8]	1.52	1,000
Dimethyl trisulfide	Swampy	10[9]	1.87	5
Dimethyl disulfide	Decaying vegetation	200[10]	1.77	20
Hydrogen sulfide	Rotten egg	1,100[11]	0.23	100
MTBE (methyl tertiarybutylether)	Sweet solvent	15,000[13]	0.94	1,000
ETBE (ethyl tertiarybutylether)	Ether	1,000[14]	1.9	100
Petroleum products	Hydrocarbon	1,000[15]		1,000

Notes: [1]Suffet *et al.* (1995); [2]Khiari *et al.* (2002); [3]Piriou *et al.* (2002); [4]Schweitzer *et al.* (1999); [5]Preti *et al.* (1993); [6]Young and Suffet (1999); [7]Burlingame *et al.* (1992); [8]Khiari *et al.* (1995); [9]Wajon *et al.* (1985); [10]Van Gemert (1977); [11]Mallevialle and Suffet (1987); [12] Whitfield *et al.* (1997); [13]Suffet (2007); [14]Van Wezel *et al.* (2009); [15]Mallevialle *et al.* (1982).

4.3 CLOSED-LOOP-STRIPPING ANALYSIS

During last two decades, CLSA-GC/MS has been the method of choice for the determination of ppt levels of volatile organics. In particular, CLSA-GC-MS allows the sub-threshold detection of odourous algal metabolites such as geosmin, methylisoborneol or 2-isopropyl- or 2-isobutyl-3-methoxypyrazine which yield intense earthy-musty odours at levels as low as 2–5 ng/L. The original design of Grob and Zürcher (1976) has been modified by several authors and a commercial apparatus is readily available (Brechbühler, Switzerland). In the closed-loop design (Figure 4.1), a sample bottle containing 1 litre of water is placed in a thermostated water bath (45°C) and connected to the closed loop. An air pump circulates the air contained in the loop at a gas flow rate of 1 L/min and generates fine bubbles through a glass frit immersed at the bottom of the bottle, in order to strip volatile compounds. A filter containing 1.5 mg of activated carbon adsorbent is inserted in the loop to adsorb the volatiles. The carbon trap is placed immediately downstream of a heater which maintains an air temperature of 55–60° C to prevent condensation of water on the trap. The system is run at equilibrium for 2 h. The loop is then disconnected and the carbon trap eluted with a small amount of solvent, usually 25 μL of carbon disulfide. Prior to extraction, the water sample is spiked with internal standards (traditionally 4 chloroalkanes, C_6, C_{10}, C_{12}, C_{16} at 200 ng/L each) and the carbon trap is spiked with an equivalent amount of two additional chloro-alkane internal standards (C_8, C_{14}) to check for recoveries. Elution of the carbon trap is the only step that requires some skill. It is carried out with four very small portions of carbon disulfide, 10, 5, 5 and 5 μL, respectively. To allow such small volumes to travel through the carbon trap, the latter is connected with short Teflon tubing to a glass receiver vial. Rotating both the trap holder and the glass receiver allows the solvent to travel 10 times through the trap. At the end, the solvent is left in the trap and an ice cube is applied along the glass receiver in order to attract, by air depression, the entire volume of solvent. The receiver is then capped with a Teflon stopper and kept at −20°C.

The very low detection limits achieved with this method stem from the high concentration factor implemented (40,000 times) and the absence of a solvent concentration step. As an illustration, a 1 ng/L concentration in the initial water sample corresponds to a concentration of 40 pg/μL in the CS_2 extract (assuming a 100% recovery) which is within reach of most modern benchtop GC-MS instruments. CLSA has been mostly useful to extract earthy-musty odourants such as geosmin, 2-MIB and the alkylpyrazines included in Table 4.1. It also works well for the C_6–C_{10} aliphatic aldehydes and probably remains the sole extraction method allowing detection at the 10 ppt odour threshold of dimethyl trisulfide. Even for compounds which are poorly extracted, the high concentration factor achieved makes it an excellent screening method for many compounds listed in Table 4.1 including iodinated THMs, dioxanes and dioxolanes, *trans*, *trans*-2,4-heptadienal, *trans*,2-*cis*-nonadienal, *cis*,3-Hexen-1-ol. It will also allow the detection of low levels of the light hydrocarbons such as gasoline and white spirit. On the other hand, halogenated phenols cannot be extracted using this method.

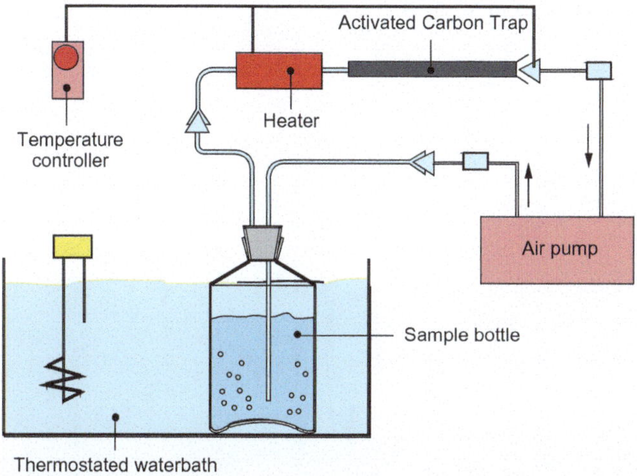

Figure 4.1 CLSA apparatus set-up.

Since the turn of the millennium, CLSA has been almost entirely replaced by SPME and SBSE for its most traditional application, i.e. the quantitative analysis of geosmin, MIB and alkylpyrazines. However, because of the wide range of odourous compounds that can be identified by this method, it remains in use in a limited number of laboratories wishing to maintain their full T&O diagnosis capability. In some cases, CLSA could not be replaced by SPME or SBSE to monitor the evolution of industrial contaminants discovered during T&O episodes. An example of this is illustrated below.

In February 1998, the Suez-Environment research centre received a sample of borehole water to examine the cause of musty-medicinal odour complaints. A previous sample collected in December 1997 had not allowed the detection of musty odourants; the presence of trifluoromethylaniline had however been reported. With this new sample, use of CLSA in combination with a LVI in GC/MS (see later) allowed the identification of 2,3,6-TCA and 2,4,6-TCA at about 1 ng/L each, easily explaining the musty odour. 2,3,6- and 2,4,6-TCA were confirmed on later samples in July and November 1998, and March 1999. Iodo-THMs and four aniline derivatives were also detected, likely explaining the medicinal odour. In October 1999, the TCAs had disappeared, but an industrial contamination by chloro-and fluoro-anilines became obvious. An example chromatogram obtained by CLSA-GC/MS showing four unusual fluorinated aromatics is shown in Figure 4.2. After considering the shutdown of the aquifer, the health authorities decided to implement weekly monitoring of different boreholes and the mixture of boreholes used to produce drinking water with a maximum contaminant level set at 30 ng/L for each individual industrial contaminant in drinking water. An example of the results from this weekly monitoring is reported in Table 4.2.

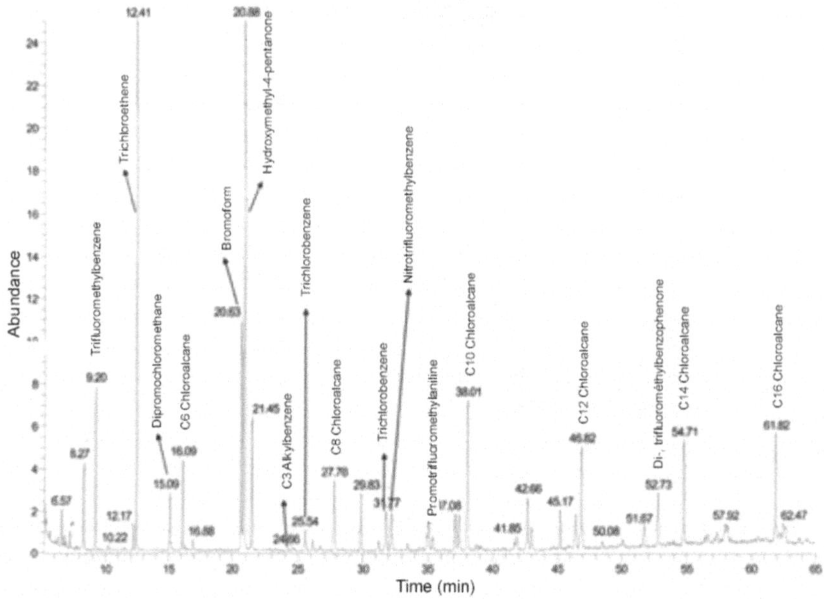

Figure 4.2 CLSA-GC/MS chromatogram showing four fluorinated organics in a contaminated borehole.

Table 4.2 Example of weekly monitoring, February (ng/L) in different boreholes.

Contaminant	Well 1	Well 2	Well 3	Well 5	Well 6
Trifluoromethylbenzene	310	210	70	<10	970
Chlorotrifluoromethylbenzene	20	10			20
Chlorotrifluoromethylbenzene	40	30			80
Trifluoromethylanisole					20
Trifluoromethylaniline			30		630
Chlorotrifluoromethylaniline	20				
Nitrotrifluoromethylbenzene	400	170			460
Trifluoromethylaniline	10		20		530
Nitrotrifluoromethylbenzene	20	10			50
Isothiocyanatotrifluoromethylbenzene	30				20
Di-, trifluoromethylbenzophenone	250	110		40	50

Few analytical standards were available when the contamination was discovered, hence the concentrations were calculated relative to a chloroalkane internal standard. This is the first time a fluorinated anisole has been reported in a ground water, but as the musty odour disappeared with 2,3,6- and 2,4,6-TCA, it is unlikely that this fluorinated anisole, having its three fluorine atoms attached to the methyl group, has a similarly low OTC than the chlorinated and brominated anisoles which have the halogenated atoms directly attached to the aromatic ring. As of 2018, monitoring of this aquifer is still on-going, although at a much lower frequency. During this interval, attempts at replacing CLSA by a more automated technique such as SPME or SBSE failed, demonstrating the value of maintaining the CLSA extraction technique.

4.4 STATIC HEAD-SPACE EXTRACTION

Static head-space extraction has been practiced since the early days of GC. In the case of water, the liquid phase is placed in a sealed vial leaving a volume of air (called the head-space) above it. After some time, each volatile analyte will reach an equilibrium between the gas and liquid phase. To speed up the time needed to reach the equilibrium, the sample vial is placed in a thermostated bath regulated at a higher than ambient temperature. An aliquot of the head space is then withdrawn with a gas syringe and injected into a heated GC injection port, usually in the split mode. This injection can be carried out manually but, since an auto sampler for head space became commercially available (in 1967 from Perkin Elmer) before any type of liquid auto sampler, the injection is almost always carried out automatically.

The equilibrium of a given analyte is controlled by its Henry's constant. The higher this constant, the higher the concentration in the gas phase. For example, geosmin has a much lower Henry's constant (6.6×10^{-5} atm m^3/mol) than chloroform (300×10^{-5} atm m^3/mol) and will not be amenable to static head space while the analysis of the latter solvent/THM has been described in standard methods using a head-space GC-ECD (ISO 10301, 1997) technique that yields a limit of detection of 0.3 µg/L. Due to insufficient sensitivity, head-space-GC/MS has long been considered inadequate for solving T&O episodes. However, progress in the sensitivity of modern GC/MS instruments makes it a viable option during episodes characterized by solvent odours when the presence of specific solvents above 0.1–0.2 µg/L is suspected.

To illustrate this, in June 2009, consumers from four cities in the Bordeaux area (France) complained of odours in their drinking water. These cities are supplied by a plant treating groundwater influenced by a surface water stream. Treatment includes coagulation, sand filtration and granular activated carbon filtration followed by disinfection with chlorine dioxide. Panelists described the odour as 'ether-like'. Three samples collected in the distribution system, and at the inlet of the plant supplying this system as well as a chlorination tank were shipped to the author's institute and analyzed by HS-GC/MS under the following conditions: water temperature, 85°C for 15 min; Injection of 2 mL of the head-space; PTV (programmed temperature vaporizer) injector temperature: programmed from −20°C (0.3 min) to 220°C at 720°C/min; Separation on a 30 m VOCOL (Supelco), 0.25 mm id, 1.5 μm column programmed from 5°C (3 min) to 100°C at 7°C/min, then 100 to 220°C at 80°C/min, held for 7 min; Column flow rate: 1.2 mL/min; MS detection in selected ion monitoring (SIM, ions at 59 and 87 m/z)/Full scan (25–450 amu) mode.

The chromatogram obtained on the distribution system sample is shown in Figure 4.3. A small peak eluting at 10.3 min that was hardly visible on the full scan chromatograms (Figure 4.3, top) appeared outstanding in the SIM mode using the ions at 59 and 87 (Figure 4.3, bottom). The identity of the compound was confirmed with an authentic standard as ethyltertiarybutylether (ETBE) which is incorporated in France as a fuel oxygenate instead of MTBE which is commonly used in other countries. Quantification results indicated a concentration of 2–2.3 μg/L for ETBE in the chlorination tank and in the distribution system. This is higher than the published OTC of 1 μg/L for ETBE (Table 4.1). The other compounds visible in the full scan mode (dichloromethane, acetic acid, toluene, tetrachloroethylene, xylene) are below their individual OTCs. This contamination was traced back to leakage from a gas station that eventually reached the stream supplying this drinking water plant. Investigation of this long-term episode, in particular to determine the best treatment technology and to more accurately define the odour detection, recognition and objection thresholds, was still on-going at the time of writing. This example indicates that head-space-GC/MS has become a viable tool for quickly diagnosing odour episodes whenever volatile compounds above 0.1–0.2 μg/L are the causative agents.

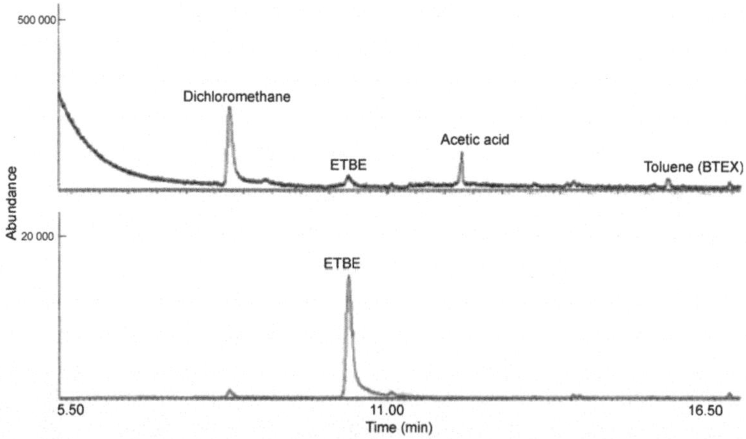

Figure 4.3 Compounds identified by static head-space-GC/MS in drinking water displaying an ether odour.

4.5 SOLID-PHASE MICRO-EXTRACTION

Solid-phase micro-extraction (SPME) has been developed by Arthur and Pawliszyn (1990) and was made commercially available three years later by the Supelco Corporation. Because of its simplicity, rapidity and sensitivity, this technique became rapidly popular and has become the method of choice for the quantification of geosmin and 2-MIB at low ppt levels in water samples.

With SPME, an extraction fibre coated at its extremity with an extraction phase (0.5 μL, 7 to 100 μm film thickness) has been inserted in the needle of a GC injection syringe. During the extraction phase the fibre is pushed outside of the needle into the headspace above an aqueous sample to extract volatile analytes (Figure 4.4, middle right), or is directly dipped into the liquid when more polar molecules are targeted. Following the extraction, the fibre is pulled back inside the needle and the syringe is inserted into a heated GC injection port and the fibre is pushed outside again to achieve thermal desorption of the adsorbed analytes (Figure 4.4, right). SPME analysis can be carried out manually with a dedicated syringe or the whole process can be automated with an autosampler. The different steps involved in SPME are represented in Figure 4.4.

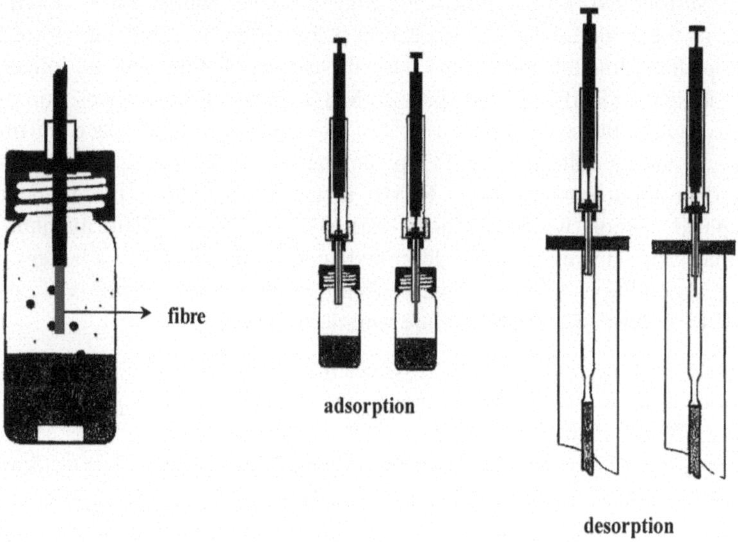

Figure 4.4 Principle of head-space SPME.

From a theoretical point of view, the amount of an analyte absorbed by the coating is directly related to its concentration in the sample

$$n = \frac{K_{fs} V_f \, C_0 V_s}{K_{fs} V_f + V_s}$$

where n is the mass of the analyte absorbed by the coating, V_f is the volume of the coating, V_s is the volume of the sample, K_{fs} is the partition coefficient of the analyte between the coating and the sample, and C_0 is the initial concentration of the analyte in the sample.

During head-space SPME, the equilibrium of the analyte between the liquid and gaseous phase, governed by its Henry's coefficient, also plays a role by controlling the initial concentration, C_0. Because of the low amount of phase used, exhaustive extraction of an analyte could only occur if K_{fs} V_f is much greater than V_s resulting in an amount extracted, $n = C_0V_s$. For most analytes, absolute recoveries are in the order of 10% or less. Therefore, calibration using standards prepared under the same conditions as the samples (water to air ratio, extraction temperature and time) is required.

The most common phases used are homogenous polymers such as PDMS (polydimethysiloxane) for the extraction of non-polar analytes, polyacrylate for more polar compounds, or polymers coated with porous particles (e.g. PDMS/divinylbenzene, PDMS/carboxen (CAR), PDMS/carboxen/divinylbenzene (DVB)) that increase the fibre capacity and selectivity.

During direct SPME where the fibre is in contact with the aqueous phase, the fibre is surrounded with a thin static aqueous layer which slows down mass transfer of the analytes into the extraction phase. Vigorous agitation is needed to speed up the extraction but in any case the extraction time will be longer than with head-space SPME. In head-space SPME, transfer of the analytes from the air to the extraction phase is usually very fast. As usual with head-space, stirring the sample or adding salts can speed up the transfer from the aqueous to the air phase. One application of direct SPME in the T&O field is the extraction of halogenated phenols which has to be carried out at acidic pH to maintain their molecular form. However, SBSE described later in this chapter, is a better option.

Typical conditions that can serve as a starting basis for the analysis of geosmin and 2-MIB in head-space-GC/MS are summarized below.

Incubation parameters: 10 mL of water in a 20 mL vial, with 100 g/L NaCl; Incubation at 65°C for 20 min with agitation at 320 rpm; Fibre type: DVB/CAR/PDMS.

GC/MS conditions: PTV injector programmed from 220 to 270°C (5 min) at 720°C/min; Column type: 30 m × 0.25 mm × 1.5 μm; Oven program: 40–230°C at 12°C/min, then held for 9.2 min; MS interface Temp: 230°C; Acquisition in SIM (*m*/*z*: 95 for MIB, 112 for geosmin) full scan mode (mass range 70–450 amu).

An example of a SIM chromatogram obtained under these conditions with 25 ng/L geosmin and MIB is shown in Figure 4.5. The limit of quantification is 2 ng/L for MIB and 1 ng/L for geosmin. With a GC run time of 25 min, the method allows the analysis of 15–20 samples per day, in addition to standard solutions and quality controls (QC). It is worth mentioning that the addition of 100 g/L sodium chloride will induce some lysis of algal cells. Consequently, if the analysis of both the total and dissolved forms of geosmin and MIB is required, the sample should be analyzed without and with prefiltration. Drinking water needs to be dechlorinated before SPME analysis to avoid significant underestimation of analytes such as geosmin, 2-MIB and MTBE (Lin *et al.*, 2003).

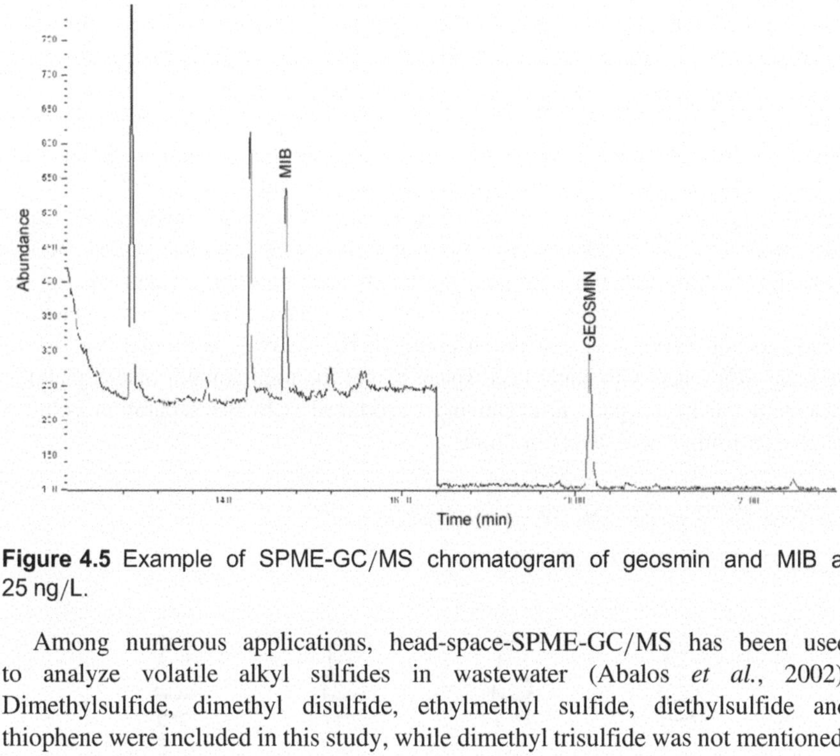

Figure 4.5 Example of SPME-GC/MS chromatogram of geosmin and MIB at 25 ng/L.

Among numerous applications, head-space-SPME-GC/MS has been used to analyze volatile alkyl sulfides in wastewater (Abalos *et al.*, 2002). Dimethylsulfide, dimethyl disulfide, ethylmethyl sulfide, diethylsulfide and thiophene were included in this study, while dimethyl trisulfide was not mentioned.

4.6 DYNAMIC HEAD SPACE EXTRACTION

With dynamic head space (DHS), the equilibrium between the aqueous and gaseous phase is displaced whether by stripping the liquid with fine bubbles of gas and adsorbing the volatiles onto a trap ('purge-and-trap' technique) or by repeatedly sending the headspace content onto an adsorbent trap ('head-space-trap'). The principle and applications of the purge-and-trap technique was described in an earlier book (Suffet *et al.*, 1995). As several brands of GC/MS instruments offer the option of adding a DHS to their other sample introduction systems and that no external purge-and-trap concentrator is needed, only the second approach will be described here.

The principle of a popular commercially available DHS accessory that can be installed on several brands of GC/MS instruments is shown in Figure 4.6.

The process involves four separate steps. The first step consists of a static head space of a liquid or solid sample. The equilibrium temperature will depend on the type of sample and analytes. For instance, measuring volatiles from solid samples may require a temperature as high as 200°C. During the second step, a two-needle device is introduced into the sealed vial. One needle is used to introduce an extraction gas that forces the content of the head space onto an adsorbent trap (Tenax TA or mixture of adsorbents) through the second needle. A dry purge of the trap is subsequently carried out (step 3). The trap is then transported by a robotic autosampler (Combipal Corporation) into a thermal desorption unit (TDU) for thermal desorption of the analytes. Analytes are cryofocused in a programmable temperature vaporizer inlet, which is then heated to release the analytes onto the column. Cryofocusing takes place in the liner of a PTV injector, packed with glass wool or Tenax TA to improve peak shape. The same adsorption trap can be reused for the whole series of samples placed on the autosampler. Although not common, the head-space equilibrium and adsorption steps 1 and 2, can be repeated several times. The main difference with classical purge-and-trap systems is that no stripping of the analytes is carried out, while the main difference with static head space is that the total amount of the analytes present in the head space is eventually introduced onto the separation column thereby ensuring lower detection limits.

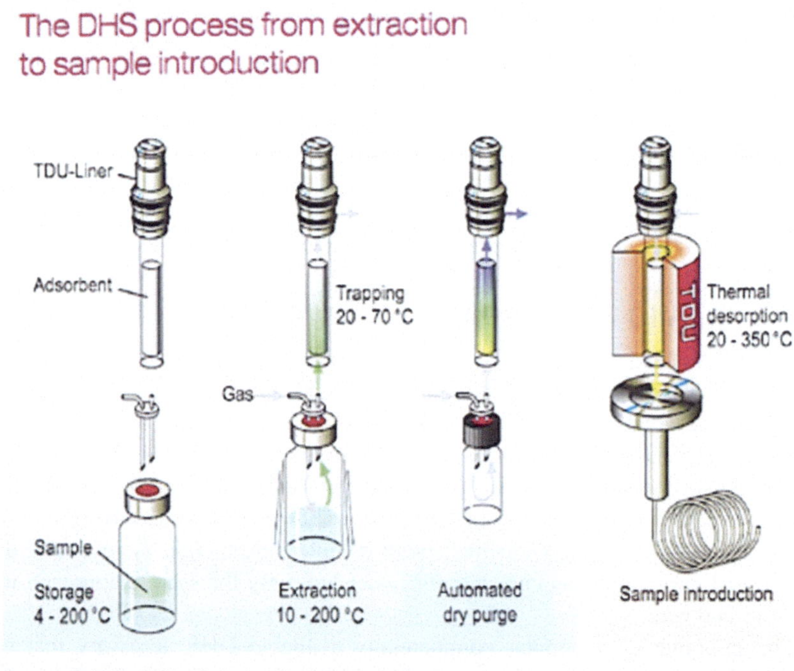

Figure 4.6 Principle of dynamic head-space.

A useful application of this technique in the T&O field concerns the determination of the gasoline additives MTBE, ETBE and tertiary amylmethylether (TAME) and the tertiary butylalcohol (TBA) degradate. Classical conditions are as follows.

DHS: Introduction of 1 g sodium chloride in 22.5 mL vials; Addition of 12 mL of the water sample and mixing until total dissolution of the salt; Addition of an internal standard (e.g. 10 μL of MTBE-D3); Addition of 200 μL of methanol; Incubation 3 min at 10°C, with agitation at 250 rpm; Purge with 150 mL of nitrogen gas at 30 mL/min; Tenax TA trap, held at 20°C.

TDU: TDU programmed from 30 to 260°C at 60°C/min, splitless desorption mode;

Cryogenic cooling at −100°C then 12°C/s up to 260°C.

GC/MS: PTV injection in split mode, injector programmed from 50 to 200°C at 20°C/min; Separation on a 30 m × 250 μm × 1.4 μm 624Sil 5 ms column programmed from 40 to 60°C at 5°C/min, then to 260°C at 25°C/min; MS interface at 220 °C, MS source at 230°C; Acquisition in SIM on ions at *m/z* 59 (300 ms dwell time), 73, 76, 87 (dwell time 100 ms each).

A chromatogram acquired under these conditions and showing the separation of MTBE, ETBE, and TBA, is shown in Figure 4.7. The limit of quantification was found equal to 25 ng/L for MTBE and ETBE and 500 ng/L for TBA. The repeatability is less than 10% at 50 ng/L for MTBE and ETBE, and 21% at 500 ng/L for TBA.

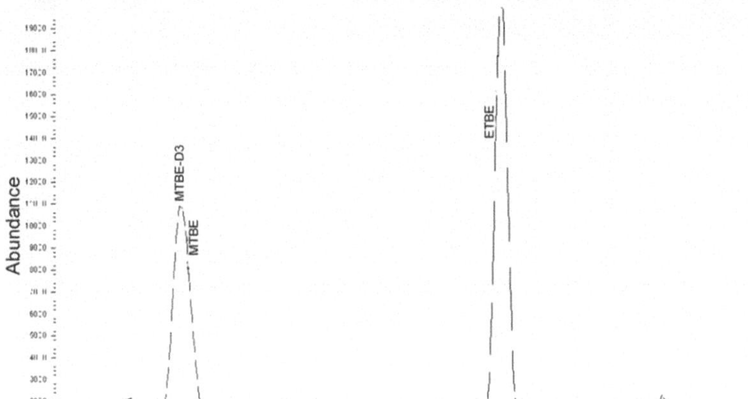

Figure 4.7 Analysis of TBA, MTBE and ETBE by dynamic head-space GC/MS.

Overall, DHS should be preferred to static head space for odourous VOCs with concentration levels ranging from 0.02 to a few μg/L.

4.7 STIR-BAR SORPTIVE EXTRACTION

Stir-bar sorptive extraction (SBSE) is one of the most recent additions to the array of sample preparation techniques for T&O compounds. The technique was initially invented by Baltussen *et al.* (1999) and was then applied by Benanou *et al.* (2004) to measure geosmin, MIB and halogenated anisoles in drinking waters, with quantification limits between 0.1 and 1 ng/L. SBSE uses a magnetic stir bar which is coated on the outside with a PDMS extraction phase. The extraction is carried out by quickly rotating the stir bar in the sample vial containing 50–100 mL of water (Figure 4.8). Agitation of the stir bar is ensured by placing the vial on a magnetic plate. Using a plate with several positions (e.g. 15) allows many samples to be extracted simultaneously.

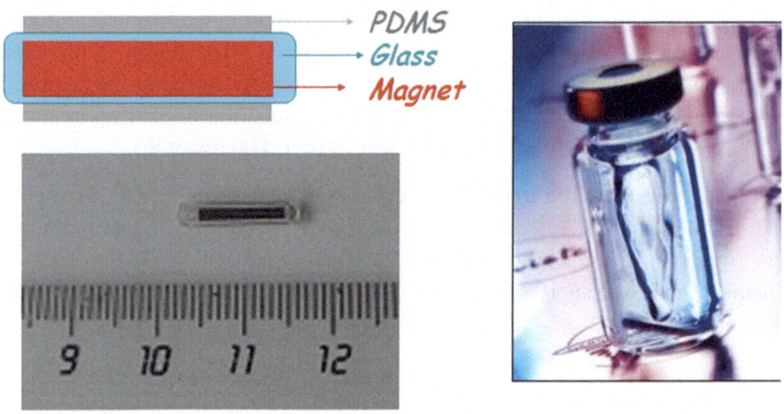

Figure 4.8 Schematic and real view of a stir bar (left) and picture of an on-going extraction (right).

The theory of SBSE is similar to that of SPME. Using the approximation that the partitioning coefficients between PDMS and water $K_{PDMS/W}$ are proportional to the octanol/water partitioning coefficients (K_{ow}), the ratio between the mass of an analyte adsorbed m_{SBSE} and its total mass initially present in the water sample m_0, the recovery equation can be written as follows (Baltussen *et al.*, 1999):

$$\frac{m_{SBSE}}{m_0} = \frac{\left(\dfrac{K_{ow}}{\beta}\right)}{1 + \left(\dfrac{K_{ow}}{\beta}\right)}$$

From this equation, it can be seen that the recovery is only governed by K_{ow} and by the phase ratio between the volume of water and the volume of PDMS, or $\beta = V_{water}/V_{PDMS}$. At $K_{ow}/\beta = 1$, the recovery is equal to 50% while when K_{ow}/β is higher than 5, the extraction is practically quantitative. Because a stir bar has much more phase than a SPME fibre, the situation is much more favorable for extraction. In the case of geosmin (log $K_{ow} = 3{,}7$ i.e. $K_{ow} = 5{,}020$), for a 50 mL water sample and a stir bar coated with 50 μL of PDMS (implying $\beta = 1{,}000$), resolution of the above equation yields a recovery of 83% at equilibrium, a result impossible to achieve with SPME where the phase ratio is much higher (e.g. 20,000 for 10 mL of water and 0.5 μL of phase). As with SPME, the time to reach equilibrium has to be determined for each target analyte. It takes about 4 h for earthy-musty compounds (Benanou *et al.*, 2004). However, sorption kinetics to the stir bar is initially rapid and shorter extraction times can be selected that still allows a good response to be obtained.

As a starting point, the following conditions can be used for earthy-musty compounds (geosmin, MIB, 2,3,4- and 2,4,6-TCA, 2,4,6-tribromoanisole, 2-methoxy-3-isobutylpyrazine). After the extraction phase, the stir bar is wiped with clean paper and submitted to thermal desorption-GC/MS.

Extraction: Introduce 100 mL water sample spiked with internal standard, e.g. 2,4,6-TCA-D5 in sample container; Add 2 g potassium carbonate and homogenize; Spike internal standard, e.g. 2,4,6-TCA-D5; Add 20 mm stir bar preconditioned for 2 h at 300°C under nitrogen; Agitate for 2 h at 680 rpm.

Thermal desorption and GC-MS: Thermal desorption is carried out in the splitless mode by programming the TDU from 30°C (1 min) to 270°C at 60°C/min, held for 8 min; Cryogenic cooling is carried out by maintaining the cooled injection system at −50°C during the desorption phase then programming it at 12°C/s to 250°C in the solvent vent mode; Chromatographic separation is carried out on a 30 m × 0.25 mm × 0.25 μm DB5-ms column, programmed from 50°C (2 min) to 200°C at 20°C/min, then to 310°C at 25°C/min, then held for 7 min; GC-Electron Impact Ionization-MS in the scan mode or SIM mode.

LODs and coefficients of variation (CV) observed with Evian water under these conditions, are shown in Table 4.3.

Table 4.3 Performance of SBSE-GC/MS for earthy-musty compounds in groundwater (LODs in ng/L).

Analyte	LOD SIM	LOD Full Scan	CV% @1 ng/L
Isobutyl methoxy pyrazine	1	5	21.49
Methylisoborneol	0.5	5	15.72
2,4,6-Trichloroanisole	0.2	2	14.98
Geosmin	0.2	1	9.43
2,3,4-Trichloroanisole	0.2	2	8.94
2,4,6-Tribromoanisole	0.2	2	16.41

Due to the higher amount of phase, lower detection limits can be obtained compared to SPME. As noted by Benanou *et al.* (2004), SBSE is time consuming but not labor intensive and a sample throughput of 15–20 samples per day can be achieved. Pre-conditioning of the stir bars is carried out in a dedicated oven that can handle 20 stir bars simultaneously.

The ability to reach low detection limits with a small sample volume and minimal sample manipulation for halogenated phenols, makes SBSE a method of choice for this class of T&O compounds. Halogenated phenols can be extracted without or with in-situ acetylation. The latter approach provides faster adsorption kinetics, better chromatographic peak shapes, and is hence recommended. Classical conditions used for a range of halogenated phenols are indicated below:

Acetylation: Add 2 g of potassium carbonate (K_2CO_3) in a 250 mL flask; Add 100 mL water sample spiked with internal standards (e.g. C^{13}-labelled pentachlorophenol) in flask and homogenize; Add 1 mL acetic anhydride (AC_2O); Add a 20 mm stir bar and agitate at 550 rpm for 1 h; Rinse the stir bar with ultrapure water and dry with clean paper.

Thermal desorption and GC-MS conditions are as described above for earthy-musty compounds.

Table 4.4 lists a selection of phenols and halogenated phenols that can be routinely quantitated using this method. Bisphenol A has been included as it may degrade into 2,4,6-trichlorophenol in distribution systems. The latter phenol can be biotransformed into the musty smelling 2,4,6-TCA. Therefore, this method is particularly adapted to the study of bisphenol A leaching from new and aged epoxy coatings, that has been shown to cause medicinal-musty odour episodes (Bruchet *et al.*, 2012).

Table 4.4 Target halogenated phenols measured by SBSE-GC/MS.

Compound	Quantitation Ion (*m*/*z*)	Ion 2 (*m*/*z*)
C13-pentachlorophenol	272	Internal standard
2-Chlorophenol	128	130
3-Chlorophenol	128	130
4-Chlorophenol	128	130
2-Isopropylphenol	121	136
3-Isopropylphenol	121	136
4-Isopropylphenol	121	136
2,6-Dichlorophenol	162	164
2,4- and 2,5-Dichlorophenol	162	164
2,4,6-Trichlorophenol	196	160
2,3,6-Trichlorophenol	196	160
C^{13}-BPA	225	Internal standard
Bisphenol A	213	228

The limit of quantification is equal to 10 ng/L for all analytes in Table 4.4. Other halogenated phenols including brominated phenols can be added to the list of target analytes. As with SPME, one advantage of SBSE is that calibration curves can be easily constructed over the entire procedure. Therefore, knowledge of absolute recoveries is not necessary to conduct quantitative work. Other advantages include the fact that SBSE is relatively insensitive to the presence of background dissolved organic carbon (DOC) as shown in Figure 4.9 where calibration curves obtained for 2,6-dibromophenol spiked in Evian ground water (DOC ~ 0.5 mg/L) and Seine river water (DOC ~ 3 mg/L) superimpose almost perfectly.

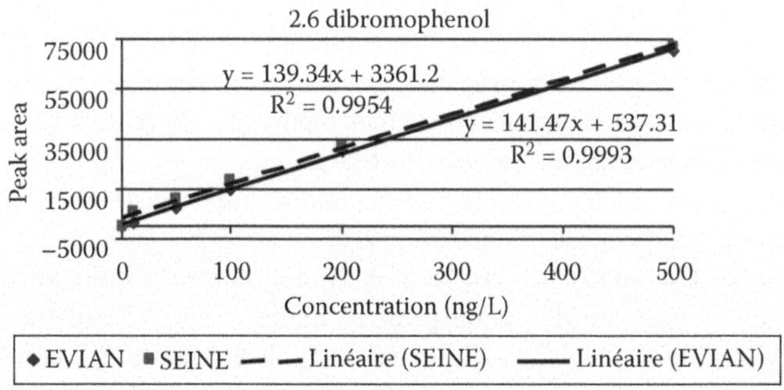

Figure 4.9 Comparison of calibration curves of 2,6-DBP in Evian and Seine river water.

Once the water has been extracted, the stir bars can be stored for at least one week at 4°C without significant loss of the analytes. This property allows extraction to be carried out in the field with a simple magnetic stirrer, followed by shipment of the bars to the laboratory where more sophisticated equipment is needed for thermal desorption-GC/MS. These advantages explain why SBSE has become quite a popular technique, particularly in Europe where accessories necessary to adapt this technique on GC/MS instruments are readily available.

It was determined that other odourants listed in the bottom part of Table 4.1 (aliphatic aldehydes, dienals, *cis*-3-Hexen-1-oland, 2-ethyl-5,5-dimethyl-1,3-dioxane (2-EDD)) could be detected below their OTCs (Bruchet, unpublished data). The quantitative aspect for these analytes however needs to be determined. On the other hand, we were not able to recover 2-ethyl-4-methyl-1,3-dioxolane (2-EMD) and dimethyltrisulfide at their OTC.

4.8 OTHER EXTRACTION TECHNIQUES

Other extraction techniques, including popular ones, have found little use in T&O investigation. This is in particular the case for solid-phase extraction (SPE) that is now more widely used than liquid-liquid extraction (LLE) for analysis of semi-volatile compounds. SPE has been used in combination with GC-plasma atomic emission detection (AED) to determine chlorophenols at sub ppb levels in drinking water (Rodriguez *et al.*, 1996). Use of an AED detector as well as the graphitized carbon cartridge employed by the authors is not common, hence a SPME or SBSE-GC/MS method is more advisable. An unusual, a combination of SPE followed by dilution of the C-18 cartridge eluent with ultrapure water and subsequent head-space-GC/MS analysis was reported (Ikai *et al.*, 2003) for geosmin and MIB. The LOD obtained with this combination was quite low, 0.1 ng/L. Although there are other ways to enhance sensitivity for T&O compounds (see next paragraph), this work indicates that SPE could be a viable option for geosmin-MIB for laboratories that do not have SPME or SBSE equipment and when sub-ppt limits of detection are not needed. SPE is, however, not recommended for the identification of unknown odourants such as during chemical spills, as it is more selective than methylene chloride LLE and as the sorbent and its packaging can introduce artifacts.

A variation of LLE, namely ultrasound-assisted dispersive-liquid-liquid microextraction has been developed to preconcentrate geosmin and 2-MIB from water and wine (Cortada *et al.*, 2011). The extraction step is quite simple but the method involves post-centrifugation prior to injection in GC and the LODs are still slightly too high, 2 and 9 ng/L, respectively. Solid phase dynamic extraction is a technique where the injection syringe needle is coated with an extraction phase. The liquid sample or head space above is repeatedly sampled through the syringe, allowing analytes to be concentrated by the extraction phase. The latter is then thermally desorbed by introducing the syringe into a heated GC injection port. This method has been applied to the determination of *trans*-2-nonenal in barley, malt and beer (Svodoba *et al.*, 2011). Despite its automated character, this method has not found application in the drinking water T&O field.

A membrane-based technique, hollow-fibre stripping analysis has been devised for extraction of geosmin, 2,3,6-TCA, *trans*-2,*cis*-6-nonadienal and 2-MIB (Zander & Pingert, 1997). Water circulates in a closed loop that includes a hollow-fibre membrane module through which volatile compounds pervaporate to a second loop where a constant gas flow drives the analytes through an activated carbon trap. Despite its good performance in terms of recovery, this modified CLSA technique has found little use. This is also the case for the CLSA/SPME combination proposed by Suffet *et al.* (2006).

4.9 ENHANCING SENSITIVITY WITH LARGE VOLUME INJECTION

Although some of the techniques described above allow sub-ppt detection limits, this may not always be sensitive enough when dealing with 2,4,6-TCA and 2,4,6-tribromoanisole which have the lowest OTCs amongst all known water odourants at 20 pg/L (Table 4.1). So far, the only way to reach such levels has been to combine an extraction technique yielding a high concentration factor with a large volume injection method, LVI-GC/MS, allowing most of the solvent extract and hence most of the analyte initially present in the water sample, to be introduced onto the GC column. Malleret *et al.* (2001) used a combination of CLSA and an on-column LVI to investigate eight earthy-musty odourants. A classical CLSA extraction was carried out with 1 liter of water, and the carbon trap was eluted with a slightly higher than usual, volume of carbon disulfide, i.e. 50 μL. A 40 μL aliquot was subsequently injected in GC/MS operated in the SIM mode.

The principle of the injection technique used by Malleret *et al.* (2001) is shown in Figure 4.10. The sample extract (up to 150 μL) is injected by an autosampler at a controlled speed through a cold on-column injector into a 12 m uncoated precolumn (UNCORET, proprietary design, Fisons Instruments, Milano, Italy). During a subsequent step, both the solvent and the most volatile solutes evaporate from the rear of the solvent injection band while heavier solutes remain throughout the precolumn. Under controlled pressure and temperature conditions the solvent vapours are vented through the solvent vapour exit until a preselected amount of final solvent volume is left inside the precolumn. The remaining solvent is then flushed into the analytical column. Both solvent effect and a 3 m retention gap allow refocussing of the deposited analytes before the latter flushing step.

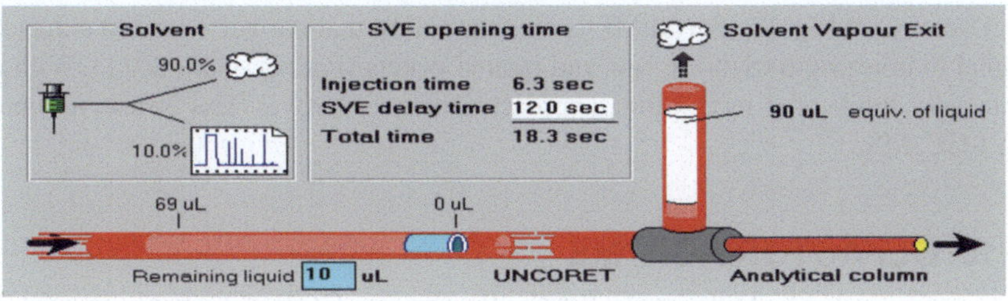

Figure 4.10 Schematic of the on-column large volume injector (from Bruchet, 2006).

A simple calculation shows that, if 20 pg/L of an analyte is present in the water sample, and if the CLSA recovery equals 50%, the above-mentioned conditions lead to an absolute injection quantity of 8 pg, which is within reach of modern GC/MS instruments in the SIM mode. As a matter of fact, the LOD observed in tap water and ground water fell in the range 13–48 pg/L for the eight target analytes investigated (Table 4.5).

This approach allowed the first identification of 2,4,6-tribromoanisole as a musty-odour causative agent in a distribution system that had been coated with a cement-based coating (Malleret & Bruchet, 2002). A large volume injector working in the on-column mode presents the advantage of avoiding losses of volatile solutes. However, such an injector has not been continuously present on the market. As an alternative, a large volume PTV injector was successfully used by Zhang *et al.* (2006) for geosmin, MIB and haloanisoles. In combination with continuous LLE, the method yielded detection limits of 35–70 pg/L for haloanisoles, 50 pg/L for geosmin, and 340 pg/L for MIB.

Table 4.5 Limits of detection by CLSA-LVI-GC/MS for eight earthy-musty odourous compounds in drinking and groundwater (pg/L).

Compound	LOD in Tap Water (S/N = 3)	LOD in Groundwater (S/N = 3)
2-Methoxy-3-isopropylpyrazine	34	17
2-Methoxy-3-isobutylpyrazine	17	16
2-Methylisoborneol	28	48
2,4,6-Trichloroanisole	24	23
2,3,6-Trichloroanisole	22	28
Geosmin	13	16
2,3,4,6-Tetrabromoanisole	15	32
2,4,6-Tribromoanisole	20	23

Source: Adapted from Malleret *et al.* (2001).

Use of CI/MS/MS after SPME is another option that has been used to enhance signal to noise ratio (S/N) in raw and treated waters (Parinet *et al.*, 2011). With a detection limit of 1 ng/L, this approach however proves less sensitive than LVI-GC/MS.

4.10 OLFACTORY-GC/MS

Since its invention by Fuller *et al.* (1964), olfactometry-GC (GC-O) has been used in the food (Acree & Teranishi, 1993) and water (e.g. Khiari *et al.*, 1999) industries and was often instrumental at pinpointing the chemicals inducing the dominant odours. The principle of GC-O is quite straightforward: the effluent from the gas chromatographic column is split into two parts, one part goes to a detector (FID, ECD, MS) and another part to a nasal cone (or sniffing port) where a trained operator smells the odour. If the system is properly designed, the retention time of the detected GC peak and the retention time of the odour perceived will perfectly match. When a mass spectrometer is used as a detector, then both the odour and mass spectra of odourants can be matched, therefore allowing identification the odour-causing compounds. This property is quite important when searching for unknown odourants as some of the extraction methods described above will often lead to the detection of more than 100 chromatographic peaks. Several GC-O devices are commercially available, however, they sometimes need adaptation to be used with MS detectors that operate under vacuum.

In particular, Hochereau and Bruchet (2004) described the design of a home-made GC-O/MS system that was used for solving T&O episodes in drinking water. The GC-Sniff/MS device (Figure 4.11) had a connection between a Voyager MD 800 GC/MS (Thermofinnigan, USA) with a sniffer 9,000 system (Brechbühler SA, Switzerland) using a home-made interface and control software. The challenge when coupling a GC-MS (that operates under a high vacuum) with a sniffing port (which is at atmospheric pressure) is to avoid the carrier gas flowing from the analytical column being entirely diverted to the MS and furthermore avoid outside air being sucked back in to the MS ion source. This difficulty was overcome by adding a restrictor that increased the pressure drop between the four-way quartz press-fit connector (universal X press tight Restek ref 20408) and the MS detector, and by carefully selecting the analytical carrier gas and make-up gas flow rates. The sniffer 9,000 interface outside the GC was manufactured from a single stainless steel tube, which was heated by direct current, giving the best possible temperature profile. This interface could hence be programmed to follow the oven temperature programme. A 1.4 m piece of empty, deactivated fused silica tubing (0.25 mm i.d) was inserted through the interface from the quartz press-fit connector to the inlet of the glass nasal cone. A make-up flow of humid air was added after the four-way connector inside the oven. This humid air was carried through the heated metal interface in a concentric manner outside the piece of fused silica tubing and was mixed with the analytes at the inlet of the nasal cone.

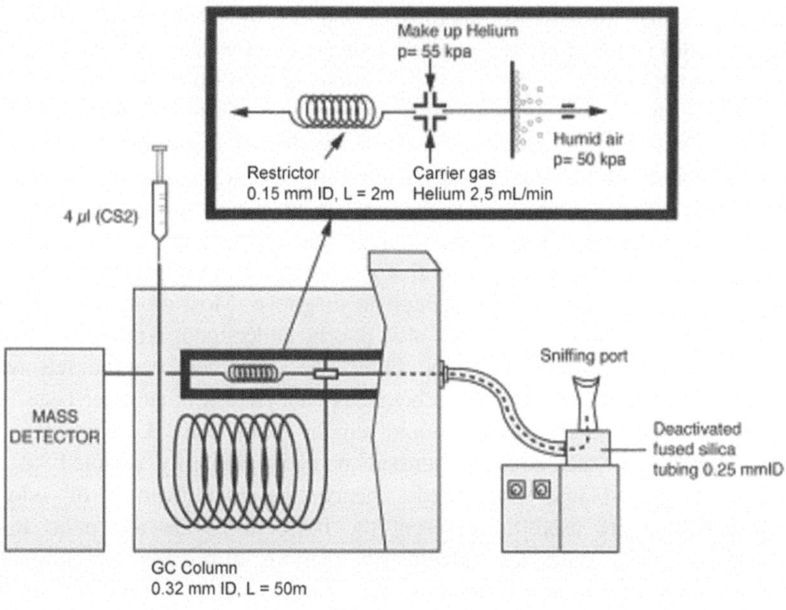

Figure 4.11 Schematic of a GC-O/MS system (Hochereau & Bruchet, 2004).

During a GC-sniff/MS run, a hand button was turned more or less as a function of the odour intensity and software was utilized to record the detection of an odour for an eluting peak and hence generate and plot an aromagram. With this system, both the mass spectrum and the odour of an analyte eluted at the same retention time, allowed the user to find the corresponding odour of a GC peak. This system has proven very useful because the human nose is extremely sensitive. Examples of odourants that could be identified with this system during long term or acute T&O episodes include 2,6-dibromophenol that caused intermittent medicinal odours (Hochereau & Bruchet, 2004). It also allowed the identification of 2-EMD and 2-EDD, the odour of which (described as olive-odour or tutti-frutti odour) caused the temporary shutdown of a major drinking water plant in South America (Bruchet *et al.*, 2005). As the human nose remains more sensitive than MS detectors, there are many cases when an odour is perceived at the olfactory port while no peak is detected by MS. However, once the retention time of the odourous chemical matching the odour of the water sample has been pinpointed, it is possible to increase the extracted volume to enhance the signal to noise ratio until the odourant is detected by the MS detector (Hochereau & Bruchet, 2004).

4.11 CONCLUSION

The array of sample preparation techniques described in this chapter allows analytical chemists to correctly diagnose a majority of T&O episodes. Statistics on the episodes submitted to our institute indicate that the causative agents can be identified in about 70% of the cases. The odour groups from the drinking water T&O wheel that are easiest to diagnose include earthy-musty (caused by geosmin, 2-MIB and halogenated anisoles), medicinal (caused by halogenated phenols or iodinated THMs), swampy (caused by sulfides) and fruity (caused by aliphatic aldehydes or alkenals). On the other hand, plastic-like odours (from the chemical odour group), fishy odours (likely caused by hard to detect volatile amines) as well as chlorinous odours that are sometimes observed in the absence of any chlorine residual, remain difficult to diagnose. Most mouthfeels such as astringent, chalky, tingling, etc., are also poorly understood. Among the tastes, the most commonly observed is the bitter one. Bitterness can be felt when drinking deionized water but many chemicals can yield a bitter after taste (e.g. halogenated phenols). The reader should remain aware that GC coupled to the sample preparation techniques described here is primarily adapted to the identification of volatile compounds, hence the identification of odours. Compounds that are used to represent the basic tastes are too polar to be separated by GC. With the advent of a more affordable generation of high-resolution liquid chromatography–mass spectrometry instruments, it will eventually become possible to identify polar compounds that contribute to tastes, mouthfeels and unknown chemical or other odours.

As far as costs are concerned, all techniques described in this chapter approximately require the same amount of capital cost. This cost is mostly determined by the purchase of a GC/MS instrument with an autosampler that can handle the different accessories necessary to conduct SPME, static or dynamic head-space extraction or the thermal desorption involved in SBSE. Two or three CLSA apparatus can be purchased for the same amount of capital as these different accessories. Thus these different techniques will differentiate themselves mostly by their automated or manual character, and by the number of samples that can be processed per day. With CLSA, two apparatus will allow extraction of 6 samples per day maximum for an 8 hour shift. As the volume of the elution solvent is quite small and its volatility is quite high, the injections however need to be carried out manually after the first extraction series. In the remaining 5 and a half hours, a maximum of 4 samples and a blank or standard can be injected when considering a 1 h chromatographic run time (including cooling of the oven). With SBSE, a magnetic plate with 15 positions would allow extraction of two series of 15, that is 30 vials, including 5 standards, a blank and 2 QCs. Starting the autosampler at noon leaves 20–21 h until the next morning, that is enough time to inject 30 vials when considering a 40 min total run time. Hence the maximum number of samples that can be processed per day is 22. For static or DHS as well as SPME, the only manual step lies in the preparation of the calibration standards, a step that will take about 1 h. For the remaining 22 h, 44 vials including 5 standards, a blank and 4 QCs can be injected if the total run time (equilibrium time for these techniques is rather short) is kept at 30 min. This would allow a maximum number of samples equal to 34 per day. In reality this rhythm cannot be maintained every day as other tasks such as data processing, stir-bar or SPME fibre conditioning, GC/MS maintenance, need to be performed. Therefore, a more reasonable estimate of the number of samples that can be processed per day is indicated in Table 4.6.

Table 4.6 Compared efficiency of various analytical techniques.

	CLSA	SPME	SBSE	Static or Dynamic Head-space Extraction
Extraction	Manual	Automatic	Manual	Automatic
Injection	Manual	Automatic	Automatic	Automatic
Number of samples per day	4*	10–15	10–15	15–20

*With two extraction apparatus.

REFERENCES

Abalos M., Prieto X. and Bayona J. M. (2002). Determination of volatile alkylsulfides in wastewater by head-space solid-phase microextraction followed by gas chromatography-mass spectrometry. *Journal of Chromatography A*, **963**, 249–257.

Acree T. E. and Teranishi R. (1993). Science. Sensible Principles and Techniques. ACS Professional Reference Book, American Chemical Society, Washington, DC.

Agus E., Lim M. H., Zhang L. and Sedlak D. L. (2011). Odourous compounds in municipal wastewater effluent and potable water reuse systems. *Environmental Science & Technology*, **45**, 9347–9355.

Anselme C., N'Guyen K., Bruchet A. and Mallevialle J. (1985). Can polyethylene pipes impart odours in drinking water? *Environmental Technology Letters*, **6**, 477–488.

Anselme C., Mallevialle J. and Suffet I. H. (1988). Effects of ozonation on tastes and odours. *Journal American Water Works Association*, **75**, 45–51.

Arthur C. L. and Pawliszyn J. (1990). Solid phase microextraction with thermal desorption using fused silica optical fibres. *Analytical Chemistry*, **62**, 2145–2148.

Baltussen E., Sandra P., David F. and Cramers C. (1999). Stir bar sorptive extraction (SBSE), a novel extraction technique for aqueous samples: theory and principles. *Journal of Microcolumn Separations*, **11**, 737–747.

Benanou D., Acobas F. and de Roubin M. R. (2004). Optimization of stir bar sorptive extraction applied to the determination of odourous compounds in drinking water. *Water Science and Technology*, **49**(9), 161–170.

Bruchet A. (2006). State of the art analytical methods for solving taste and odour episodes. *Water Science and Technology: Water Supply*, **6**(3), 157–165.

Bruchet A., N'Guyen K., Mallevialle C. and Anselme C. (1989). Identification and behaviour of iodinated haloform medicinal odour. In: Organic Micropollutants in the Aquatic Environment, G. Angelettiand and A. Bjørseth (eds), Springer, Dordrecht, the Netherlands.

Bruchet A., Hochereau C. and Campos C. (2005). An acute taste and odour episode solved by olfactory GC-MS. *Water Science and Technology*, **55**(5), 223–230.

Bruchet A., Elyasmino N., Decottignies V. and Noyon N. (2012). Leaching of Bisphenol A and F from New and Old Epoxy Coatings: Laboratory and Field Studies. IWA World Water Congress and Exhibition, Busan, Korea, 16–21 September.

Burlingame G. A., Muldowney J. J. and Maddrey R. E. (1992). Cucumber flavor in Philadelphia's drinking water. *Journal American Water Works Association*, **86**(8), 92–97.

Burttschell R. H., Rosen A. A., Middleton F. M. and Ettinger M. B. (1959). Chlorine derivatives of phenol causing taste and odour. *Journal American Water Works Association*, **51**(2), 205–214.

Cancho B., Ventura F. and Galceran M. T. (1999). Solid-phase microextraction for the determination of iodinated trihalomethanes in drinking water. *Journal of Chromatography A*, **841**, 197–206.

Cortada C., Vidal L. and Canals A. (2011). Determination of geosmin and 2-methylisoborneol in water and wine samples by ultrasound-assisted dispersive liquid-liquid microextraction coupled to gas chromatography-mass spectrometry. *Journal of Chromatography A*, **1218**, 17–22.

Fuller G. H., Steltenkamp G. A. and Tisserand G. A. (1964). The gas chromatograph with human sensor: perfumer model. *Annals of the New York Academy of Sciences*, **116**, 711–724.

Gerber N. N. (1968). Geosmin, from microorganisms, is *trans*-1,10-dimethyl-trans-9-decalol. *Tetrahedron Letters*, **25**, 2971–2974.

Gerber N. N. and Lechevalier H. A. (1965). Geosmin, an earthy-smelling substance isolated from Actinomycetes. *Applied Microbiology*, **13**, 935–938.

Grob K. and Zürcher F. (1976). Stripping of trace organic substances from water. Equipment and procedure. *Journal of Chromatography*, **117**, 285–294.

Hansson R. C., Henderson M. J., Jack P. and Taylor R. D. (1987). Iodoform taste complaints in chloramination. *Water Research*, **21**, 1265–1271.

Hayes K. P. and Burch M. D. (1989). Odourous compounds associated with algal blooms in South Australian waters. *Water Research*, **23**, 115–121.

Hochereau C. and Bruchet A. (2004). Design and application of a GC-SNIFF/MS system for solving taste and odour episodes in drinking water. *Water Science and Technology*, **49**(9), 81–87.

Ikai Y., Honda S., Yamada N., Onuma S., Tomita B.-I., Kawamura N. and Miyazaki Y. (2003). Determination of geosmin and 2-methylisoborneol in water using solid phase extraction and head-space-GC/MS. *Journal of the Mass Spectrometry Society of Japan*, **51**, 174–178.

ISO (1997). Standard 10301 – Water Quality-Determination of Highly Volatile Halogenated Hydrocarbons-Gas Chromatographic Methods. ISO, Geneva, Switzerland.

Khiari D., Brenner L., Burlingame G. A. and Suffet I. H. (1992). Sensory gas chromatography for evaluation of taste and odour events in drinking water. *Water Science and Technology*, **25**(2), 97–104.

Khiari D., Suffet I. H. and Barrett S. E. (1995). Extraction and identification of chemicals causing grassy odours in flavor profile analysis (FPA) reference standards. *Water Science and Technology*, **31**(11), 93–98.

Khiari D., Young C. C., Amah G., Ye Q., Atasi K., Huddleston J. and Suffet I. H. (1999). Factors affecting the stability and behaviour of grassy odours in drinking water. *Water Science and Technology*, **40**(6), 287–292.

Khiari D., Barrett S., Chinn R., Bruchet A., Piriou P., Matia L., Ventura F., Suffet I. H., Gittleman T. and Luitweiler P. (2002). Distribution System Generated Taste-and-Odour Phenomena. AWWA Research Foundation, No. 90897, American Water Works Association, Denver, CO.

Lin T.-F., Liu C.-L., Yang F.-C. and Hung H.-S. (2003). Effect of residual chlorine on the analysis of geosmin, 2-MIB and MTBE in drinking water using the SPME technique. *Water Research*, **37**, 21–26.

Malleret L. and Bruchet A. (2002). A taste and odour episode caused by 2,4,6-tribromoanisole. *Journal American Water Works Association*, **94**, 84–96.

Malleret L., Bruchet A. and Hennion M.-C. (2001). Picogram determination of 'earthy-musty' odourous compounds in water using Modified Closed-Loop-Stripping Analysis and Large Volume Injection GC/MS. *Analytical Chemistry*, **73**, 1485–1490.

Mallevialle J. and Suffet I. H. (1987). Identification and Treatment of Tastes and Odours in Drinking Water. AWWA Research Foundation and Lyonnaise Des EauxCooperative Research Report, American Water Works Association, Denver, Colo., USA.

Mallevialle J., Lefebvre M., Rousseau C. and Saglie C. (1982). Pollution with petroleum products: qualitative and quantitative analysis at low concentrations. *Revue Française des Sciences de L'Eau*, **1**, 125–138.

Montiel A., Ouvrard J., Rigal S. and Bousquet G. (1987). Etude de l'origine et du mécanisme de formation de composés sapides responsables de goûts de moisi dans les eaux distribuées. *TSM-l'Eau*, **February**, 73–83.

Nyström A., Grimvall A., Krantz-Rülcker C., Sävenhed R. and Akerstrand K. (1992). Drinking water off-flavour caused by 2,4,6-trichloroanisole. *Water Science and Technology*, **25**(2), 241–249.

Parinet J., Rodriguez M., Serodes J. and Proulx F. (2011). Automated analysis of geosmin, 2-methyl-isoborneol, 2-isopropyl-3-methoxypyrazine, 2-isobutyl-3-methoxypyrazine and 2,4,6-trichloroanisole in water by SPME-GC-ITDMS/MS. *International Journal of Environmental and Analytical Chemistry*, **91**, 505–515.

Pirbazari M., Borow H. S., Craig S., Ravindran V. and McGuire M. J. (1992). Physical chemical characterization of five earthy-musty-smelling compounds. *Water Science and Technology*, **25**(2), 81–88.

Piriou P., Soulet C., Acero J. L., Bruchet A., Von Gunten U. and Suffet I. H. (2002). Understanding medicinal taste and odour formation in drinking waters. *Water science and Technology*, **55**(5), 85–94.

Preti G., Gittelman T. S., Staudte P. B. and Luitweller P. (1993). Letting the nose lead the way. Malodourous components in drinking water. *Analytical Chemistry*, **65**, 699A–702A.

Rodriguez I., Turnes M. I., Mejuto M. C. and Cela R. (1996). Determination of chlorophenols at the sub-ppb level in tap water using derivatization, solid-phase extraction and gas chromatography with plasma atomic emission detection. *Journal of Chromatography A*, **721**, 297–304.

Rosen A. A., Mashni C. I. and Safferman R. S. (1970). Recent developments in the chemistry of odour in water: the cause of earthy/musty odour. *Water Treatment Examination*, **19**, 106–109.

Schweitzer L., Noblet J., Ruth E. and Suffet I. H. (1999). The formation, stability, and odour characterization of 2-ethyl-4-methyl-1,3-dioxolane (2-EMD). *Water Science and Technology*, **40**(6), 293–298.

Suffet I. H. (2007). A re-evaluation of the taste and odour of methyltertiary butyl ether (MTBE) in drinking water. *Water Science and Technology*, **55**(5), 265–273.

Suffet I. H., Mallevialle J. and Kawczyski E. (1995). Advances in Taste-and-Odour Treatment and Control. American Water Works Association, Denver, CO.

Suffet I. H., Bruchet A. and Young C. C. (2006). Development of a closed-loop stripping analysis using solid-phase microextraction to analyse geosmin and MIB in drinking water. *Water Science and Technology: Water Supply*, **6**(3), 167–174.

Svodoba Z., Mikulikova R., Belakova S., Benesova K., Marova I. and Nesvadba Z. (2011). Optimization of modern analytical SPME and SPDE methods for determination of trans-2-nonenal in barley, malt and beer. *Chromatographia*, **73**(1), 157–161.

Van Gemert L. J. (1977). Compilation of Odour Threshold Values in Air and Water. Boelens Aroma Chemical Information Service, The Netherlands.

Van Wezel A., Puijker L., Vink C., Versteegh A. and de Voogt P. D. (2009). Odour and flavour thresholds of gasoline additives (MTBE, ETBE and TAME) and their occurrence in Dutch drinking water collection areas. *Chemosphere*, **76**, 672–676.

Wajon J. E., Alexander R. and Kagi R. I. (1985). Dimethyl trisulfide and objectionable odours in potable water. *Chemosphere*, **14**, 85–89.

Watson S. B., Brownlee B., Satchwil T. and Hargesheimer E. H. (2000). Quantitative analysis of trace levels of geosmin and MIB in source and drinking water using headspace SPME. *Water Research*, **34**, 2818–2828.

Whitfield F. B., Hill J. L. and Shaw K. J. (1997). 2,4,6-tribromoanisole: a potential cause of mustiness in packaged food. *Journal Agricultural Food Chemistry*, **45**, 889–893.

Young C. C. and Suffet I. H. (1999). Development of a standard method – analysis of compounds causing tastes and odours in drinking water. *Water Science and Technology*, **40**(6), 279–285.

Zander A. K. and Pingert P. (1997). Membrane-based extraction for detection of tastes and odours in water. *Water Research*, **31**(2), 301–309.

Zhang L., Hu R. and Yang Z. (2006). Routine analysis of off-flavour compounds in water at sub-part-per-trillion level by large-volume-injection GC/MS with programmable temperature vaporizing inlet. *Water Research*, **40**, 699–709.

Chapter 5

Advances in sensory measurement determinations

Andrea M. Dietrich[1] and Pinar Ömür-Özbek[2]

[1]Department of Civil and Environmental Engineering, Virginia Polytechnic Institute and State University, Blacksburg, VA, USA
[2]Civil and Environmental Engineering, Colorado State University, Fort Collins, CO, USA

5.1 INTRODUCTION

This chapter provides descriptions for sensory tests that have been used in the water industry since the 20th century (e.g. flavour profile analysis (FPA), threshold odour number (TON), flavour rating assessment (FRA), and triangle tests and thresholds). The chapter then introduces the newer methods gaining usage in the water industry since the start of 21st century, though these tests have been used in the food and beverage industry for a much longer time. These tests include the attribute rating test (ART), total intensity of odour (TIO), 2-of-5 test, ranking test, and strategies for tracking and interpreting consumer feedback.

5.2 TYPES OF SENSORY TESTS

Sensory evaluation is defined as 'a scientific method used to evoke, measure, analyze, and interpret those responses to products as perceived through the senses of sight, smell, touch, taste, and hearing' (Stone & Sidel, 1993). There are two general categories of sensory tests: affective and analytical. Affective tests

 Taste and Odour in Source and Drinking Water: Causes, Controls, and Consequences
Editor(s): Tsair-Fuh Lin, Susan Watson, Andrea M. Dietrich and I. H. (Mel) Suffet
doi: 10.2166/9781780406664_0143

frequently use untrained subjects to measure both preference and acceptance. Analytical tests are more sophisticated than affective tests. The analytical tests primarily use trained subjects to measure defined characteristics including specific tastes and odour (T&O) and their intensities. Applications of both affective and analytical tests are employed to produce high quality drinking water that will be accepted and desired by consumers (Dietrich, 2006).

This section briefly summarizes the types of sensory tests used in the evaluation of foods and beverages. The reader can find more details concerning procedures and applications for sensory tests in *Sensory Evaluation Techniques* by Meilgaard *et al.* (2007). Expanded descriptions on the types of tests and their appropriate use for drinking water are available in *Practical Taste-and-Odour Methods for Routine Operations: Decision Tree* by Dietrich *et al.* (2004a). *Sensory Evaluation of Food: Principles and Practice* by Lawless and Heymann (2010) is also recommended as a source of information about the field and practice of sensory science. Practical information on strengths and challenges of performing effective sensory tests of drinking water with human panelists is addressed in a review article by Burlingame *et al.* (2017).

5.2.1 Affective tests

5.2.1.1 Acceptance tests

Acceptance tests provide subjective data that describe the degree to which a sample is acceptable based on its aesthetic qualities. These are also described as 'liking' tests. In an acceptance test, the panelist rates a single sample using a pre-determined scale. Category or hedonic scales are used because they are simple and readily understood.

5.2.1.2 Preference tests

Preference tests determine which sample a panelist prefers. Preference test by definition involves two or more samples as the panelist must choose which sample is preferred compared to other sample(s).

5.2.2 Descriptive test

Descriptive tests are a type of analytical test in which trained analysts identify the T&O characteristics of samples. Typically, 4–15 panelists are involved in determining the characteristics of a sample.

5.2.2.1 Attribute rating

These tests are used for rating previously identified 'attributes' or characteristics of a sample, such as musty, sweet, or salty. A variety of scales can be used to rate the attribute. A category scale is used to describe the attribute semi-quantitatively, such as weak or strong intensity. A ratio scale involves comparison a sample to a control, and then describes the sample relative to the control, such as 'the same as' or 'twice as strong as the control'. Rating by comparison allows the panelists to match each sample to standards arranged in order from weak to strong.

5.2.2.2 Descriptive analysis

These tests attempt to describe multiple characteristics of a sample and in some cases, to describe a sample in its entirety. Descriptive analysis combines qualitative identification of the characteristics with a quantitative assessment of their intensity. A panel is trained in advance to perform descriptive analysis and then works together to characterize individual samples.

5.2.3 Discrimination tests

Discrimination tests are designed to select the differences between samples. For many tests, between 30 and 100 panelists are recommended in order for the test to be statistically valid. The term 'forced-choice' is often used to describe discrimination tests as the panelist is forced to choose a sample according to the instructions of the test, even if the panelist perceives no differences.

There are many types of discrimination tests. When two samples are evaluated, it is commonly referred to as a 'paired comparison'. The 'triangle test' employs three samples – two of which are identical and one of which is different. The triangle test is the most widely used discrimination test. The 'duo-trio test' is similar to the triangle test. One of the three samples is a reference sample, and the panelist must determine which of the other two samples most closely matches the reference sample. The '2-of-5 test' is a forced choice test and requires the panelist to 'sort' the samples. Five samples are presented to the panelist: two are the same, and the other three are the same and different from the group of two. The task of the panelist is to sort the samples correctly into the correct groups of two and three samples. Another group of discrimination tests is the ranking test. For a series of samples, the panelist must arrange the samples in the order from the one with the least amount of an attribute to the one with the greatest amount of the attribute.

5.2.4 Threshold tests

Threshold tests are designed to measure the human sensitivity to a particular substance. There are several types of thresholds including detection, recognition, and dilution thresholds. To perform detection and recognition thresholds, panelists are given samples of increasing concentration and asked to identify the sample that is different (detection) or the one in which a specific taste or odour is recognized (recognition). Dilution thresholds determine the relative strength or persistence of a sample's taste or odour in relation to a control. The test is performed by continuous dilution of the initial sample until a just perceivable taste or odour can be detected.

5.3 ESTABLISHED METHODS TO ASSESS DRINKING WATER TASTES AND ODOURS

This section briefly summarizes the procedures and applications of four methods widely employed to assess the aesthetic quality of drinking water.

5.3.1 Flavour profile analysis

Flavour profile analysis (FPA) is a descriptive test for flavour analysis in the drinking water industry that can evaluate diverse tastes and odours in a water sample. The method is published as Standard Method 2170 (APHA, 2012) and its widespread applications are described in a report (Desrochers, 2008). The technique uses a trained panel to both describe different T&O in a water sample and to rate the intensities of each taste or odour present in the sample (Cairncross & SjöstrÖm, 1950; Krasner *et al.*, 1985). The method involves intensive training and standardization of panelists in order to achieve reliable and reproducible results. During the training, which typically lasts two days, panelists taste solutions of different concentrations of sucrose, sodium chloride, citric acid, and caffeine or quinine to familiarize them with sweet, salty, sour and bitter tastes, respectively. Select concentrations of sucrose, sodium chloride, citric acid, and caffeine or quinine are associated with known values of the FPA scale. The FPA scale is from 0 (no odour or taste) to 12 (strong taste or odour) (e.g. the sugar content of jelly). Known odours are also presented to the panelists, usually at an intensity known to have a weak–moderate odour, with the exception of *n*-hexanal, which is an international odour standard that has concentrations established for many FPA intensities on the 0–12 scale (Ömür-Özbek & Dietrich, 2008). Exposing panelists to a variety of standards for both different T&O and different FPA ratings allows them to develop the ability to provide both descriptive and intensity data for flavour (APHA, 2012; Schweitzer & Suffet, 1999).

Once trained, a panel of 4–7 members characterizes and assigns intensities to T&O in a water sample. Typically, the testing is performed at 45°C but may be performed at 25°C. When assessing tastes or odours, each panel member will make an individual assessment of a sample. Each assessment consists of one or more descriptors and a corresponding intensity rating on the 0–12 scale. Each panelist reports her or his individual results and the full panel then discusses all the results. If at least 50% of the panelists agree on a descriptor, an average for its intensity is calculated. An intensity of zero is used in the average for panel members who did not identify a particular descriptor (Table 5.1). Since an FPA panel can often identify multiple taste or odour components in one water sample, a 'profile' of the flavour is generated. Once trained and practiced, an FPA panel can efficiently process 6–8 samples within one hour.

Table 5.1 Example data for an FPA panel of 5 panelists describing water impacted by a cyanobacteria bloom.

Odour Descriptor	Individual Panelists					Consensus
	A	B	C	D	E	
Chlorinous	2	4	2	0	2	2
Musty	8	2	8	6	6	6
Medicinal	0	0	0	2	0	0[1]

[1]Exclude data in which only one or a few panelist identifies.

FPA can be applied to calculate a taste or odour threshold concentration (OTC) for an individual tastant and odourant when FPA intensity data are plotted against the log of the concentration using a Weber-Fechner plot, which is a form of a dose-response curve (Figure 5.1). Odour intensity is rated using FPA data with the 0, 2, 4, 6, 8, 10, 12 scale. The OTC is the concentration at which the FPA intensity is two (Krasner, 1988; Rashash *et al.*, 1997). Several publications are valuable resources that provide FPA data and/or Weber-Fechner plots for intensity versus concentration for diverse substances in water, these include: nonadienal (Burlingame *et al.*, 1992); fifteen drinking water odourants (Rashash *et al.*, 1997); iodophenols (Dietrich *et al.*, 1999); iodinated thrihalomethanes (Cancho *et al.*, 2001); aldehydes (Khiari *et al.*, 2002); halogenated anisoles (Díaz *et al.*, 2005); disinfectants in water (Krasner & Barrett, 1984; McDonald *et al.*, 2009); and geosmin and 2-MIB (Piriou *et al.*, 2009; Rashash *et al.*, 1997; Worley *et al.*, 2003; Yu *et al.*, 2014).

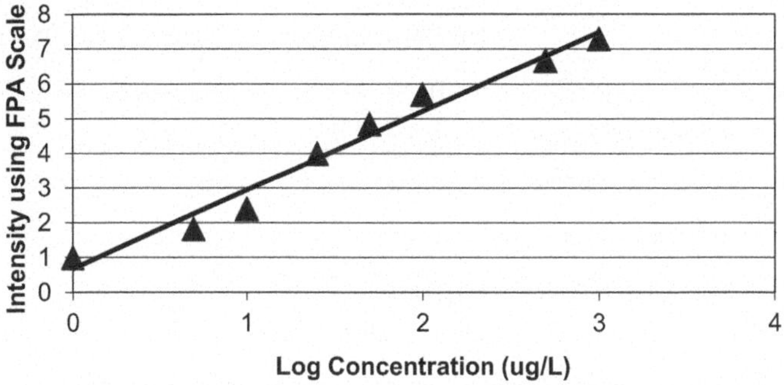

Figure 5.1 Weber-Fechner plot for n-hexanal. The odour threshold concentration is 4 ug/L.

FPA is also a valuable tool for assessing the individual T&O in source waters and drinking waters. A study in China applied FPA to survey 111 drinking water treatment plants and found that 80% of the source waters had odour problems but only 45% of the treated drinking waters had odour problems. The major odour categories were earthy/musty and swampy/septic (Sun *et al.*, 2014). FPA was instrumental in identifying fishy and musty as the major odour issues for source and drinking waters in Southern Taiwan; experimental data also demonstrated that the musty ratings from FPA were well correlated with GC/MS measurements for 2-MIB (Lin *et al.*, 2002). Likewise, FPA in conjunction with GC/MS was instrumental in identifying 2-MIB as the cause of musty odours in source waters throughout China (Yu *et al.*, 2009). In Singapore, FPA coupled to GC/MS and GC/O is instrumental in identifying odourous compounds in wastewater before becoming reuse water and possibly drinking water source (Agus *et al.*, 2012). FPA is also recommended for determining odourus compounds that leach from materials into drinking water as assessed by the Utility Quick Test (Schweitzer *et al.*, 2004).

A variation of FPA is often used in the water and food/beverage industry. This is quantitative descriptive analysis (QDA) (Stone *et al.*, 1974; Stone & Sidel, 1993). Instead of just reporting an average for the intensity of a taste or odour as is done in FPA, QDA reports a mean and standard deviation for the intensity of individual T&O detected in a water sample (Table 5.2). Many researchers and utilities are trained to use FPA, but actually process their data in the manner of QDA. Using QDA, panelists focus on a few descriptors pre-determined to be targeted for the samples. Instead of panelists generating a consensus opinion, a mean and standard deviation are calculated from the data of individual panelists.

Table 5.2 Application of quantitative descriptive analysis (QDA).

Odour Descriptor	**Odour Intensity of 50 μg/L ETBE[1] in Water**		
	No Disinfectant	**0.7 mg/L Cl_2**	**2.0 mg/L Cl_2**
Chlorinous	0	3.7 ± 0.6	3.8 ± 1.3
Sweet/bitter/burning/chemical	3.4 ± 1.5	2.5 ± 1.3	1.5 ± 0

[1]Ethyl-t-butyl ether is a compound that leaches from PEX-b pipes into tap water (Durand & Dietrich, 2007).

5.3.2 Threshold odour number

The threshold odour number (TON) is a dilution threshold test and has been used in the drinking water industry for over sixty years. The TON, which is Standard Method 2150B (APHA, 2012), measures the extent of dilution of a water sample with odour-free water in order to reduce the odour to a barely perceptible level. The test can be performed at either 40°C or 60°C with a volume of 200 mL for each water sample. Dilutions of the sample water are prepared in Erlenmeyer flasks, shaken and then sniffed until the dilution has no perceptible odour. The TON value is calculated by dividing the total volume (odour-free water volume + sample volume = 200 mL) by the volume of the odourous sample. For example, if 50 mL sample was diluted to 200 mL and an odour was just perceptible, then TON = 200/50, or 4.

The TON method has been employed for aesthetic guidelines in several countries (e.g. Taiwan, 2017; USEPA, 1979). Advantages of the method are that the endpoint is the same for all samples: a just-perceptible odour. The TON process requires little training and can be used to compare samples. The method has some major disadvantages and is therefore less desirable for routine monitoring and drinking water evaluation. The major disadvantage is that consumers do not receive and drink 'diluted water'. Therefore, the method has no direct relationship to water quality at the tap and does not correlate with the level of acceptability by the consumers. Other disadvantages include: TON does not describe individual odours or their intensities; TON measures the most dominant odour after dilution; and the procedure is cumbersome and time consuming to perform properly because many dilutions need to be prepared and smelled (AWWA, 2002; Dietrich & Burlingame, 2015; Dietrich *et al.*, 2004a).

5.3.3 Flavour rating analysis

The flavour rating analysis (FRA) is an affective test; it is described in detail in Standard Method 2160 C (APHA, 2012). FRA is designed to evaluate the acceptability of drinking water for daily consumption and was initially developed to assess the acceptability of the mineral content of a drinking water. This method provides an assessment of the acceptability of a drinking water based on its flavour. Panelists ingest a water sample and then assess its flavour using nine descriptors listed below: (1) I would be very happy to accept this water as my everyday drinking water; (2) I would be happy to accept this water as my everyday drinking water; (3) I am sure that I could accept this water as my everyday drinking water; (4) I could accept this water as my everyday drinking water; (5) Maybe I could accept this water as my everyday drinking water; (6) I don't think I could accept this water as my everyday drinking water; (7) I could not accept this water as my everyday drinking water; (8) I could never drink this water; (9) I can't stand this water in my mouth and I could never drink it.

The FRA can be used by an individual to describe the assessment of a water sample, or it can be used by a panel to provide multiple ratings that are then averaged to provide a single value for overall acceptability. FRA is applied by several drinking water providers to assess the acceptability of their water for consumers. Seattle Public Utilities (Seattle, Washington; USA) uses panels to assess the water flavour; if the water is rated from 1–5 then its flavour will be acceptable to their consumers when they drink the water (Dietrich *et al.*, 2004a). For the calendar year 2012, the Seattle Public Utilities successfully met its flavour goal for its drinking water, the water was rated a 3 or better on the FRA scale (Seattle Public Utilities, 2012).

5.3.4 Determining thresholds

An important property of a substance is its sensory threshold, or the minimum amount required to produce a perceptible human response. ASTM E679-91 – Standard Practice for Determination of Odour and Taste Thresholds by a Forced-Choice Ascending Concentration Series Method of Limits – is a commonly applied threshold test for estimating both detection and recognition thresholds (ASTM, 1991). Typically, eight concentrations are tested with 50–100 panelists participating in the threshold determination. Test concentrations should be selected to bracket the expected threshold level. The lowest concentration should be below the threshold and the highest concentration should be detectable by most people with normal sensitivity.

In performing the threshold determination, test concentrations are presented in an ascending concentration series in a forced-choice format with one sample containing the tastant or odourant and the other two samples being blanks. Each panelist evaluates the three samples at each concentration, then must choose a sample that is different (e.g. forced choice). The individual usually evaluates all samples in one session. An individual's threshold is calculated as the geometric mean of the highest concentration for which their response was incorrect and the lowest concentrations for which their response was correct and continued to be correct (Table 5.3). The best estimate threshold (BET) for a group or population is estimated as the geometric mean of these individual thresholds. The procedure recognizes that the degree of training of the panelists with the test substance may influence the threshold obtained.

Table 5.3 Example data for determining individual thresholds for panelists.

Panelist	Test Concentrations (μg/L)								Individual Threshold (μg/L)
	0.025	0.1	0.4	1.6	6.4	25.6	102.4	409.6	
1	–	–	–	+	–	+	+	+	12.8
2	+	–	–	–	–	+	+	+	12.8
3	–	–	+	+	+	+	+	+	0.2
4	–	+	–	–	+	+	+	+	3.2
5	–	–	–	+	+	+	+	+	0.8
6	–	+	–	–	–	–	–	+	204.8

Notes: A multiplier of 4 was applied to determine test concentrations, with the initial concentration selected to be 0.025 μg/L. + symbol indicates the panelist responds correctly in choosing the sample with the substance; – symbols indicates that the panelist selects the blank. Shaded boxes indicate the two concentrations used to calculate the individual panelist's geometric mean. The individual threshold is calculated when the panelist continues to have correct responses; the occasional correct response is considered a random correct response (e.g. panelist 2, 0.025 μg/L).

Variations to ASTM E679-91 exist and are routinely applied to generate valid data. Although all sample concentrations are usually evaluated by a panelist in one session, this approach is not possible if the test substance has a lingering effect or temporarily alters the human senses such as for metals like copper (Cuppett *et al.*, 2006; Hong *et al.*, 2010) or iron (Ömür-Özbek & Dietrich, 2011) which cause oral lipid oxidation. In these cases, a few hours between testing each concentration the panelists need to be refreshed. Manganese does not cause lipid oxidation; taste tests with Mn salts can be done with a few minute rest between samples as long as the panelists are given time to treat each sample individually and not become fatigued (Sain & Dietrich, 2014). To increase the accuracy of an individual test, a 1-of-5 forced choice test may be performed instead of the 1-of-3 (i.e. triangle test). The threshold method was used to determine taste thresholds for sulfate and nitrate (López *et al.,* 2017) and manganese (Sain *et al.*, 2014) in water, an odour threshold for methyl-*t*-butylether (MTBE) in water (Stocking *et al.*, 2001), and individual odour thresholds for individual *cis*-and *trans* isomers for components of a spill of crude methylcyclohexane methanol (MCHM) that disrupted the drinking water quality for an entire city for months (Gallagher *et al.*, 2015; Phetxumphou *et al.*, 2016).

The reader should keep in mind that different methods can be used to present samples or evaluate the data. Threshold methods are not truly capable of establishing single numbers and only estimating threshold ranges. Although not a part of the ASTM method, many statistical alternatives calculate group threshold range using logistical regression and confidence intervals (Abbott, 1925; Gallagher & Cuppett, 2007; Gallagher *et al.*, 2015; Lawless, 2010; Peng *et al.*, 2012).

It is important to note that a threshold is not a constant value for a given tastant or odourant. Thresholds can be viewed as constantly changing values on a sensory continuum from not perceptible to easily perceptible. An individual's thresholds will change from day to day based on psychological and physiological factors including mood, hunger, and satiety (Burlingame *et al.,* 2017; Meilgaard *et al.*, 2007). This is perhaps why the ASTM method uses the term BET when describing the numerical results for applying this method.

5.4 RECENT AND EVOLVING METHODS FOR SENSORY ANALYSIS OF DRINKING WATER

Methods to analyze tastes and odours in drinking water are continuously evolving to incorporate new developments in sensory science and to meet the needs of the drinking water industry. Over time, the established methods previously described have become part of the drinking water profession. The TON (Standard Method 2150B) first appeared in the 9th edition of *Standard Methods for the Examination of Water and Sewage* (APHA, 1946). FPA (Standard Method 2170) first appeared in the 19th edition of *Standard Methods for the Examination of Water and Sewage* (APHA, 1995). Since its introduction, the flavour profile method has gained rapid acceptance worldwide as a valuable tool for assessing the flavour of drinking water (Desrochers, 2008) and was revised in the 22nd edition of *Standard Methods for the Examination of Water and Wastewater* (APHA, 2012).

This next section describes methods which have recently been applied in the drinking water industry. Overall, these three methods are rapid and efficient; training and implementation require less than an hour. Thus, these practical methods can be used for routine applications. These methods are expected to gain acceptance and to eventually become standard methods published in *Standard Methods for the Examination of Water and Wastewater.*

5.4.1 Attribute rating test

The attribute rating test (ART) is a quick and practical descriptive method designed for detection and rating of individual odourants with respect to a target concentration (Dietrich *et al.*, 2004a). The target concentration is predetermined and should be near the odour level that would result in customer complaints if exceeded (Bae *et al.*, 2002; Dietrich *et al.*, 2004a, b). Although developed for odourants, the method is readily adaptable to tastants. Training for this method requires about an hour and implementation requires less than an hour. The method is designed to be performed by one analyst, although if more than one person performs the test, the statistical confidence in the results increases.

In this method, a standard is prepared for the target odourant and concentration. All analysts who perform this test are prescreened to detect the odourant at the desired concentration and that the analysts are not anosmic or insensitive. The analyst first sniffs odour free water to clear the nose, then smells the standard to perform a calibration. Next the analyst smells a water sample to rate the intensity of the specific odour as being either 'not detected' or a value of 0, 'less than the standard' which is a value of 1, 'equal to the standard' which is a value of 2, or 'greater than the standard' which is a value of 3. Data are recorded and if multiple analysts participate, data are compared and averaged. The method is useful for routine monitoring for an odourant in raw or processed water at the water treatment facility, analysis of water samples from consumer premises, tracking an odourant in a watershed or distribution system, or evaluation of odourant-removal capability by water treatment processes.

The ART is successfully employed to monitor the amount of powdered activated carbon (PAC) necessary to remove geosmin from a lake water source to below the target concentration of 15 ng/L. As shown in Figure 5.2, daily monitoring of the intensity of the earthy geosmin odour in raw and full-scale treated water demonstrated geosmin was successfully being removed by PAC and the water distributed to consumers had either no or a slight geosmin odour below the target concentration for this water treatment plant (Dietrich *et al.*, 2004b). The operators and staff who performed the ART reported the method was easy to perform and having the reference standard provides confidence that the geosmin odour was recognized at the target. In another treatment study, bench-scale research to evaluate the removal of 2-MIB and geosmin by PAC successfully employed the ART as a monitoring technique and demonstrated that this sensory technique agreed with chemical monitoring data produced by GC/MS (Bae *et al.*, 2002).

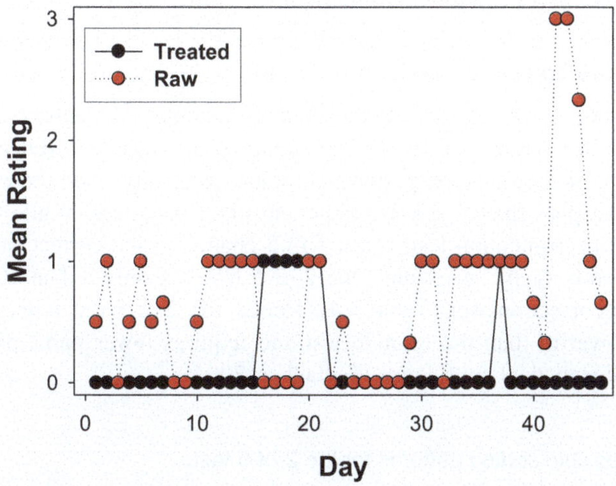

Figure 5.2 Plot of the geosmin odour intensity in raw and treated water from a water treatment plant applying powdered activated carbon to remove geosmin from source water. Reference concentration = 15 ng/L geosmin standard. The *y*-axis is defined as: 0 = not detected; 1 = <15 ng/L; 2 = 15 ng/L; 3 = >15 ng/L geosmin.

5.4.2 Total intensity of odour

Total intensity of odour (TIO) is a technique for deriving a numerical strength or intensity of a sample's overall perceived odour (Dale *et al.*, 2011; Jacobsen-Garcia *et al.*, 2017). A single numerical rating is assigned to a water sample by comparison to odour reference standards. This method requires minimal training, is economical and efficient, and does not require advanced laboratory techniques. The method is similar in approach to FPA, although instead of describing each odour and its intensity, an overall odour intensity is determined for the whole water sample. This method is not a dilution method, as dilution may change the impact of chemicals on odour intensity when multiple odours are present.

TIO is designed to be conducted by at least four panelists who individually rate the odour intensity of a water sample maintained at a temperature of 25°C. The odour reference standards are prepared from hexanal at concentrations corresponding to values between 0–12 on the FPA scale. The panelists compare the intensity of the odour in the water samples to the intensity of the reference standards. The panel discusses its findings and reaches a consensus opinion that describes the overall odour intensity of the water sample.

This method can be used to evaluate odour intensity of drinking water sources, any sampling points within the water treatment train, and finished drinking water samples; it can be an indicator of potential consumer complaints. TIO is currently under review to become a Standard Method.

5.4.3 2-of-5 test

The 2-of-5 test is an easily implemented discrimination method designed to determine if the odours of two water samples are different (Dietrich *et al.*, 2004a). The 2-of-5 test can be performed by just one analyst and therefore is very practical to be performed by water treatment plant operators. While the triangle test, another discrimination test, has a 1-in-3 chance of the correct sample being chosen randomly or by 'guessing', the 2-of-5 test has only a 1-in-10 chance of guessing the correct answer (Table 5.4). Hence, the 2-of-5 test is more sensitive and more powerful than the triangle test and requires fewer participants for the same level of statistical confidence (Gallagher, 2004).

Table 5.4 Possible sorting patterns for the 2-of-5 test.

Combination	Possible Sorting					Response
	Group of 2		Group of 3			
1	T1	T2	C1	C2	C3	CORRECT
2	C3	T2	C1	C2	T1	Incorrect
3	C3	T1	C1	C2	T2	Incorrect
4	C2	C3	C1	T1	T2	Incorrect
5	C1	C3	C2	T1	T2	Incorrect
6	C1	C2	C3	T1	T2	Incorrect
7	C2	T2	C1	C3	T1	Incorrect
8	C2	T1	C1	C3	T2	Incorrect
9	C1	T1	C2	C3	T2	Incorrect
10	C1	T2	C2	C3	T1	Incorrect

Notes: The three control samples are labeled C1, C2, and C3, the two test samples are labeled T1 and T2. Only one of the ten possible combinations correctly sorts the control and water samples.

In the 2-of-5 procedure, five flasks are used: two contain the test sample and the other three the control water. The choice of the control water is determined by operational goals for aesthetic testing. For example, the control water can be odour free water if the goal is to produce drinking water that is free of odours. The control can be water from the point of entry to the distribution system if the goal is to determine if water from a certain location in the distribution system has developed odours not present in water leaving the treatment plant.

One person can effectively perform the 2-of-5 test, although more than one person is preferred. Five flasks are labeled by affixing labels to the bottom of each flask so that the analyst cannot see them without turning the flasks over. Two of the five flasks are labeled the test water (labeled T1 and T2), and three of the five flasks are labeled the control water (labeled C1, C2, and C3; Table 5.4). The flasks can be either glass Erlenmeyer flasks or odour-free plastic cups The analyst places the corresponding water in each flask, then places the five flasks on a rotating plate, twirls the plate, then walks away for about 2 min. The twirling without the analyst looking 'randomizes' the order of the flasks. The analyst evaluates the odours in each flask, allowing ~30 s between samples, then sorts them into a group of two and a group of three based on their odour characteristics. The probability of the analyst sorting correctly by chance alone is only 10% (1 out of 10). If the samples were correctly sorted, the two waters are considered to be noticeably different and the analyst describes the odour by selecting a term from a list of the common odour attributes.

The 2-of-5 odour test can be adapted to taste testing if the water is safe to ingest. The test is used to evaluate drinking water under a variety of situations, including routine monitoring of raw or finished water; evaluation of odour problems in the distribution system; evaluation of treatments to remove odours; and tracking odour problems in watersheds. The Philadelphia Water Department used this method in the field for routine use to approve water for distribution to consumers after superchlorination of newly installed water mains (Dietrich *et al.*, 2004a). The control water was taken from the distribution system at a point before the location of the new water main. The sample water was taken from after the new water main. If the field crew sorted the control and the test waters correctly, then flushing of the water main was continued until no detectable difference was found in water before and after the water main. Implementation of the 2-of-5 test system was effective for distributing water with a typical amount of chlorinous odour to the customers impacted by the new water main.

5.4.4 Ranking tests

A ranking test requires panelists to rank the samples in order from the least to the strongest intensity and may be applied for hedonic scaling to determine the acceptability of water or for intensity scaling to rate a specific attribute (Lawless & Heymann, 2010; Meilgaard *et al.*, 2007). Advantages are that the test is relatively easy to perform, requires minimum instructions, and the data are straight-forward to interpret. The ranking test is designed to provide a comparison of different waters and not to describe a single water sample. It can demonstrate which samples cannot be discriminated and which can be discriminated from one another and thus is a powerful tool for assessing differences, such as variations in chlorine concentrations in the distribution system or taste intensity of different blended water samples.

An example scenario and data are presented in Tables 5.5 and 5.6, respectively, where the goal is to determine if consumers can detect a difference in the taste of varying blends of two water sources, a treated surface and a treated ground water. Six different blends were prepared and panelists tasted each of the six samples and ranked them in order of taste, with the water with the least taste given a '1' and the water with the most taste given a '6'. The ranking data for each of the six blends were statistically analyzed using the Friedman's T Statistic (Meilgaard *et al.*, 2007) and found to be statistically different (calculation not shown). Since overall the blended waters were significantly different, a Fisher's least-significant difference (LSD_{rank}) value was calculated to determine which specific pairs of waters are different as:

$$LSD_{rank} = z_{\alpha/2}\sqrt{\frac{bt(t+1)}{6}} = 13.73 \approx 14$$

where b = number of panelists = 7; t = number of samples = 6; $z_{\alpha/2} = 2.447$ for this data set.

Table 5.5 Six water blends for the example scenario that portrays the ranking test.

Water Type	Blended Water Samples					
	A	B	C	D	E	F
Surface water (%)	90	80	70	30	20	10
Ground water (%)	10	20	30	70	80	90

Table 5.6 Ranking data for each of the six water blends in the example scenario.

Panelist	Rankings for Blended Waters					
	A	B	C	D	E	F
1	1	2	3	4	5	6
2	2	3	1	4	5	6
3	1	2	3	4	5	6
4	2	3	1	5	4	6
5	2	3	1	4	5	6
6	3	1	2	4	6	5
7	2	1	3	5	4	6
Rank sum	13	15	14	31	33	41

The value $LSD_{rank} \approx 16$ is the required minimum value between the Rank Sums for a pair of blended waters for the difference to be significant. The 'Rank Sums' is simply the sum of the individual panelist rankings for each water sample. As can be seen in Table 5.6, the difference between the Rank Sums for any combination of pairs of waters A, B, and C is less than 16, therefore, these three waters have the same taste. Likewise, the difference between the Rank Sums for any combination of pairs of waters D, E, and F is also less than 16, therefore, these three waters have the same taste. For any pair-wise comparison of waters A, B, or C with waters D, E, or F, the Rank Sums are always greater than or equal to 16 and therefore there is a statistical difference in the taste of the water if the blend has more ground water than surface water.

The ranking test was applied to determine if the taste of drinking water improved when reverse osmosis treated water or electrodialysis-reversal treated water were blended with a conventionally treated drinking water of ~1100 mg/L total dissolved solids (TDS; Devesa *et al.*, 2010; García *et al.*, 2015). Panelists were asked to rank a series of blended water samples in order of preference. The ranking test data demonstrated there were differences in the taste of the blended waters compared with the conventionally treated river water. The blended waters were also subjected to an affective test in which panelist rated the likeability of the water. There was agreement between the ranking data and the rating data, trained and untrained panelists, and also chemical data for TDS; blended waters with lower TDS values also had preferred ranking scores and higher likeability. These studies demonstrated the effectiveness of the ranking test for evaluating the taste of drinking water.

5.5 CONSUMER FEEDBACK

5.5.1 Consumer complaint surveillance

Many water utilities consider consumer feedback important and respond effectively to consumer complaints (Lauer, 2004). Systematically tracking consumer complaints is a valuable, but underutilized, tool for assessing water quality. Consumers are present everywhere, everyday throughout the water distribution network to drink and use water 24 hours a day, 7 days a week. A Consumer Complaint Surveillance system is a valuable means to augment, but not replace, chemical and microbiological water quality monitoring (Gallagher & Dietrich, 2014; Speight *et al.*, 2010; PWD and CH2M Hill, 2013; USEPA, 2008). Surveillance in this case refers to a methodology that relies on detecting water quality problems based on consumer input, such as an increase in complaints about aesthetic issues.

Consumer Complaint Surveillance can take on many forms, although electronic and not paper records of the data are highly recommended. A good approach is to employ the 'funnel-filter-focus' technique (Whelton *et al.*, 2007). This approach funnels all consumer complaints to a central location such as a call center, then filters to just the water quality complaints (e.g. separate from billing or non-water quality), and finally focuses on categorizing the complaints into standard terms, such as illnesses, tastes and odours of the T&O wheel, colors of the water, and presence of particles. The type of data to be entered should include: date and time of complaint; customer name and address; GIS information for the exact location; nature of the complaint, including placing into subcategories; and resolution of complaint (PWD and CH2M Hill, 2013).

The power of a Consumer Complaint Surveillance system is found in maintaining long-term records of complaints and systematically examining them for trends and changes. The frequency for total number of complaints or number of complaints in specific categories (e.g. rusty particles, chemical odours, earthy-musty odours) is examined to determine if there is cause for concern or if the pattern is typical and atypical for that utility. This can be achieved through developing exceedance plots (Gallagher & Dietrich, 2014) which demonstrate the likelihood that the number of complaints exceeds a certain number in a defined time period. Triggers can be established to determine when the number of complaints exceeds expected number of complaints. Establishing triggers can be done manually (USEPA, 2011). Preferably, instead of triggers, changes above background should be monitored through event detection algorithms and event detection systems that integrate data, algorithms, and management (PWD and CH2M Hill, 2013; USEPA, 2010). Event-detection systems development is in an early stage for assessing chemical water quality using on-line monitors (Rosen & Bartrand, 2013). At present there are few event detection algorithms and no event detection systems for aesthetic issues in drinking water. When developed, these will need to take into account the number of consumers in the distribution system, the temporal-spatial locations of the complaints (Proulx *et al.*, 2007), the nature of the sensory complaint (Dietrich *et al.*, 2014), and statistical evaluation of historical data as total complaints and in select categories.

The difficulties in developing Consumer Complaint Surveillance systems in the drinking water industry are also related to understanding consumer perception, consumer preferences, and how to categorize consumer complaints. Many of these issues have been previously addressed for improving a consumer surveillance system (Dietrich, 2006). That article incorporated the water industry's approach to: (1) identifying the sensory issue; (2) identifying the cause of the sensory issue; and (3) treating to control chemical or biological cause of the sensory issue. Specific suggestions were:

- Comprehensively identify the chemicals that cause T&O in drinking water. The drinking water T&O wheel that represents the best version of our current understanding was expanded in Chapter 2. The drinking water wheel in distribution systems (Khiari *et al.*, 2002) should be expanded.
- Identify specific causes and sources of tastant or odourant (e.g. microbial produced, leached from materials, etc.) so that they can be treated and controlled.

- Understand population variations in describing and providing feedback about specific tastants and odourants so that consumer complaints can be properly categorized. Water treatment plant personnel, who are more likely to be familiar with and accepting of odours in source and drinking water, can better describe odours than consumers who are less familiar and expect their drinking water to be clear and without unfamiliar odours (Dietrich *et al.*, 2014; Phetxumphou *et al.*, 2017; Webber *et al.*, 2015).
- Understand population variations in sensory perception of specific tastants and odourants as well as determine human preferences for different water qualities (e.g. mineral levels, chlorine levels) and relate to demographic and geographical factors (Piriou *et al.*, 2015; Yu *et al.*, 2014). While utilities will not be able to adjust water quality to match the variability in the population, understanding the range of concentrations that humans can detect will allow utilities to understand sensitive and insensitive human responses.
- Investigate and understand the sensory behavior of mixtures of tastes and odourants. To date, most tastants and odourants are investigated individually for their T&O properties and thresholds. Synergisms and antagonisms are common and should be understood. For example, 1 mg/L free chlorine is acknowledged to be antagonistic to geosmin and 2-MIB and suppresses their odours (Worley *et al.*, 2003). Minor components in mixtures can contribute substantial odours if they possess high volatility (Phetxumphou *et al.*, 2016).
- Evaluate water quality changes from various treatments and devices used to improve aesthetic water quality.
- Educate consumers about the value of water (Dietrich & Burlingame, 2015; Roy *et al.*, 2015).

Consumer Complaint Surveillance can serve many overlapping purposes, including being part of a Contaminant Warning System that is designed to detect serious threats to drinking water through intentional or unintentional contamination (PWD and CH2M Hill, 2013; USEPA, 2008). The global changes in climate are expected to result in more extreme weather events, and these weather events have been demonstrated to impact water quality and water aesthetics (Newcombe *et al.*, 2012; Stanford *et al.*, 2013). In a 2012 survey of utilities, 56% report that they are likely or very likely to implement Consumer Complaint Surveillance if available (Ömür-Özbek, 2012). Thus, harnessing consumer complaints as a water quality indicator is a challenge that should be undertaken by the water industry in the near future.

5.5.2 Setting guidance levels for individual odourants

Setting internal guidance for a country or within a utility or geographic region is one approach to providing aesthetic guidance levels for problematic and frequently occurring odourants. This has been done for geosmin (earthy) and 2-MIB (musty), the worldwide cyanobacterial metabolites that plague utilities and consumers because they are odourous at 1–10 ng/L (Piriou *et al.*, 2009; Rashash *et al.*, 1997; Yu *et al.*, 2014). For reasons of consumer satisfaction with their drinking water, The Japanese Ministry of Health, Labour and Welfare established national standards of 10 ng/L for geosmin and 2-MIB (Kishida *et al.*, 2013, 2015). In Australia, the operational guidelines use 10 ng/L for geosmin and 2-MIB as actions levels for increased monitoring and treatment in order to maintain consumer satisfaction (Australia Drinking Water Guidelines, 2011). In China, regional FPA studies were performed that demonstrated the variability and sensitivity of local populations; this variability is being used to establish regional standards for 2-MIB and geosmin (Yu *et al.*, 2014). The Philadelphia Water Department established odour-specific guidance for geosmin and 2-MIB at 10 ng/L in response to contaminant levels at which increased consumer complaints occurred (Suffet *et al.*, 2008). Likewise, the Metropolitan Water District of Southern California notifies consumers and its member agencies when the geosmin or 2-MIB concentrations are in the 10–20 ng/L range (Taylor *et al.*, 2006). Odours besides those causing earth-musty odours are also regulated. The State of California (California Code of Regulations, 2006) enforces a standard of 5 μg/L MTBE due to the sweet solvent odour of the groundwater contaminant. In rethinking and replacing the odour standard of TON, a proposal is to consider establishing guidance for individual odourants that are problematic in drinking water (Dietrich & Burlingame, 2015).

5.6 SUMMARY

A targeted goal of water treatment and distribution is to provide safe and palatable drinking water that consumers will perceive as valuable and acceptable for all their needs. Sensory tests are valuable tools for the drinking water industry to use for assessing source and drinking water quality and also for determining sensory thresholds and achieving consumer satisfaction.

REFERENCES

Abbott W. S. (1925). A method of computing the effectiveness of an insecticide. *Journal of Economic Entomology*, **18**, 265–267.

Agus E., Zhang L. and Sedlak D. L. (2012). A framework for identifying characteristic odour compounds in municipal wastewater effluent. *Water Research*, **46**, 5970–5980.

American Public Health Association (APHA), American Water Works Association (1946). Standard Methods for the Examination of Water and Sewage, 9th edn. American Public Health Association, Washington, DC.

American Public Health Association (APHA), American Water Works Association, and Water Pollution Control Federation (1995). Standard Methods for the Examination of Water and Wastewater, 19th edn. American Public Health Association, Washington, DC.

American Public Health Association (APHA), American Water Works Association, Water Environment Federation (2012). Standard Methods for the Examination of Water and Wastewater, E. W. Rice, R. B. Baird, A. D. Eaton and L. S. Clesceri (eds), 22nd edn, American Public Health Association, Washington, DC.

American Standards for Testing Materials (1991). Standard Practice for Defining and Calculating Individual and Group Sensory Thresholds Form Forced-Choice Data Sets of Intermediate Size. Annual Book of Standards, American Society for Testing of Materials, Philadelphia, PA, Vol. **15**, pp. 67–74.

American Water Works Association Taste and Odour Committee (2002). Options for a taste and odour standard. *Journal American Water Works Association*, **94**, 80–87.

Australian Government (2011). Australian Water Quality Guidelines 6. Volume 1. Natural Resource Management Ministerial Council, Commonwealth of Australia, Canberra.

Bae B.-U., Kim Y.-I., Dugas D. W., Burlingame G. A. and Dietrich A. M. (2002). Demonstration of new sensory methods for drinking water taste-and-odour control. *Water Science and Technology: Water Supply*, **2**(5–6), 241–247.

Burlingame G. A., Muldowney J. J. and Maddrey R. E. (1992). Cucumber flavour in Philadelphia's drinking water. *Journal American Water Works Association*, **94**, 92–97.

Burlingame G. A., Doty R. L. and Dietrich A. M. (2017). Humans as sensors to evaluate drinking water taste and odor: a Review. *Journal of the American Water Works Association*, **109**, E13–E24.

Cairncross S. E. and SjöstrÖm L. B. (1950). Flavour profiles: a new approach to flavour problems. *Food Technology*, **4**, 308–319.

California Code of Regulation (2006). Title 22, Division 4, Environmental Health Chapter 15. Domestic Water Quality and Monitoring Regulation. Article 16. Secondary Standards. R-21-03, California Code of Regulation, May 2, 2006.

Cancho B., Fabrellas C., Díaz A., Ventura F. and Galceran M. T. (2001). Determination of the odour threshold concentrations of iodinated trihalomethanes in drinking water. *Journal of Agricultural and Food Chemistry*, **49**, 1881–1884.

Cuppett J. D., Duncan S. E. and Dietrich A. M. (2006). Evaluation of copper speciation and water quality factors that affect aqueous copper tasting response. *Chemical Senses*, **31**, 689–697.

Dale M. S., Desrochers R. D., Labernik S. M. and Moylan M. S. (2011). Threshold Odour Number Versus Flavour Profile Analysis: A Case Study in the Endeavor to Measure Consumer Acceptance. Presented at the 9th IWA Symposium on Off-Flavours in the Aquatic Environment, Aberdeen, Scotland, August 2011.

Desrochers R. (2008). Sensory analysis in the water industry. *Journal American Water Works Association*, **100**, 50–54.

Devesa R., García V. and Matía L. (2010). Water flavour improvement by membrane (RO and EDR) treatment. *Desalination*, **250**, 113–117.

Díaz A., Fabrellas C., Ventura F. and Galceran M. T. (2005). Determination of the odour threshold concentrations of chlorobrominated anisoles in water. *Journal of Agriculture and Food Chemistry*, **53**, 383–387.

Dietrich A. M. (2006). Aesthetic issues for drinking water. *Journal of Water and Health*, **4** (Supplement 1), 11–16.

Dietrich A. M. and Burlingame G. A. (2015). Critical review and rethinking of USEPA secondary standards for maintaining consumer acceptability of organoleptic quality of drinking water. *Environmental Sciences & Technology*, **49**, 708–720.

Dietrich A. M., Mirlohi S., DaCosta W. F., Dodd J. P., Sauer R., Homan M. and Schultz J. (1999). Flavour profile analysis and GC-MS detection of phenolic iodinated disinfection byproducts in drinking water for the USA space program. *Water Science and Technology*, **40**(6), 45–51.

Dietrich A. M., Hoehn R. C., Burlingame G. A. and Gittelman T. (2004a). Practical Taste-and-Odour Methods for Routine Operations: Decision Tree. AWWA Research Foundation, Denver, CO.

Dietrich A. M., Whelton A., Hoehn R., Anderson R. and Wille M. (2004b). Attribute rating test for sensory analysis. *Water Science and Technology*, **49**(9), 61–67.

Dietrich A. M., Phetxumphou K. and Gallagher D. L. (2014). Systematic tracking, visualizing, and interpreting of consumer feedback for drinking water quality. *Water Research*, **66**, 63–74.

Durand M. L. and Dietrich A. M. (2007). Contributions of silane cross-linked PEX pipe to chemical/solvent odours in drinking water. *Water Science and Technology*, **55**(5), 153–160.

Gallagher D. L. (2004). Statistical comparison of the triangle test and the two-of-five test for taste and odour evaluation. *Water Science and Technology*, **49**(9), 107–114.

Gallagher D. L. and Cuppett J. (2007). Evaluation of threshold limit methods for sensory data. *Water Science and Technology*, **55**(5), 67–75.

Gallagher D. L. and Dietrich A. M. (2014). Statistical approaches for analyzing customer complaint data to assess aesthetic episodes in drinking water. *Journal of Water Supply: Research and Technology-AQUA*, **63**, 358–367.

Gallagher D. L., Phetxumphou K., Smiley E. and Dietrich A. M. (2015). Tale of two isomers: complexities of human odour perception for cis- and trans-4-methylcyclohexane methanol from the chemical spill in West Virginia. *Environmental Science & Technology*, **49**(3), 1319–1327.

García V., Fernández A., Medina M. E., Ferrer O., Cortina J. L., Valero F. and Devesa R. (2015). Flavour assessment of blends between desalinated and conventionally treated sources. *Desalination and Water Treatment*, **53**, 3466–3474.

Hong J.-H., Duncan S. E. and Dietrich A. M. (2010). Effect of copper speciation at different pH on temporal sensory attributes of copper. *Food Quality and Preference*, **21**, 132–139.

Jacobsen-Garcia A., Dale M. and Krasner S. (2017). Total intensity of odour: a new method to evaluate odours. *Journal American Water Works Association*, **109**, E42–E48.

Khiari D., Chinn R., Barrett S., Matia L., Ventura F., Suffet I., Gittelman T. and Leutweiler P. (2002). Distribution Generated Taste and Odour Phenomena. AWWA Research Foundation, American Water Works Association, Denver, CO.

Kishida N., Konno Y., Nemoto K., Amitani T., Maki A., Fujimoto N. and Akiba M. (2013). Recent trends in microorganism-related off-flavour problems in drinking water treatment systems in Japan. *Water Science and Technology: Water Supply*, **13**(5), 1228–1235.

Kishida N., Sagehashi M., Takanashi H., Fujimoto N. and Akiba M. (2015). Nationwide survey of organism-related off-flavour problems in Japanese drinking water treatment plants (2010–2012). *Journal of Water Supply: Research and Technology-AQUA*, **64**, 832–838.

Krasner S. W. (1988). Flavour-profile analysis – an objective sensory technique for the identification and treatment of off-flavours in drinking-water. *Water Science and Technology*, **20**(8–9), 31–36.

Krasner S. W. and Barrett S. E. (1984). Aroma and Flavour Characteristics of Free Chlorine and Chloramines. Proceedings of the AWWA Water Quality Technology Conference, Denver, CO, 2–5 December 1984.

Krasner S. W., McGuire M. J. and Ferguson V. B. (1985). Tastes and odours: the flavour profile method. *Journal American Water Works Association*, **77**(3), 34–39.

Lauer W. C. (2004). Water Quality Complaint Investigator's Field Guide. American Water Works Association, Denver, CO, USA.

Lawless H. T. (2010). A simple alternative analysis for threshold data determined by ascending forced-choice methods of limits. *Journal of Sensory Studies*, **25**, 332–346.

Lawless H. T. and Heymann H. (2010). Sensory Evaluation of Food: Principles and Practice, 2nd edn. Springer, New York.

Lin T.-F., Wong J.-Y. and Koa H.-P. (2002). Correlation of musty odour and 2-MIB in two drinking water treatment plants in South Taiwan. *Science of the Total Environment*, **289**, 225–235.

López P., Pérez-Rodríguez I., Estrany F. and Devesa R. (2017). Moderate levels of sulfate improve the taste of water, nitrate is not relevant. *Journal of and Water Supply: Research and Technology – AQUA*, **66**, 598–605.

McDonald S., Lethorn A., Loi L., Jol C., Driessen H. and Heitz A. (2009). Determination of odour threshold concentration ranges for some disinfectants and disinfection by-products for an Australian panel. *Water Science and Technology*, **60**(10), 2493–2506.

Meilgaard M. C., Carr B. T. and Civille G. V. (2007). Sensory Evaluation Techniques, 4th edn. CRC Press, Boca Raton, FL.

Newcombe G., Chorus I., Falconer I. and Lin T.-F. (2012). Cyanobacteria: impacts of climate change on occurrence, toxicity and water quality management. *Water Research*, **46**, 1347–1358.

Ömür-Özbek P. (2012). Global Taste and Odour Survey of Water Utilities: Final Report. American Water Works Association, Denver, CO, USA.

Ömür-Özbek P. and Dietrich A. M. (2008). Developing *n*-hexanal as an odour reference standard for sensory analysis of drinking water. *Water Research*, **42**, 2598–2604.

Ömür-Özbek P. and Dietrich A. M. (2011). Retronasal perception and flavour thresholds of iron and copper in drinking water. *Journal of Water and Health*, **9**(1), 1–9.

Peng M., Jaeger S. and Hautus M. (2012). Determining odour detection thresholds: incorporating a method-independent definition into the implementation of ASTM E67. *Food Quality and Preference*, **25**, 95–104.

Phetxumphou K., Dietrich A. M., Shanaiah N., Smiley E. and Gallagher D. L. (2016). Subtleties of human exposure and response to chemical mixtures from spills. *Environmental Pollution*, **214**, 618–626.

Phetxumphou K., Raghuraman A. and Dietrich A. M. (2017). Implementing the drinking water taste-and-odor wheel to improve consumer lexicon. *Journal of American Water Works Association*, **109**, E453–E463.

Piriou P., Devesa R., de Lalande M. and Glucina K. (2009). European reassessment of MIB and geosmin perception in drinking water. *Journal of Water Supply: Research and Technology – AQUA*, **58**, 532–538.

Piriou P., Devesa R., Puget S., Thomas-Danguin T. and Zraic F. (2015). Evidence of regional differences in chlorine perception by consumers: sensitivity differences or habituation? *Journal of Water Supply: Research and Technology – AQUA*, **64**, 783–792.

Proulx F., Rodriguez M. J., Sérodes J. and Bouchard C. (2007). A methodology for identifying vulnerable locations to taste and odour problems in a drinking water system. *Water Science and Technology*, **55**(5), 177–183.

PWD and CH2M HILL (2013). Philadelphia Water Department Contamination Warning System Demonstration Pilot Project: Customer Complaints Surveillance Guidance. White paper submitted to USEPA as part of the Water Security Initiative Grant Awarded to Philadelphia Water Department. www.ch2mhill.com/iws

Rashash D. M. C., Dietrich A. M. and Hoehn R. C. (1997). Flavour profile analysis of selected odourous compounds. *Journal American Water Works Association*, **89**, 131–142.

Rosen J. and Bartrand T. (2013). Using online water quality data to detect events in a distribution system. *Journal American Water Works Association*, **105**, 22–26.

Roy S., Phetxumphou K., Dietrich A. M., Estabrooks P., You W. and Davy B. M. (2015). An evaluation of the readability of drinking water quality reports: a national assessment. *Journal of Water and Health*, **13**, 645–653.

Sain A. E. and Dietrich A. M. (2014). Rethinking aesthetic guidelines for manganese and iron in drinking water. *Journal of Water Supply: Research and Technology-AQUA*, **64**, 775–782.

Sain A. E., Griffin A. and Dietrich A. M. (2014). Assessing taste and visual perception of Mn (II) and Mn(IV) in drinking water. *Journal American Water Works Association*, **106**, 32–40.

Schweitzer L. and Suffet I. H. (1999). Exposure assessment of the taste and odour standards used in the method of flavour profile analysis. *Water Science and Technology*, **40**(6), 209–215.

Schweitzer L., Tomboulian P., Atasi K., Chen T. and Khiari D. (2004). Utility quick test for analyzing materials for drinking water distribution systems for effect on taste and odour. *Water Science and Technology*, **49**(9), 75–80.

Seattle Public Utilities (2012). Annual Report Card. http://www.seattle.gov/util/AboutUs/Management/AnnualReportCard/index.htm

Speight V., Grayman W. and Khanal N. (2010). Syngeries between security research and distribution system water quality management. *Journal American Water Works Association*, **102**, 30–33.

Stanford B. D., Wright B., Routt J. C., Debroux J. F. and Khan S. J. (2013). Water Quality Impacts of Extreme Weather-related Events (WaterRF #4324). Water Research Foundation, Denver, CO, USA.

Stocking A. J., Suffet I. H., McGuire M. J. and Kavanaugh M. C. (2001). Implications of an MTBE odour study for setting drinking water standards. *Journal American Water Works Association*, **93**, 95–105.

Stone H. and Sidel J. L. (1993). Sensory Evaluation Practices, 2nd edn. Academic Press, San Diego, CA.

Stone H., Sidel J. L., Oliver S., Woosley A. and Singleton R. C. (1974). Sensory evaluation by quantitative descriptive analysis. *Food Technology*, **28**, 24–31.

Suffet I. H., Burlingame G. A. and Mackey E. (2008). Controlling taste and odour events for the city of Philadelphia, USA: a case study. *Water Science and Technology: Water Supply*, **8**(2), 135–141.

Sun D., Yu J., Yang M., An W., Zhao Y., Lu N., Yuan S. and Zhang D. (2014). Occurrence of odour problems in drinking water of major cities across China. *Frontiers in Environmental Science and Engineering*, **8**, 411–416.

Taiwan Environmental Protection Administration (2017). Drinking Water Quality Standards, https://oaout.epa.gov.tw/law/EngLawContent.aspx?lan=E&id=171&KW=drinking+water+quality (accessed 25th December 2018).

Taylor W. D., Losee R. F., Torobin M., Izaguirre G., Sass D., Khiari D. and Atasi K. (2006). Early Warning and Management of Taste-and-Odour Events. AWWA Research Foundation, Denver, CO, USA.

USEPA (1979). National secondary drinking water regulations. Final rule. United States environmental protection agency. *Federal Register*, **44**(140), 42195.

USEPA (2008). Water Security Initiative: Interim Guidance on Developing an Operational Strategy for Contamination Warning Systems, EPA-817-R-08-002. United States Environmental Protection Agency, Washington, DC.

USEPA (2010). Water Quality Event Detection Systems for Drinking Water Contamination Warning Systems, EPA/600/R-10/036. United States Environmental Protection Agency, Washington, DC.

USEPA (2011). Alarm Estimation Tool Microsoft Excel File. United States Environmental Protection Agency. Washington DC. http://water.epa.gov/infrastructure/watersecurity/techtools/

Webber M. A., Atherton P. and Newcombe G. (2015). Taste and odour and public perceptions: what do our customers really think about their drinking water? *Journal of Water Supply: Research and Technology-AQUA*, **64**, 802–811.

Whelton A., Dietrich A. M., Gallagher D. L. and Roberson J. A. (2007). Using customer feedback for improved water quality and infrastructure monitoring. *Journal American Water Works Association*, **99**, 62–76.

Worley J. L., Dietrich A. M. and Hoehn R. C. (2003). Dechlorination techniques to improve sensory odour testing of geosmin and 2-MIB. *Journal American Water Works Association*, **95**, 109–117.

Yu J. W., Zhao Y. M., Yang M., Lin T.-F., Guo Z. H., Gu J. N., Li S. and Han W. (2009). Occurrence of odour-causing compounds in different source waters of China. *Journal of Water Supply: Research and Technology-AQUA*, **58**, 587–594.

Yu J., An W., Cao N., Yang M., Gu J., Zhang D. and Lu N. (2014). Quantitative method to determine the regional drinking water odorant regulation goals based on odor sensitivity distribution: illustrated using 2-MIB. *Journal of Environmental Sciences*, **26**, 1389–1394.

Chapter 6

Removal of odourants from drinking water

Chao Chen[1], Arash Zamyadi[2], Tsair-Fuh Lin[3] and Daniel Gallagher[4]

[1]State Joint Key Lab of Environmental Simulation and Pollution Control, School of Environment, Tsinghua University, Beijing, China
[2]Department of Civil, Geological and Mining Engineering (CGM), Polytechnique Montréal, University of Montreal, Montreal, Canada
[3]Department of Environmental Engineering, National Cheng Kung University, Tainan, Taiwan
[4]Department of Civil and Environmental Engineering, Virginia Polytechnic Institute and State University, Blacksburg, VA, USA

6.1 INTRODUCTION

6.1.1 Overview of treatment of odourants

Treatment of odourants in drinking water is an important issue for water professionals and the general public. Although unpalatable odour in finished water may not be directly linked to health risks, consumers may question the unpleasant odour with regard to safety and quality of the water supply, leading to a loss of confidence in the public water supply. Complaints about unpleasant odour in drinking water have been reported in London as early as 1913, the smell was attributed to invasion by an organism with no pathogenic significance (Anonymous, 1913). However, before 1970 only about 10 papers relevant to this

Editor(s): Tsair-Fuh Lin, Susan Watson, Andrea M. Dietrich and I. H. (Mel) Suffet
doi: 10.2166/9781780406664_0167

topic were published. Only after the late 1970s, more scientific reports were published about the drinking water odour issues. Figure 6.1 shows the numbers of scientific papers published from 1913 based on the search of two keywords, 'drinking water' and 'odour'. Among the 1,383 papers published for drinking water and odour, 57% (788 papers) were identified with an additional keyword – 'treatment'. In the last 20 years, scientific papers published relevant to odour and drinking water totalled about 50 papers each year, while about 28 contained the additional keyword treatment, suggesting that treatment is always an important topic for the research of odour in drinking water.

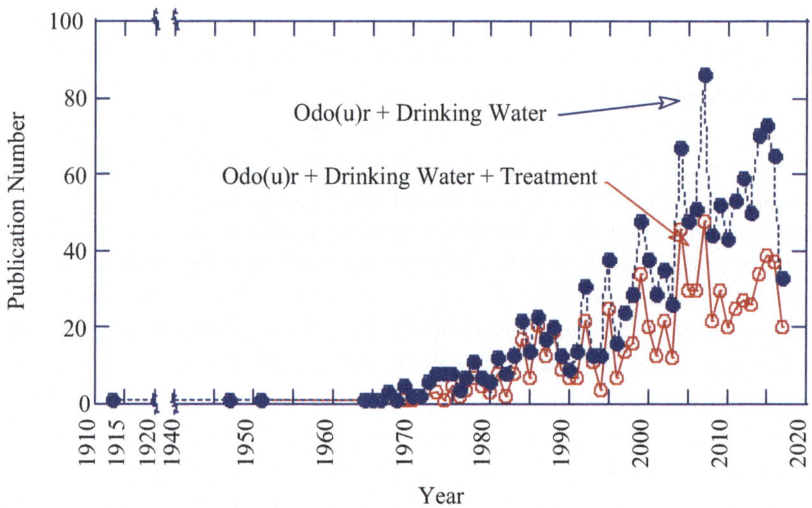

Figure 6.1 Publications about odour in drinking water searched from Scopus, from 1913 to October 22, 2017.

Figure 6.2 shows the word-cloud plots based on the keywords extracted from the 788 and 280 papers published for all years and for the last 10 years (2008 to Oct 2017), respectively, searched from Scopus database. Geosmin is the most popular chemical studied, with about 141 and 63 publications in all years and in the last 10 years, respectively. 2-Methylisoborneol (2-MIB), another important odourant, was the second most studied, and appeared in 106 and 46 publications for all years and the last 10 years, respectively. For the all year case, the most common treatment processes linked together with odour are: disinfection (117 out of 788 papers), oxidation (80), adsorption (78), filtration (65), ozonation (56), activated carbon treatment (55), biodegradation (39), hydrogen peroxide (37), bacteria (36), coagulation (34), powdered activated carbon (PAC) (29), ultraviolet

(UV) radiation (23), and microfiltration (19). For those published in the years 2008–2017, they are: activated carbon (57), ozone (45), disinfection (39), adsorption (34), oxidation (31), filtration (24), coagulation (22), PAC (20), hydrogen peroxide (18), chemical treatment (17), biodegradation (14), UV (12), bacteria (12), granular activated carbon (GAC) (10), and membrane (10). From the keywords, it is clearly demonstrated that when facing odour issues in drinking water, oxidation and adsorption processes, more specifically advanced oxidation including ozone, hydrogen peroxide and UV, and activated carbon adsorption are the most common processes studied or used (Suffet *et al.*, 1995). Besides oxidation and adsorption, conventional water treatment processes, including coagulation and filtration are also well studied, mainly for the removal of odourant-laden cyanobacteria and microorganisms (Henderson *et al.*, 2008). Biodegradation is another important process to remove odourant, mainly happening in filtration processes (Ho *et al.*, 2005). In the last 10 years, membrane processes have become another research topic for the removal of odourants from drinking water (Dixon *et al.*, 2011).

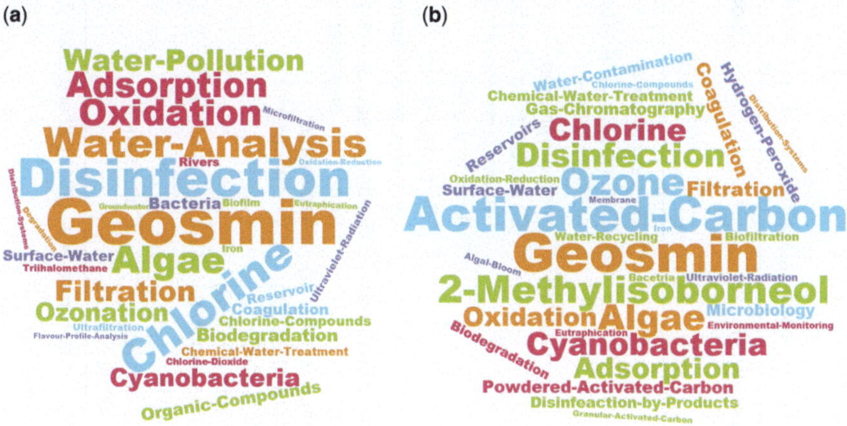

Figure 6.2 Word-cloud analysis of the keywords appearing in publications about odour in drinking water treatment, searched from Scopus, (a) from 1913 to October 22, 2017, and (b) from 2008 to 2017.

The chemicals commonly causing odour problems in water supply can be linked to three sources (Table 6.1): microbiological sources (mainly cyanobacteria and actinomycetes), human activities, and those relevant to water treatment. In the table, water treatment methods commonly used to remove the odourants are also listed and will be discussed in the following sections of this chapter.

Table 6.1 Compounds causing taste and odour issues in water supply and their sources.

Odour Type	Compound(s)	Source	Treatment Methods (Discussed in this Chapter)
Earthy	Geosmin	Microorganisms Cyanobacteria: *Anabaena*, *Oscillatoria;* Actinomycetes:	Activated carbon treatment
Musty	2-MIB	Microorganisms; Cyanobacteria: *Anabaena*, *Oscillatoria*, *Phormidium*, *Planktothrix*, *Pseudanabaena;* Actinomycetes:	Activated carbon treatment
Woody, Tobacco	β-Cyclocitral	Microorganisms; Cyanobacteria: *Microcystis*	Oxidation
Swampy, Rotten, Garlic, Seaweed	Volatile Sulfur Compounds (VSC): Hydrogen sulfide (H_2S) Methyl mercaptan (MeSH) Dimethyl sulfur (DMS) Dimethyl disulfide (DMDS) Dimethyl trisulfide (DMTS) Ethyl thiol (EtSH)	Decomposed or living cyanobacteria; Anaerobic degradation of substrate in sediments; Wastewater	Oxidation
Chlorine Medicinal	Chlorine, Chloramines Aldehydes, Chlorophenols,	Chlorination	–

Sources: Suffet *et al.* (1995), Watson (2003), Newcombe *et al.* (2010), Zhang *et al.* (2010), Chiu *et al.* (2016).

6.1.2 Treatment of Geosmin and 2-MIB

Geosmin and 2-MIB are the two most commonly detected odourants in water supply systems (Suffet *et al.*, 1999), their forms (dissolved or cell-bound phases) and removal in water treatment processes are discussed here further. Geosmin has been reported to be mostly cell bound, 95–99% for cultured *Anabaena circinalis* (now called *Dolichospermum circinale*) and 60–85% in *Phormidium amoenum* (Li *et al.*, 2012), and 85–97% in water dominated with *Anabaena spiroides* in the Yanghe Reservoir, China (Li *et al.*, 2010). Unlike geosmin, reports indicated that much of 2-MIB may be present in dissolved phase. For laboratory cultures, only 20–40% were cell bound for *Phormidium sp.* (Li *et al.*, 2012), and >90% were in dissolved phase for raw water of the Feng-Shen Reservoir, Taiwan (Tung *et al.*, 2008). In addition, Chiu *et al.* (2016) have reported that *Pseudanabaena galeata* is dominant in Tsai-Pu-Wo reservoir, Taiwan, with about 50% of 2-MIB being cell bound, while for three other Taiwanese reservoirs, the ratios between dissolved and total 2-MIB may change from time to time, 0–53% and 71–99% in different seasons in different reservoirs.

Removal of cyanobacterial cells, intact and without damage, is an important step to remove intracellular 2-MIB and geosmin (Chow *et al.*, 1999). Methods to remove cyanobacterial cells in conventional processes are coagulation together with sedimentation or flotation. As indicated in Henderson *et al.* (2008), removal of cyanobacterial and algal cells by sedimentation is around 63–94%, and when operated with polymer and an additional filtration process, the removal can be improved to ~99%. For dissolved air flotation (DAF), the removal is typically better than that of sedimentation. With appropriate operation, DAF can remove 96–99.8% of cells. Similar observation was reported in a water treatment plant (WTP) in Kinmen, Taiwan; DAF removed 77–79% of cyanobacterial cells, and when including rapid and slow sand filtration processes, ~99.4% removal was achieved for the entire treatment train (Hsieh *et al.*, 2014). However, for both sedimentation and flotation, the removal is strongly dependent on coagulant doses and types applied, cyanobacterial properties, such as morphology, sizes and colonies, and water matrix (Henderson *et al.*, 2008).

Pre-oxidation, using ozone, chlorine or permanganate, is a common practice used in many water utilities (Chen & Yeh, 2005; Henderson *et al.*, 2008; Plummer & Edzwald, 2002) for the improvement of particle separation, odour control and ammonia removal. However, oxidant may lyse cyanobacteria cells and cause the release of intracellular metabolites (Daly *et al.*, 2007; Fan *et al.*, 2016; Lin *et al.*, 2009). Figure 6.3 shows the removal of 2-MIB in a conventional WTP in Matzu, Taiwan, in which *Pseudanabaena* spp. was the dominant cyanobacteria in the source water. The treatment processes include pre-chlorination, coagulation with aluminium sulfate and sedimentation, rapid sand filtration, and post-chlorination. As shown in the figure, about 82% of 2-MIB was cell bound in the raw water. However, after chlorination, almost all the 2-MIB (~92%) became dissolved in water, indicating that a majority of the cyanobacteria cells were ruptured, leading to an increase of dissolved 2-MIB in the water. The figure also demonstrates that, in the treatment processes, total 2-MIB concentrations were almost constant throughout the treatment processes. Since 2-MIB is biodegradable, with appropriate bacteria community, it has been reported that 2-MIB could be biologically removed (McDowall *et al.*, 2009); however, no 2-MIB removal was found through the conventional treatment processes. As pre-chlorination was applied in some WTPs, residual chlorine was expected in the subsequent processed water and, therefore, biodegradation is suppressed in the treatment processes. Besides biodegradation, oxidation by chlorine and permanganate are the only two processes in conventional WTP that may remove/degrade 2-MIB. As indicated by Lalezary *et al.* (1986), chlorine and permangnate can only degrade ~10% of 2-MIB at a dose of 10 mg/L, suggesting that 2-MIB is very resistant to chlorine and permanganate oxidation. Therefore, it is reasonable to see that the total 2-MIB concentrations remained almost unchanged in the processed water of the WTP in Matzu.

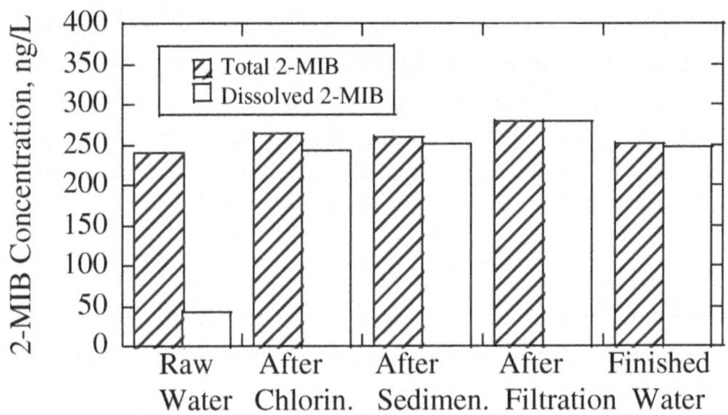

Figure 6.3 Removal of 2-MIB in a conventional water treatment plant, Matzu, Taiwan.

The removal of 2-MIB and geosmin in different water treatment trains are listed Table 6.2. The data were collected from four WTPs in Taiwan and off-shore islands, Kinmen and Matzu from 2009–2017. The treatment trains of the WTP represent four types of processes, including two conventional and two advanced. Different treatment processes for each train (plant) are shown in the footnote of Table 6.2. Conventional Train A removed very little 2-MIB (Table 6.2). Since flotation and slow sand filtration were present in Conventional Train B, the train removed 77% of 2-MIB on average, probably due to better removal of cyanobacteria cells and biological degradation in the processes. For the two advanced trains, >95% of 2-MIB could be removed, indicating that advanced treatment processes are appropriate for 2-MIB control.

Table 6.2 Removal of 2-MIB and geosmin in different water treatment trains.

Type of Treatment Processes	Number of Samples	>2-MIB Concentration						>Geosmin Concentration					
		Raw Water (ng/L)		Finished Water (ng/L)		Removal (%)		Raw Water (ng/L)		Finished Water (ng/L)		Removal (%)	
		Mean ± SD	Range	Mean ± SD	Range	Range	Mean	Mean ± SD	Range	Mean ± SD	Range	Range	Mean
Train A (Plant L)	7	8.8 ± 7.5	3.8–23.6	10.6 ± 10.7	4.2–31.7	0–23	4	–	–	–	–	–	–
Train B (Plant T)	12	44.2 ± 32.8	3.1–112.1	8.7 ± 11.1	ND–35.3	45–100	77	14.6 ± 13.7	ND–43.2	2.3 ± 2.5	ND–7.5	0–100	79
Train C (Plant T)	6	21.8 ± 16.2	6.2–45.1	ND	ND	100	100	5.3 ± 4.4	ND–11.4	ND	ND	100	100
Train D (Plant R)	17	34.7 ± 40.2	2.6–171.5	1.5 ± 3.9	ND–14.7	33–100	95	6.4 ± 3.6	ND–14.6	1.8 ± 2.4	ND–7.7	0–100	75

Notes: Train A: pre-chlorination, coagulation, sedimentation, rapid filtration and post-chlorination; Train B: pre-chlorination, coagulation, flotation, rapid filtration, slow filtration and post-chlorination; Train C: pre-chlorination, coagulation, flotation, rapid filtration, slow filtration, ultrafiltration, nanofiltration, and post-chlorination; Train D: pre-chlorination, coagulation, flotation, rapid filtration, slow filtration, activated carbon filtration, ultrafiltration, reverse osmosis, and post-chlorination; ND = not detected.

6.2 ACTIVATED CARBON ADSORPTION

6.2.1 General knowledge about activated carbon adsorption

Activated carbons (ACs) are carbonised 'vegetable' substances that have been subjected to a thermal treatment known as 'activation'. This process increases their porosity and makes them particularly effective for adsorption applications. Coconut shell, wood, peat and coal are commercially used to produce activated carbon on the basis of cost and availability. Two activation techniques are used to produce activated carbon: (1) physical process which includes carbonisation by activation at 800–1,000°C with steam and carbon dioxide present; and (2) chemical process during which a chemical dehydrating agent is used followed by activation at 400–600°C. The number of pores in a single gram of activated carbon can provide an internal surface area of up to 2,500 m^2. Produced activated carbons are classified by their pore size: (a) micropores <2 nm, (b) mesopores 2–50 nm, and (c) macropores >50 nm. Coconut-based activated carbons are mainly microporous while wood-based carbons are mostly meso- and macroporous. Coal-based carbons are primarily within the category of micro- and mesopores. The chemical structure of target contaminant(s) during water treatment would dictate the pore type of carbon to be used.

Contaminant adsorption steps on activated carbon include: (1) contaminant transport in the solution, (2) contaminant transport by diffusion, and (3) contaminant transport in the pore and adsorption. Carbon adsorption capacity varies depending on target contaminants. Thus adsorption isotherms have been developed for different contaminants and carbon types (Crittenden *et al.*, 2012). The adsorption isotherm describes the relationship between the equilibrium concentration of adsorbate (target contaminant) and adsorbent (activated carbon) in water.

Notably, adsorption isotherms depend on type and concentration of contaminant and activated carbon, water quality and presence of other substances competing for adsorption sites with the target contaminant, and water temperature. The three most common mathematical expressions used to relate the adsorption isotherm are (Crittenden *et al.*, 2012): (1) Freundlich equation; (2) Langmuir equation; and (3) Brunauer-Emmett-Teller (BET) equation. Freundlich equation for adsorption isotherms is an empirical equation (which is verifiable by experience) that describes adsorption data:

$$q_e = KC_e^{1/n} \tag{6.1}$$

and it can be linearized as follows:

$$\log q_e = \log K + \frac{1}{n} \log C_e \tag{6.2}$$

where q_e is equilibrium loading on the carbon (mg chemical/g carbon), C_e is equilibrium concentration in the water (mg chemical/L), K is a constant at unit concentration (mg/g) $(L/mg)^{1/n}$, and $1/n$ is strength of adsorption (dimensionless). K and $1/n$ reference values are published for several compounds by health authorities and regulatory agencies like United States Environmental Protection Agency (USEPA).

Langmuir equation for adsorption isotherms is a mechanistic model which takes into account several adsorption sites on carbon surface and describes the adsorption data as follows:

$$q_e = \frac{q_{\max} b C_e}{1 + b C_e} \tag{6.3}$$

and similar to the previous equation it can be linearized as:

$$\frac{1}{q_e} = \frac{1}{q_{\max} b C_e} + \frac{1}{q_{\max}} \tag{6.4}$$

where $q_{\max}$ equals ultimate adsorption capacity (mg chemical/g carbon), and b is relative energy of adsorption which needs to be evaluated for different water conditions.

BET equation for adsorption isotherms is developed to take into consideration of several layers of adsorption on carbon surface. BET provides a more sophisticated equation:

$$\frac{J}{J_o} = \frac{AC}{(C_s - C)\left[1 + (A - 1)\dfrac{C}{C_s}\right]} \tag{6.5}$$

where J is the adsorption capacity of contaminant, and J_0 is the capacity of the 1st layer, C_s is concentration of saturation and A is the equation constant. The main challenge with BET is that C_s is mostly unknown and trial-and-error assays are required to estimate the C_s and use the linearized form of this equation.

6.2.2 Kinetics of adsorption by activated carbon

Generally, the WTPs apply activated carbon adsorption process in two modes: PAC or GAC. The former is used as a pre-treatment reagent and added before mixing well or with the coagulant. The typical dosage of PAC is 5–15 mg/L in WTPs and the dosage can be elevated to 40–80 mg/L under the extreme condition to address the chemical spills or odourant breakout. The PAC adsorption efficiency will be greatly impaired in the settling tank due to the insufficient mixing and the interference of flocs. In some cases, the water professionals add PAC in the water intake to extend the contact time as long as possible which will allow PAC to perform to its maximum adsorption capacity when the total contact time is over 2 h (Zhang & Chen, 2009; Zhang *et al.*, 2011).

One group of PAC, called superfine PAC (SPAC, particle size <1 μm) has recently been proposed for the removal of geosmin and 2-MIB in drinking water

(Matsui *et al.*, 2013). SPAC was considered to significantly enhance the adsorption capacity and adsorption kinetic. The increased adsorption rate was attributed to the smaller transport distance (or smaller particle size) and more external surface area. However, the increased capacity may be linked to more sites being accessible in the same time period, compared with PAC, and change of surface hydrophobicity when milling was conducted (Matsui *et al.*, 2015).

GAC is used as the adsorption treatment process after the conventional process of coagulation, sedimentation and sand filtration. The operation cost of GAC filter generally is lower than PAC to address the water frequently contaminated by odourants or other organic matter. However, the investment of GAC filter is much higher than PAC addition facility. More importantly, GAC filter is generally applied with empty bed contact time (EBCT) of 5–15 min. Therefore, it is not GAC but PAC that is the more reasonable choice to address seasonal odourant breakout or sporadic chemical spill with high and flexible load (Zhang & Chen, 2009; Zhang *et al.*, 2011).

The adsorption kinetics is a very important characteristic for odourant adsorption which determines the contact time of activated carbon in real application. The kinetics of several frequently detected odourants, i.e. 2-MIB, geosmin and β-cyclocitral, in source water is presented in Figure 6.4.

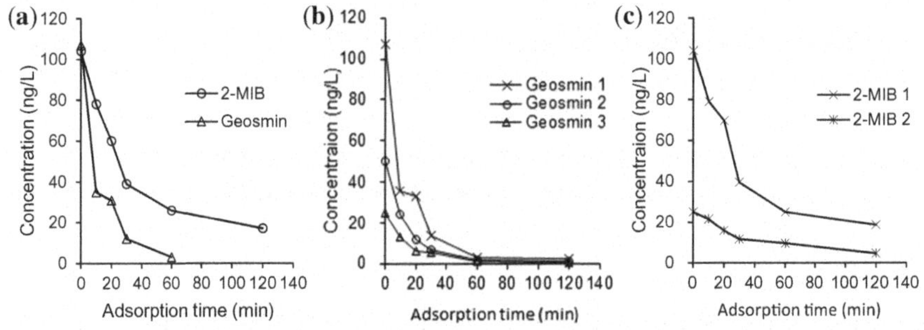

Figure 6.4 Adsorption kinetics of 2-MIB and geosmin by PAC in distilled water. (a) Comparison of 2-MIB and geosmin; (b) geosmin adsorption kinetic with different initial concentrations; (c) 2-MIB adsorption kinetic with different initial concentrations. (*Note*: PAC was prepared by milling the Calgon F400 GAC and sieved by the 200 mesh (78 μm). The PAC dosage was 10 mg/L.) (*Data source*: Li, 2009).

As mentioned before, 2-MIB and geosmin are the most frequently detected odourants in source water (Li *et al.*, 2007; Ömür-Özbek, 2012; Suffet *et al.*, 1996). With 30 min adsorption, 10 mg/L of PAC could remove 100 ng/L of geosmin to slightly higher than the odour threshold concentration (OTC, 10 ng/L) in distilled water. Compared with geosmin, 2-MIB is more recalcitrant to PAC adsorption (Figure 6.4(a)). 10 mg/L of PAC can hardly remove 100 ng/L of 2-MIB below its OTC of 10 ng/L. The result indicates 30 min of adsorption can only treat 20 ng/L of 2-MIB to below OTC. Higher dosage or longer contact time is needed to eliminate the odour of 2-MIB of 100 ng/L.

Besides 2-MIB and geosmin, β-cyclocitral has also been widely detected in eutrophic water sources. It is not a big nuisance with a relatively high OTC of 3 μg/L. It is secreted by the M*icrocystis aeruginosa* and other algae and originated from the cleavage of β-carotene (Zhang *et al.*, 2010). PAC can remove β-cyclocitral quite efficiently. As shown in Figure 6.5, 10 mg/L of PAC could reduce 43 μg/L of β-cyclocitral to 6 μg/L in 30 min, approaching its OTC. If the β-cyclocitral load was high in the source water, increasing PAC dosage would enhance the removal. As shown in Figure 6.5(b), 10 mg/L of PAC could decrease 200 μg/L of β-cyclocitral to 100 μg/L after 2 h and 30 mg/L of PAC could further decrease to 20 μg/L after 1 h.

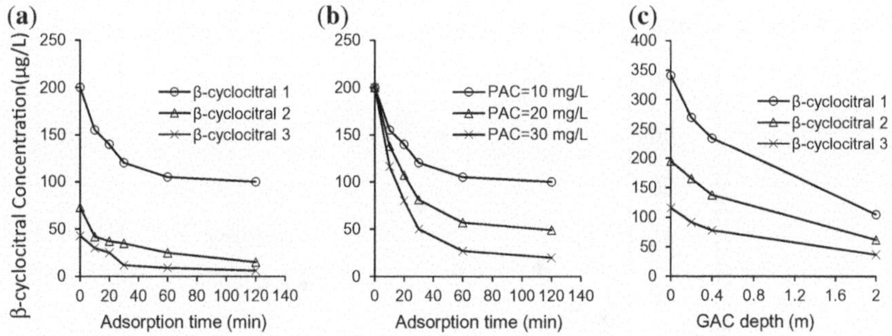

Figure 6.5 Adsorption kinetics of β-cyclocitral by PAC in distilled water, where (a) is for different initial β-cyclocitral concentrations and PAC dose = 10 mg/L; and (b) is for initial β-cyclocitral concentration = 200 μg/L and different PAC dosages; and (c) concentration changes of β-cyclocitral along the depth of GAC. (*Data source*: Li, 2009).

It is worth noting that GAC did not perform well for odourant β-cyclocitral removal. GAC filter with media depth of 2 m could only decrease 200 μg/L of β-cyclocitral to 62 μg/L or decrease 116 μg/L of β-cyclocitral to 37 μg/L.

6.2.3 Adsorption isotherm of activated carbon

The adsorption capacity is also one of the important characteristics for odourant adsorption which determines the dosage of activated carbon in real applications. The adsorption capacity of several frequently detected odourants, i.e. 2-MIB, geosmin and β-cyclocitral in distilled water and source water is presented below.

As mentioned before, Freundlich and Langmuir isotherms are the most widely used for evaluating adsorbent adsorption capacity. The former presents higher suitability for most situations and the latter fits well for monolayer adsorption. Adsorption isotherm is determined by the properties of odourant with the same adsorbent and the capacity could be calculated with certain equilibrium concentration. The lower the equilibrium concentration, the less capacity the activated carbon can present.

According to Figure 6.6, with an OTC of geosmin or 2-MIB of 10 ng/L, the adsorption capacity will be 38 ng/mg or 15 ng/mg, respectively, with sufficient contact time of 24–48 h. 2-MIB is more recalcitrant to PAC adsorption than geosmin which could be explained by their different structure. It is noted that 10 mg/L of PAC could address 390 ng/L of geosmin or 160 ng/L of 2-MIB under the contact time of ~24 hr according to this figure. However, the PAC contact time in real WTP will be limited within 2 h. Thus, the carbon adsorption capacity will be quite low in real applications.

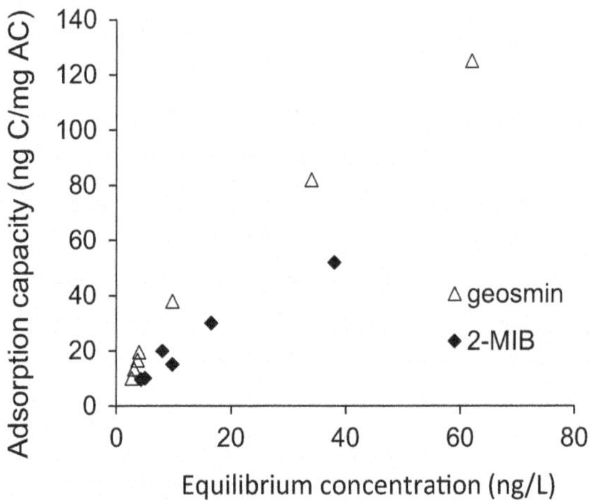

Figure 6.6 Adsorption capacities of 2-MIB and geosmin by powdered activated carbon (PAC) in distilled water. (*Data source*: Li, 2009).

Another restraint for PAC capacity comes from the interference or competitive adsorption from the natural organic matter (NOM) in source water. Figure 6.7 presents the adsorption capacity of several common odourants under the conditions of distilled water and real source water. Generally, the carbon adsorption capacity for geosmin or 2-MIB will decrease 20% to 50% in range of a few to 50 ng/L in the presence of NOM in source water as compared with those without NOM. The carbon adsorption capacity of β-cyclocitral will decrease 10% to 30%. The reduction proportion will be more prominent when the equilibrium concentration is lower.

The mechanisms of competitive adsorption can be explained from three aspects. First, some NOM molecules will occupy some adsorption sites, which inhibits the target odourant adsorption. Second, the large NOM molecules will block the diffusion pathway for odourant to reach the adsorption sites. Third, some NOM molecules will interact with odourants and make them un-adsorbable by activated carbon.

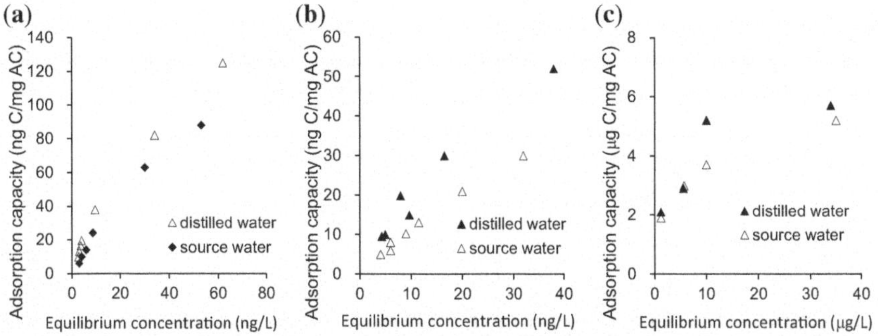

Figure 6.7 Adsorption capacities of some odourants by PAC in source water: (a) 2-MIB; (b) geosmin; (c) β-cyclocitral. (*Data source*: Li, 2009).

The carbon adsorption capacity under the distilled water could be compared between different investigations with respect to the type of ordourant; the capacity under the real source water will differ greatly. At least double the PAC dosage calculated from literature will be needed to address a real odourant breakout episode if there is no onsite experimental data available at the time. In addition to the aforementioned odourants, dozens of odourants have been identified and the list is still increasing (Suffet *et al.*, 1999; Li, Wang *et al.*, 2015; Li, Zamyadi *et al.*, 2015). Therefore, it will be helpful to establish the prediction model of different odourants based on their physico-chemical properties instead of performing individual experiments. For example, Li *et al.* (2016) tested the adsorption characteristics of 5 odourants at ng/L level, i.e. geosmin, 2-MIB, β-cyclocitral, 2-isopropyl-3-methoxypyrazine (IPMP), 2-isobutyl-3-methoxypyrazine (IBMP), and 2,4,6-trichloroanisole (TCA). They found carbon adsorption capacities were related with their basic physico-chemical properties of adsorbate, including octanol/water partition coefficient, molecular weight, polarizability, and aqueous solubility. The mechanism behind the correlations was adsorbent–adsorbate interaction forces represented by those properties. For earthy/musty odourants, hydrophobic interaction seemed to be dominant in adsorption while dispersion force shows its influence when a benzene ring was present.

Bunmahotama *et al.* (2015) developed a new model approach to predict the adsorption isotherms of low-molecular-weight nonpolar organic compounds (LMWNPOCs) onto activated carbons. The model is based on the Polanyi-Dubinin (PD) equation, with the limiting pore volume of adsorbent estimated from the pore size distribution (PSD) data, and the adsorption affinity of adsorbate described by the molecular connectivity index. They used the adsorption results of 34 LMWNPOCs from literature to test their model and found that the model fit the experimental data well, with only 39.2% of errors. Then, they used this model to predict the adsorption capacity of different LMWNPOCs on different brands of activated carbon products. Thus, it is possible to predict the fate of odourants in the activated carbon adsorption process. The approach was further extended to adsorption of polar organic compounds and dyes onto different activated carbons (Bunmahotama *et al.*, 2018).

As the adsorption of organic compounds on activated carbons is strongly influenced by NOMs in natural water, Bunmahotama *et al.* (2017) further proposed a model to predict the adsorption isotherms of 2-MIB and three other organic pollutants onto activated carbons in the presence of NOM. Similar to that proposed by Bunmahotama *et al.* (2015), the adsorption capacity could be predicted by the PD equation and PSD data. The competitive effect was described by the ideal adsorbed solution theory, equivalent background compound model approximation. The proposed model was able to predict the adsorption isotherms for 2-MIB onto different activated carbons in natural water, with only one set of isotherm in deionized water and one set in natural water needed to obtain the parameters for the prediction. The model approach made the prediction of adsorption capacity feasible for 2-MIB onto different activated carbons in the same natural water. This will certainly reduce the effort required for water utilities if change of activated carbon types is needed. However, the model was only limited to thermally activated carbons with known pore size information.

6.2.4 Activated carbon application in WTPs

Activated carbon is a strong barrier against T&O compounds in water and it is commonly used by water utilities. Activated carbon is available in various particle sizes (grain size) depending on the intended application. PAC can be mixed with the intake water, mainly at the early stages of the treatment, as demonstrated in Option 1 in Figure 6.8, and it is available in sizes ranging between 2 and 200 μm. GAC, available in sizes between 0.1 and 4 mm, is used within a contactor similar to filter beds (Option 2). In advanced oxidation processes (AOP) such as combination of UV and H_2O_2, GAC could be used to quench the H_2O_2 residual without any impact on GAC adsorption capacity (Faruqi *et al.*, 2018; Li, Wang *et al.*, 2015; Li, Zamyadi *et al.*, 2015). Efficient application of PAC depends on contact time and adequate mixing of carbons. While efficiency of GAC is a function of the GAC contactor depth, particle size and loading rate described as EBCT. The latter parameter has a significant effect on the performance of GAC adsorbers (Faruqi *et al.*, 2018; Schideman *et al.*, 2012) and is calculated as the total volume of bed divided by the liquid flow rate. Similar to filtration beds, GAC contactors are equipped with backwash systems to prevent clogging of the bed (Figure 6.9).

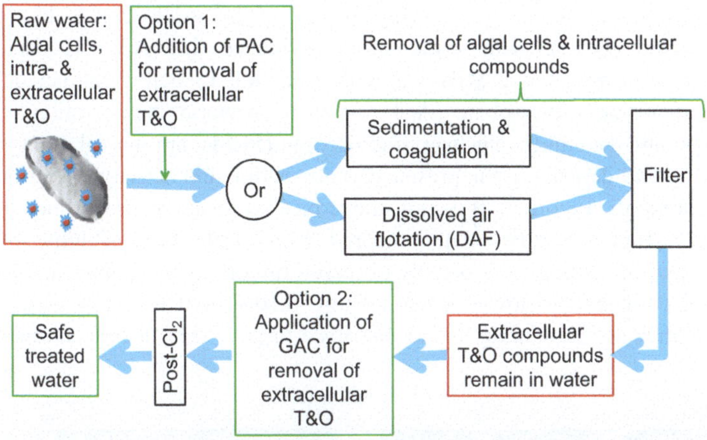

Figure 6.8 Schematic of conventional water treatment process for removal of algal T&O compounds from water with Option 1 for addition of PAC and Option 2 for application of GAC contactors.

Figure 6.9 Full-scale GAC contactor and air scour supply for backwash.

Activated carbon is only capable of removing dissolved compounds, including T&O compounds. Therefore, the ratio of intra/extra-cellular compounds would strongly influence the performance of PAC and GAC for removal of target contaminants. GAC particle could act similar to filter media and remove cells; however unsuccessful removal of cells in prior processes and their accumulation over GAC would lead to filter clogging.

Recently, there has been a growing interest in the application of biologically active carbon (BAC), with some strong evidence that BAC is capable of controlling geosmin and 2-MIB via biodegradation to below human detection limits. Adenosine triphosphate (ATP) analysis, bacterial enumeration by flow cytometry and scanning electron microscopy (SEM) are useful analyses for quantifying and describing the presence, composition and activity of biofilms and their constituents. Figure 6.10 demonstrates biofilm forms on the surface of GAC as it ages (Figure 6.10(b)–(d)) compared with virgin GAC (Figure 6.10(a)). Stronger biofilm formation is usually observed on top layers (Figure 6.10(b)) of a GAC bed compared to lower levels (Figure 6.10(c) and (d)); top layers are in contact with water containing higher dissolved organic carbon concentrations.

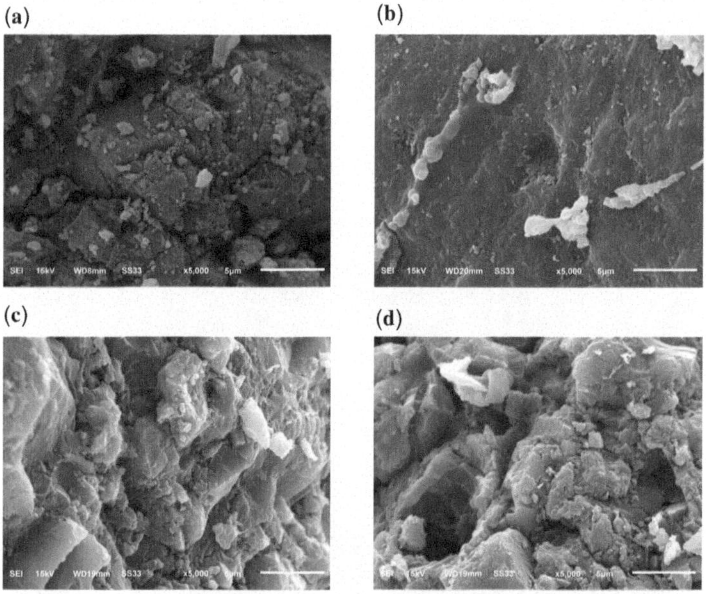

Figure 6.10 Scanning electron microscopy (SEM) data for (a) virgin GAC, (b) GAC from top of the bed, (c) GAC from middle of the bed, and (d) GAC from bottom of the bed (Zamyadi *et al.* 2015a, b).

Equilibrium adsorption experiments and batch kinetic tests may be used to determine kinetic parameters associated with the adsorption of T&O compounds onto activated carbon in batch bench-scale settings, or kinetic adsorption experiments in short-bed adsorbers (SBA). Rapid small-scale column tests, whilst useful in assessing virgin GAC performance, are unsuitable for use in assessing the remaining capacity of aged GAC.

The remainder of this section focuses on modelling removal of algal metabolites by activated carbon and future direction. Using an intraparticle diffusion constant (IDC) model, Greenwald *et al.* (2015) demonstrated the importance of understanding the kinetics of adsorption for predicting GAC bed life. Their results show a positive correlation between rapid sorption kinetics and long GAC bed life, and that IDCs and breakthrough curves possess similar relationships to GAC pore volume. Adsorption isotherms may be used to determine the most efficient GAC for MIB and geosmin removal; however, the characteristics of the test water should be carefully selected to exemplify the target water (Mackenzie *et al.*, 2005).

The SBA/homogeneous surface diffusion model trialled by Ho and Newcombe (2010) was found to be ineffective in predicting the performance of aging GAC for removal of T&O compounds, predicting 90% breakthrough after a six-month operation period. Pilot scale experiments found breakthrough after a six-month operation period to be at most 30% (with pre-chlorination). This discrepancy has been attributed to lack of consideration of biological activity of the GAC, and the combined effects of using small volumes of GAC with high filtration rates to determine the equilibrium and kinetic parameters in SBAs. However, unexpectedly, large laboratory column tests using aged GAC (bed height 15 cm, EBCT 15 min) successfully predicted performance in pilot scale GAC columns, and showed that high adsorption efficiency (up to 80%) persisted in six-month old sterilised GAC (Faruqi *et al.*, 2018).

For batch sorption tests, rapid kinetic tests and isotherm tests were carried out to determine the kinetics of adsorption onto GAC and predict the initial loading of the contactor. Rapid small-scale column tests were also used to predict the performance of virgin GAC, but were not effective at predicting the capacity of aged GAC. Pilot-scale columns run in parallel with full-scale columns providing important information about the effectiveness of scale-up procedures.

Performance of aging GAC is difficult to predict and may be costly and time-consuming to replicate in laboratory experiments. Schideman *et al.* (2012) built upon the breakthrough prediction method developed by Gillogly *et al.* (1999) to create a model for breakthrough at various EBCTs, though requiring short-term bench-scale experimental data using only a few different EBCTs, saving time and costs associated with conducting multiple column tests.

$$\frac{C}{C_0} = e^{-\phi \times \mathrm{EBCT}} \tag{6.6}$$

At breakthrough, C/C_0 is the fraction of a contaminant in the effluent (C) compared with the influent concentration of the same contaminant (C_0), and EBCT is measured in minutes. ϕ is a fitting coefficient, and must be determined by fitting an exponential function to experimental data. Models developed using data from three measured EBCTs in the range 0–7 min may be sufficient to predict performance at a wider range of EBCTs. When experimental EBCTs are small (e.g. in the range 0–3.5 min), experimental data from only two EBCTs may be sufficient.

A recent pilot study in Ontario, Canada, was conducted to verify the performance of virgin and aged GAC at different depths for the removal of geosmin and MIB (Li, Wang *et al.*, 2015; Li, Zamyadi *et al.*, 2015; Zamyadi *et al.*, 2015a, b). Five different bases of GAC were used for this study: (1) catalytic bituminous coal; (2) non-catalytic coal; (3) acid-washed lignite coal; (4) coconut; and (5) bituminous coal. Pilot-scale glass columns (a display of 22 columns) containing GAC were used for this test (Figure 6.11(a)). The set up was installed in a treatment plant in Lake Ontario. The 16 columns allowed for the testing of the GAC in duplicate. Also, another six columns were used to conduct similar tests on aged GAC removed from full-scale contactors. The columns were each 2.5 cm in diameter, and 128 cm in depth to simulate the GAC depth of the full-scale plant with sampling port at pre-determined depths (Figure 6.11(b)). Water samples were dosed with 100 ng/L of geosmin and MIB and directed through the columns at EBCT of 4.1 min (similar to the full-scale plant operational EBCT) for predetermined bed volumes. Geosmin and MIB concentrations were measured along the depth of the column (Figure 6.11(b)).

(a) (b)

Figure 6.11 (a) 22 GAC column pilot set-up, and (b) columns sampling ports at different depths. (Photo credit: Arash Zamyadi).

Data from the pilot study showed that all GAC types and ages, even virgin GAC, were less efficient for removal of MIB compared with geosmin. All tested types of GAC provided similar performance for removal of geosmin, while catalytic bituminous coal-based GAC provided the best MIB removal. For virgin GAC, independent of the carbon type, the top 25 cm of GAC was the most effective for removal of geosmin (Figure 6.12(a)), while the top 75 cm of virgin GAC removed the greatest percentage of MIB (Figure 6.12(b)). However, significant reduction in removal of both geosmin and MIB was observed with two-year old GAC (Figure 6.12). Biofilm formation was observed on aged GAC surface but the ATP test did not confirm its biological activity.

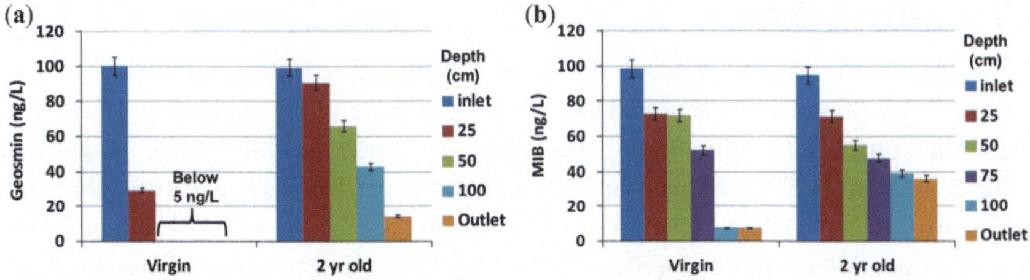

Figure 6.12 (a) Geosmin and (b) MIB removal using virgin GAC and carbon aged in full-scale contactors.

The results of the pilot study and presence of inactive biofilm on GAC demonstrate a need to review the performance of virgin and aged GAC and BAC for removal of algal T&O compounds using other published studies. Figure 6.13 shows the variation in capacity of GAC for removal of geosmin and MIB with respect to age and type of GAC. Adsorption capacity decreases with time, with four-year old catalytic GAC achieving less than 20% removal of both MIB and geosmin. However, some variations should be noted. BAC achieved more than 30% greater MIB removal than a comparable (non-biologically active) GAC, and four-year old BAC from one study performed almost as well as a comparable carbon at one-and-a-half-year old. It should be noted that the data presented are from several studies, each with unique source water characteristics. This can significantly impact results, as dissolved organic matter in source water decreases GAC capacity (Summers *et al.*, 2013).

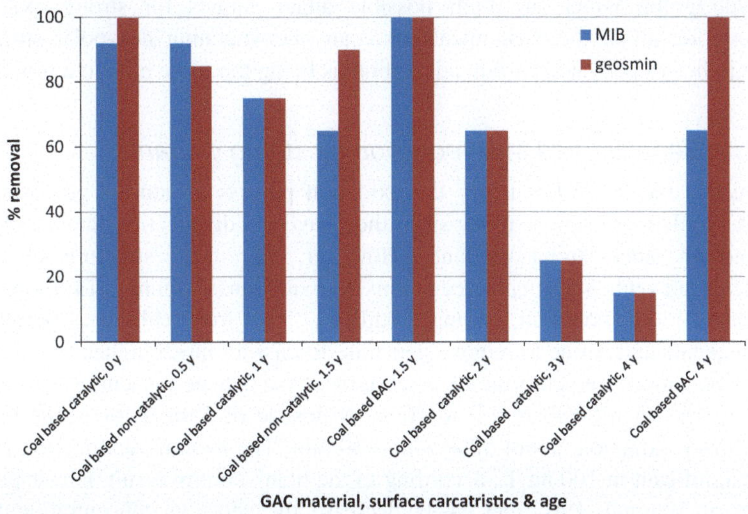

Figure 6.13 Variation in capacity of GAC and BAC for removal of geosmin and MIB with respect to age and type of GAC ('y' stands for year).

6.3 OXIDATION

Odourous chemicals are among the most troublesome water quality issues in WTPs. They have very low OTCs which challenge the capability of water-treatment processes. Chemical oxidation is one of the basic treatment technologies in WTPs. The common oxidants used in WTP are free chlorine, potassium permanganate ($KMnO_4$), chlorine dioxide and ozone. The choice of oxidants depends on the properties of oxidants and odourants as well as the bulk water quality. Although AOPs have been investigated at bench scale, real application in WTPs is still seldom performed due to the practical and economic concerns in this emerging technology.

In this section, the efficiency of the chemical oxidation process on two categories of the most common odourous chemicals, i.e. microbial metabolite-related odourants and volatile sulfur chemicals are discussed.

6.3.1 Microbial metabolite-related odorous chemicals

Algae, actinomycetes and other microorganisms can produce some odourous chemicals during their metabolism. The most well known microbial metabolite-related odourous chemicals are 2-MIB and geosmin. In addition, IPMP, IBMP, 2,4,6-TCA and β-cyclocitral are also reported frequently all over the world.

These aquatic microbial metabolism-related odourous chemicals, except for β-cyclocitral, have fairly stable structures which hinders the attack of common oxidants. β-cyclocitral has one –C=C– bond in the hexal ring and also one aldehyde group which are both possible attack targets for strong oxidants. Furthermore, all of these chemicals have an overwhelming non-polar structure which indicates activated carbon adsorption as being the more effective process.

6.3.1.1 Geosmin and MIB reduction by strong oxidants

It is reasonable for WTPs to try the oxidation process to address geosmin and 2-MIB problems in source water since they have the disinfection facility of free chlorine or other strong oxidants. However, the stable structure of these chemicals prevents significant reduction by oxidation. Fxigure 6.14 shows the performance of common strong oxidants, i.e. free chlorine, potassium permanganate and ozone to remove geosmin. Even with much higher dosage than normal operation, free chlorine or potassium permanganate presents very limited removal (less than 20%) in 1 h reaction. A dosage of 3 mg/L of ozone brings much better reduction, about 50% within 30 min. The natural vaporization effect is not significant in 100 ng/L, according to the blank control result. However, the dosage of 10 mg/L PAC can easily decrease the geosmin concentration from 100 ng/L to below OTC within 30 min. The performance of oxidants on 2-MIB removal is quite the same as that on geosmin.

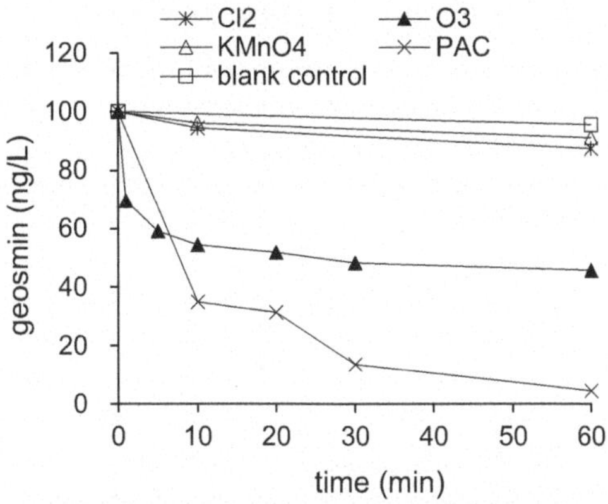

Figure 6.14 Geosmin removal in distilled water by free chlorine, ozone and potassium permanganate oxidation vs. PAC adsorption. (*Note*: dosages of free chlorine, ozone, potassium permanganate and PAC are 5, 3, 5 and 10 mg/L, respectively.) (*Data source*: Li, 2009).

6.3.1.2 IPMP, IBMP and TCA reduction by strong oxidants

The experiment results showed that free chlorine cannot effectively oxidize these odourous chemicals (Figure 6.15). The changes of these three odourous chemicals of no more than 20% was still within standard derivation of detection. As for $KMnO_4$, the results show that there is almost no reaction when applying $KMnO_4$ to remove IPMP, IBMP or TCA. The changes of these three odourous chemicals were no more than 10% which was within the standard derivation of detection.

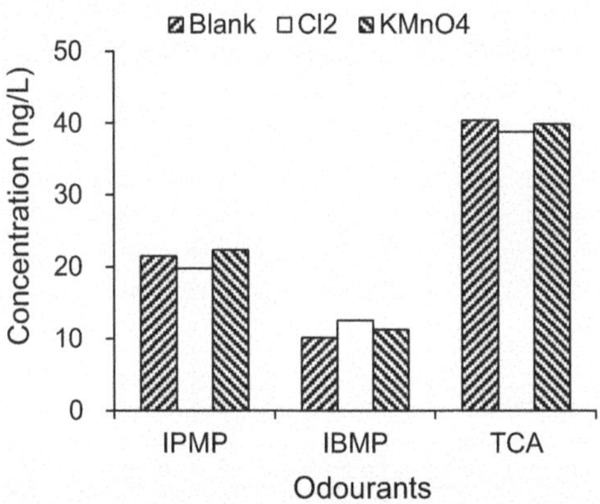

Figure 6.15 Performance of free chlorine and potassium permanganate to oxidise IPMP, IBMP or TCA. (*Note*: Dosages of free chlorine and potassium permanganate are 4 and 4 mg/L, respectively.) (*Data source*: Li, 2016).

6.3.1.3 β-Cyclocitral reduction by strong oxidants

As mentioned before, it is possible for strong oxidants to attack β-cyclocitral since this chemical has two unsaturated bonds. Some common oxidants were applied to remove this chemical, as shown in Figure 6.16.

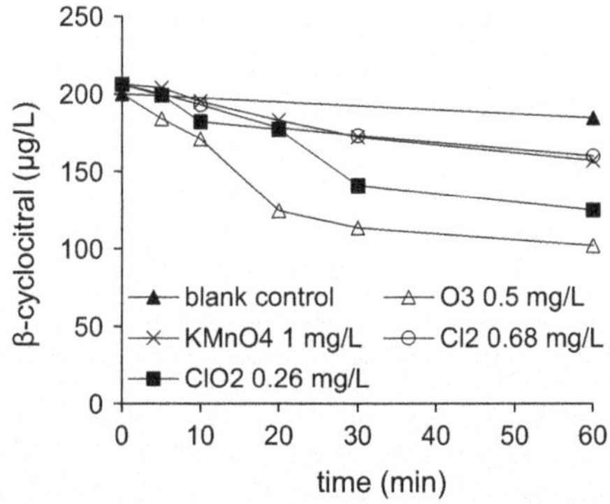

Figure 6.16 β-Cyclocitral removal by chlorine dioxide, free chlorine, ozone and potassium permanganate with the same molar dosage. (*Data source*: Li, 2009).

The removal of β-cyclocitral by different oxidants depends on their oxidation potential. Potassium permanganate only reduced about 20% of β-cyclocitral after 60 min reaction. Free chlorine at 0.68 mg/L gave almost the same removal performance as $KMnO_4$ dose = 1 mg/L. Chlorine dioxide (0.26 mg/L) and ozone (0.5 mg/L) performed better and give about 40% and 50% removal, respectively.

Higher reduction could be obtained if higher dosage of oxidants was applied. Figure 6.17 plots show the increase of removal with the increase of potassium permanganate dosage. However, only 50% removal (initial β-cyclocitral = 200 ng/L) could be achieved at high dosage of 5 mg/L $KMnO_4$ and long reaction time of 60 min.

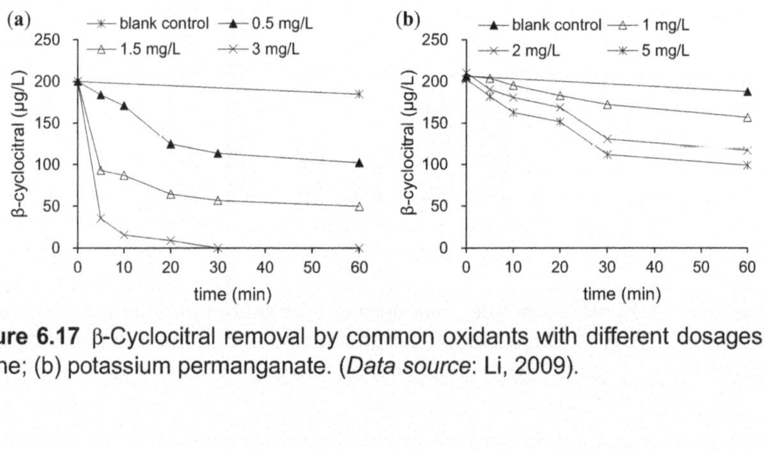

Figure 6.17 β-Cyclocitral removal by common oxidants with different dosages: (a) ozone; (b) potassium permanganate. (*Data source*: Li, 2009).

If higher removal is required, higher dosage of ozone could be a solution. Figure 6.17 showed 0.5, 1.5, 3.0 mg/L of ozone could result in 40%, 70% and even 100% reduction of β-cyclocitral in about 30 min. However, these efficiencies were obtained under a pure water atmosphere. The ozone consumption by other reductive chemicals must be accounted for in real water treatment. Therefore, the ozone dosage could be much higher which will raise other problems, such as high cost and elevated risk of aldehyde and bromate as ozonated disinfection byproducts.

6.3.2 Volatile sulfur chemicals

6.3.2.1 Profile of VSCs as odourous chemicals

Volatile sulfur chemicals (VSCs) are notorious for their stink, rotten egg or swampy smell (Suffet *et al.*, 1999). The main members of this family include H_2S, MeSH, DMS, DMDS, DMTS, EtSH, among others. These chemicals are often found in wastewater treatment, sludge treatment, composting and other anaerobic situations. Sometimes, VSCs can also enter the drinking water system and there are more problems associated with their flavor than their toxicological effect (Zhang *et al.*, 2010). The basic odour profiles of different VSCs are shown in Table 6.3.

Table 6.3 Basic odour profiles of different VSCs.

Compound	CAS	Type of Odour	Odour Threshold Concentration
H_2S	77830-6-4	Rotten eggs	0.00062 ng/m^3 in air
MeSH	74-93-1	Rotten onions	0.00015 ng/m^3 in air
EtSH	75-08-1	Garlic	1 μg/m^3 in air
DMS	75-18-3	Rotten cabbage or algae	0.0083 ng/m^3 in air
DMDS	624-92-0	Pulp mill or rotten cabbage	0.0092 ng/ m^3 in air; 0.2–5 μg/L in water
DMTS	3658-80-8	Rotten onions	10 ng/L in water
Dimethyl Tetrasulfide $CH_3S_4CH_3$	5756-24-1	Garlic or seaweed	–

The VSC precursors exist ubiquitously in the environment, including the organic sulfur compounds such as sulfur-containing protein and also the inorganic sulfate. The organic sulfur could be digested into methionine or cysteine which will be further degraded by bacteria into VSCs (Bentley & Chasteen, 2004; Juttner,

1984). The sulfate can also be transformed into the acidic, corrosive and toxic H_2S by the sulfate reducing bacteria, which is one of the main threats to sewage collection pipes (Pikaar *et al.*, 2014).

6.3.2.2 Oxidation process for VSCs

It is very clear that VSCs are reductive chemicals formed in an anaerobic atmosphere, which means they can easily be oxidized by strong oxidants. The common oxidants, chlorine, chlorine dioxide, potassium permanganate and ozone, could be applied to eliminate the VSCs from the water (Table 6.4).

Researchers tested the efficiency of common oxidants used in water industry. The oxidants, chlorine, chlorine dioxide, potassium permanganate and ozone, reacted with VSCs very quickly and eliminated the odour within 5 min (Li, 2009). Hydrogen peroxide and monochloramine did not perform well. However, the efficiencies of chlorine and chlorine dioxide were almost ruined by the interference of other reductive chemicals, such as organic matter and ammonia in concentrations of mg/L.

The reactions between potassium permanganate and three kinds of VSCs (DMDS, DMTS and EtSH) could be described by pseudo second-order model:

$$\frac{d[\mathrm{VSC}]}{dt} = -k[\mathrm{KMnO_4}][\mathrm{VSC}] \tag{6.7}$$

The kinetics rates were calculated as 0.03, 0.007, 0.025 L/(min.mg) at 20°C, respectively. The products of VSCs included suloxide, sulfone, sulfinic acid, sulfonic acid and eventually sulfate (Li, 2009). These results helped the emergency response in eliminating the strong off-flavor caused by VSCs in practice. It was reported that potassium permanganate oxidation was the effective technology for VSC removal when treating bio-waste polluted water (Zhang *et al.*, 2010).

6.3.2.3 Application of oxidation process of VSCs

The application of oxidation process for effective VSC removal depends on the type of oxidants. Well known disinfectants applied in WTPs can be used as oxidants for VSC removal. However, new oxidants might be used if the disinfectants are easily consumed by other accompanying contaminants. For example, the organic matter in natural water, usually present at mg/L level of concentration, may compete with VSC, usually at μg/L level, for the reaction with free chlorine, chlorine dioxide or ozone. Moreover, the capacity of a reagent-addition facility is designed for normal situations. It is usually insufficient to address any unusual situations. Therefore, it is necessary for WTPs, in which water intake is under threat of VSC contamination, to have an emergency oxidant-addition system in reserve.

Table 6.4 Performance of common oxidants to VSCs.

Oxidants	**Cl_2**	**$KMnO_4$**	**ClO_2**	**O_3**	**H_2O_2**	**NH_2Cl**
Oxidation potential (eV)	1.36	1.51 (acidic) 0.56 (basic or neutral)	1.57	2.07	1.78	
VSC Removal	Excellent for MeSH, DMS, DMDS, DMTS removal	Excellent for MeSH, DMS, DMDS, DMTS removal	Excellent for MeSH, DMS, DMDS, DMTS removal	Excellent for MeSH, DMS, DMDS, DMTS removal	Not good for DMS, DMDS, DMTS but MeSH removal	Almost no effect on VSCs
Reaction kinetic	Fast	Fast	Fast	Fast	Fairly fast	Slow
Interference	Sensitive or easily quenched by NH_4^+, organics	Not sensitive to NH_4^+, organics	Sensitive to organics	Sensitive to organics	Not sensitive to NH_4^+, organics	–
Application	Available, consider the interface	Best available technology (BAT)	Available, not easily accessed in emergency	Available, not easily accessed in emergency	Not available	Not available

As mentioned before, potassium permanganate could be used as the effective technology for VSC removal in WTPs. One major disadvantage of potassium permanganate is the strict requirement for accurate dosing, which is not easily implemented (Zhang *et al.*, 2010). It is well known that residual potassium permanganate will bring a pink color to the finished water. On the other hand, a dosage lower than the stoichiometric value will produce a reductive atmosphere, which will transform the high valent manganese, i.e. Mn(VII), into the soluble Mn(II). Since manganese is also included in the drinking water standard in many countries, over dosage of potassium permanganate will result in pink water and excess Mn in the form of $KMnO_4$, while a lower dosage will result in insufficient oxidation of VSCs and an excess of Mn in the form of Mn^{2+}.

The addition system includes the weighing, dissolution and metric pump parts (Figures 6.18 and 6.19). Besides the addition system, an accurate online water quality surveillance system is also needed to adjust the reagent dosage according to the fluctuation of source water and finished water quality.

Figure 6.18 The permanent $KMnO_4$ addition system in one large WTP in Wuxi, China. Blue devices are stirring devices, white ones are the high-accuracy weighing and feeding devices for $KMnO_4$, lower tank is the dissolution tank for $KMnO_4$. (Photo Credit: Wuxi Waterworks Company).

Figure 6.19 The mobile particle-form permanganate addition system showing (left) dissolution tank, (middle) vacuum feeding part and (right) metric pump part. Permanganate and PAC, in the form of fine particles, can be added using this system. (Photo Credit: Beijing General Municipal Engineering Design & Research Institute Company).

6.4 CASE STUDIES FOR ROUTINE ODOUR REMOVAL

6.4.1 Case study I: MIB removal by activated carbon adsorption in Beijing, China

Over half of the water supply in Beijing relies on the Miyun Reservoir. Resulting from climate change, inlet water flux decrease and eutrophication, the consequent algal odour problem has been aggravated in every autumn since 2000. The 2-MIB concentration was detected to be as high as 160 ng/L in 2005, which exceeded the 2-MIB removal capacity of WTPs in Beijing.

Although conventional process and a GAC filter have been applied in the No. 9 WTP (Figure 6.20), the finished water still exhibits strong earthy odour due to the limited contact time for adsorption. The maximum GAC capacity for 2-MIB removal was determined as 40 ng 2-MIB/L according to the operation record. Consequently, additional PAC has been added in the water intake in Miyun Reservoir. The long transport time, 16 h from the water intake to the WTP, could make full use of PAC adsorption capacity. The range of PAC dosage was from 5 to 25 mg/L, which was adjusted to guarantee a 2-MIB concentration in the GAC filter influent of no more than 40 ng/L. Pre-chlorination was suspended to avoid breaking algal cells, releasing more odourous chemicals and also interfering the PAC adsorption.

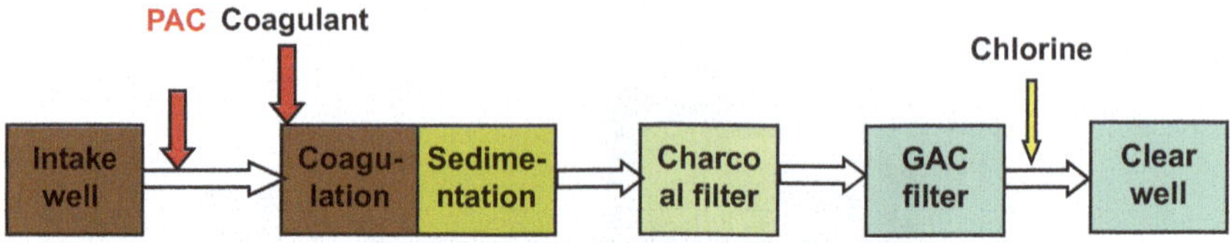

Figure 6.20 Flowchart of No. 9 water treatment plant in Beijing.

Moreover, an online water quality surveillance system was established at the water intake, including temperature, pH, dissolved oxygen (DO), nitrogen, phosphate, organic matter and algal count. The engineers in Beijing Water Company found the DO level to be a good indicator for the odour problem. When DO in the source water turns lower in late August due to the consumption of decayed algae, the peak of odourous chemicals would occur a few weeks later. Good preparations can now be made before the odour problem happens. With the application of PAC adsorption, the Beijing Water Company now can tackle the off-flavor problem caused by eutrophication in the source water.

6.4.2 Case study II for routine odour removal

The objectives of this case study was to demonstrate the accumulation and breakthrough of T&O compounds and their producing cells inside full-scale water treatment plants and during sludge/spent water recycling, and evaluate the performance of different treatment processes for their removal. This presentation reports on intensive water samplings which were conducted in 11 Australian and 4 North American full-scale WTPs: at raw water, after sludge supernatant reuse, after pre-oxidation, post addition of PAC, post clarification, post filtration/membrane, post AOPs, post GAC, post oxidation, and from the sludge supernatant in sludge treatment basins. Here we focus on the results from one plant, highlighting the breakthrough of cells and their operational issues.

Figure 6.21 demonstrates high algal cell numbers (up to 3×10^6 cells/mL) at the source (Figure 6.21(a)) and water intake (Figure 6.21(b)) in one of the studied plants. The maximum geosmin and 2-MIB concentrations found in the samples taken from the water intake of the studied plant were 114 and 40 ng/L, respectively. Breakthrough of these high cell numbers into the studied plant caused several treatment challenges including difficulties with coagulant mixing as shown in Figure 6.22. Operational solutions to handle issues similar to Figure 6.22 are limited, time consuming and costly. It has been suggested that online monitoring of cyanobacterial cells and real time detection of breakthrough events could help operators adjust the treatment process on time, for example increase the coagulant dose. However, cell numbers at the source and the water intake could limit the efficiency of online measurement tools.

Figure 6.21 (a) Harmful blooms at the source and (b) breakthrough of bloom into treatment plants.

Figure 6.22 Breakthrough of bloom into treatment plants and causing major mixing and coagulation issues.

Interestingly even with very high cell number at the water intake the studied plant, the system was able to achieve over 95% removal of cells and T&O compounds. Figure 6.23 demonstrates successful accumulation and removal of cells in the DAF system. However, the remaining cells (5% or less) in the clarified water ranged from 10,000 to 40,000 cells/mL causing filter clogging and breakthrough. Also, due to accumulation of cells, the maximum concentration of T&O compounds in the clarification sludge were over 9,400 ng/L. Furthermore, accumulation of geosmin- and MIB-containing cells and cell lysis over the filters led to detection of 23 ng/L of geosmin and 42 ng/L of MIB in treated water.

Figure 6.23 Breakthrough of bloom into treatment plants and cell accumulation DAF clarification sludge.

Monitoring in full-scale plants provided the opportunity to rank the performance of studied treatment processes for removal of dissolved geosmin and MIB: GAC > PAC > filtration > clarification > chlorination. These results demonstrate the scale of algal/cyanobacterial T&O compounds monitoring and treatment challenges in full-scale plants. In short, a multi-barrier treatment approach could significantly reduce the risk of breakthrough of T&O compounds into treated water. Furthermore, review of these recent monitoring results in full-scale treatment plants clearly outlines the knowledge gaps in this field.

6.5 CASE STUDIES FOR EMERGENCY RESPONSE

6.5.1 Case study III: VSC removal by $KMnO_4$ in odorous water problem in Wuxi, China

An odourous drinking water crisis occurred in the city of Wuxi, China, between May and early June of 2007. During this crisis, the drinking water delivered to roughly 2 million people became septic and colored. The crisis was initially attributed to a bloom of the cyanobacterium *Microcystis aeruginosa* in Lake Taihu, the source of drinking water for the city (Guo, 2007). The chemicals responsible for the foul smell in both the source and the tap water were later identified as VSCs (Yang *et al.*, 2008; Zhang *et al.*, 2010). The production of VSCs from cyanobacteria or algae has been reported for many years.

6.5.1.1 Screening of odourous chemicals

Initial efforts to address the water quality crisis focused on identification of the contaminants responsible for the odour problem. MeSH, DMS, DMDS, DMTS, β-cyclocitral, and toluene were identified as the dominant volatile chemicals (Table 6.5). Liquid-liquid extraction (LLE) pretreatment and GC/MS analysis were used to detect semi- or non-volatile chemicals including indole and its derivates as well as some aldehydes and ketones (Table 6.5). In addition to the presence of these individual contaminants, poor intake water quality was demonstrated by the high chemical oxygen demand (with permanganate as the oxidant, here denoted as COD_{Mn}) and turbidity, high concentrations of NH_4^+ and low DO level. Indeed, the concentration of COD_{Mn} and turbidity almost doubled after May 28, while the DO was depleted quickly to near zero and the ammonia also increased by 10 to 20-fold when compared with the normal conditions before May 28 (Zhang *et al.*, 2010).

The olfactory characteristics of thiols and thioethers matched those odours detected in the water, and their concentrations were dominant among odourants, indicating that they were the immediate cause of the offensive T&O in the tap water. MeSH (204 μg/L) and DMS (93.9 μg/L) were the dominant VSCs in the first assay on May 30 and presumably contributed most to the odour. Small concentrations of H_2S and DMDS were also detected at that time, but no DMTS was found. The second assay was conducted on June 2, when DO level had risen to ~5 mg/L, at which time the both MeSH and DMS were found to have almost entirely disappeared and been replaced by the more oxidized DMDS (46.1 μg/L) and DMTS (17.2 μg/L). β-cyclocitral was also detected in source and tap water during the crisis, but contributed little to the odour due to its lower concentration and high OTC. Geosmin and 2-MIB, the most common odourous chemicals in algae-laden lake water, were detected, but not above the drinking water standard of 10 ng/L, indicating that these compounds were not responsible for the odour of the water during the crisis.

VSCs, such as MeSH, DMS, and DMDS can be produced in the decomposition pathway of sulfur-containing materials, algae cell from algae mat in the lake or some amino acids, predominantly methionine. A substantial cyanobacterial bloom associated with hot and dry weather since April 2007 had resulted in the accumulation of a great deal of organic matter and sulfur-containing compounds in the water and sediments of Lake Taihu. The death and decay of cyanobacterial blooms is often associated with high concentrations of VSCs and accompanied by drastic increases in organic matter, nitrogen, and phosphorus.

VSCs may also originate from proteins present in domestic waste. However, the amount of sulfur-containing materials in domestic wastewater near the intake was negligible when compared with the massive mat of cyanobacteria in the lake. Thus, domestic wastewater should not be regarded as the main source of VSCs.

Table 6.5 Main compounds detected in the source water at the intake during the crisis and their basic characteristics.

Compound	Concentration (μg/L)	
	May 30	**June 2**
VSCs		
H_2S	+	ND
MeSH	204/+++	ND
DMS	93.9/+++	0.01/+
DMDS	2.51/++	46.1/+++
DMTS	ND	17.17/+++
$CH_3S_4CH_3$	ND	+
Other odourants		
β-cyclocitral	8.14/++	21/++
Toluene	0.46/++	0.44/++
2-MIB	ND	ND
Geosmin	ND	ND
Semi- or non-volatile chemicals		
Polysulfides (S_6, S_8)	+	NT
3,5-Di-*t*-butyl phenol	+	NT
Indole & derivatives	++	NT
1-Hexanal	+	NT
1-Octanal	+	NT
3-Hexanone	+	NT
Cyclohexanone	+	NT
Microcystin-LR	7.59	0.73
Microcystin-RR	9.43	0.60

Notes: Some of the detected chemicals were not quantified due to the lack of standards for pure chemicals. The comparative abundances of these chemicals were assessed by their peak area from several analyses of purge and trap (P&T)-GC/MS, solid-phase microextraction-GC/MS and LLE-GC/MS. + = trace; ++ = some; +++ = abundant; ND = not detected; NT = not tested. Microcystin-LR and RR were determined by LC/MS. (*Data source*: Zhang *et al.*, 2011).

6.5.1.2 Emergency water treatment process

Based on odourant identification, an emergency water treatment process was developed to address the odour problem within one day (Zhang *et al.*, 2010). In this process, $KMnO_4$ was added to the intake of the influent pipeline at 3–5 mg/L to oxidize VSCs during the 6 h transit to the treatment plant. After the source water reached the WTP, PAC (30–50 mg/L) was added into the mixing well to adsorb pollutants and decompose the residual $KMnO_4$ (Figure 6.24). The doses of $KMnO_4$ and PAC were adjusted according to the values measured by the online oxidation-reduction potential (ORP) meter. The Wuxi Water Company implemented the treatment process on the morning of June 1 and the quality of the treated water met all 106 items of the Chinese Standard for Drinking Water Quality (GB5749-2006) (MOHC, 2006), which is similar to the standards of the WHO. In addition, microcystins were detected at levels far below the criteria (1 μg/L) in the treated water.

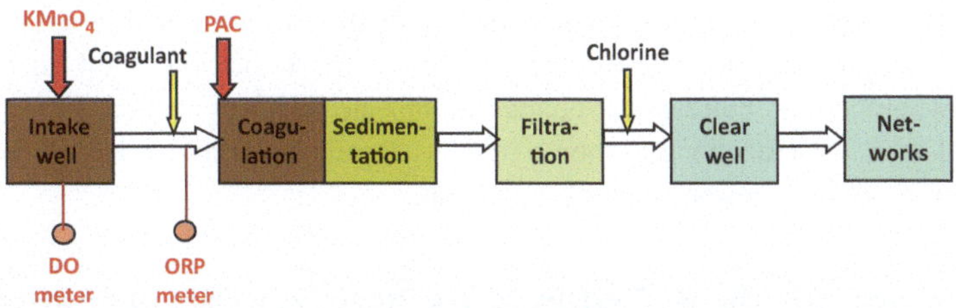

Figure 6.24 Flow chart of the emergency water treatment process applied in Wuxi, China.

6.5.2 Case Study IV: VSC removal by ozone

Some cities along the Pearl River suffered off-flavor problems caused by the wastewater discharge from an up-stream community (Figure 6.25). VSCs were detected at μg/L concentrations as the dominant odourous chemicals in the contaminated source water.

Figure 6.25 The contaminated water source during the flooding season. (Photo Credit: Chao Chen).

A pilot test of an ozone contactor followed by a GAC filter was set up at a WTP (Li, 2009). Ethyl mercaptan (C_2H_5SH) was added (100–200 μg/L) as a typical VSC species to the effluent of sand filter to simulate the odourous source water. The ozone dosages were 1–7 mg/L. The results showed that ozonation alone can effectively remove ethyl mercaptan. However, its efficiency was compromised by other reductive chemicals in water. The dosage of ozonation was over 40 times that of the ethyl mercaptan concentration. About 90% removal of ethyl mercaptan could be achieved at 15 min contact time with this dosage. In this study, GAC filter without ozonation had limited efficiency for removing ethyl mercaptan. Thus, when ethyl mercaptan concentration in influent is over 100 μg/L, pre-oxidation with ozone or other oxidants should be combined with GAC to remove the odours caused by ethyl mercaptan.

6.5.3 Case Study V: Crude (4-methylcyclohexyl)methanol (MCHM) spill into the Elk River in West Virginia

The ability of a conventional drinking water plant to remove contaminants by a moderately soluble (e.g. ~1,000 mg/L) mixture of industrial chemicals was tested on January 9, 2014 in Charleston, West Virginia, USA. Approximately 37,800 L of the industrial chemical mixture crude MCHM leaked from a Freedom Industries chemical storage tank into the Elk River (Foreman *et al.*, 2015; Gallagher *et al.*, 2015). Crude MCHM is used to process coal. The intake for the West Virginia American Water drinking water treatment plant (AWDWTP) was located on Elk River approximately 2.4 km downstream of the spill. Crude MCHM, which has a distinct sweet, liquorice odour, was processed through the AWDWTP and distributed to over 300,000 consumers. By noon on January 9, 2014, AWDWTP and consumers were aware of the odourous contamination; consumers complained to their AWDWTP and their government which prompted the county government to issue public health announcements (Bahadur & Samuels, 2015; Gallagher *et al.*, 2015). Since few physicochemical and toxicological data were available for crude MCHM at the time of the spill, a state of emergency was declared about 6 pm on January 9, 2014 and consumers were told to 'Do not use' the water for drinking, bathing or washing. This order lasted for 4–10 days for different sections of the distribution system. During the time of the 'Do not use' orders and for months after, consumers continued to complain of sweet, liquorice smelling of drinking water especially when using hotter water for showering or dish washing (Gallagher *et al.*, 2015; Phetxumphou *et al.*, 2016; Schade *et al.*, 2015).

At the time of the spill, few data were available concerning components of crude MCHM. Subsequent research demonstrated that there were at least ten components in the crude MCHM including several *cis*/*trans* isomer pairs (Gallagher *et al.*, 2015). The major odourants in the crude MCHM were *trans*-4-MCHM and *trans*-methyl-4-methylcyclohexanecarboxylate (MMCHC); their physicochemical and sensory properties are summarized in Table 6.6. *trans*-4-MCHM constituted 60.3% of the organic chemicals in crude MCHM (Phetxumphou *et al.*, 2016) and has a low air OTC of 0.06 ppbv for a predominantly liquorice odour (Gallagher *et al.*, 2015). *trans*-4-MMCHC constituted only 0.3% of crude MCHM (Phetxumphou *et al.*, 2016), has an air OTC of 0.021 ppbv with a predominantly sweet, fruity odour and a Henry's Law Constant approximate 50-fold higher than *trans*-4-MCHM. An analysis of the relative exposure of consumers to these two chemicals during showering demonstrated that *trans*-4-MCHM contributed about 82% of the odour because of its high concentration and low odour threshold and that *trans*-4-MMCHC also contributed substantially to the odour due to its high volatility even though its concentration was only 0.3% of the spilled crude chemicals (Phetxumphou *et al.*, 2016). Both ingestion and inhalation are routes for human exposure to contaminated water. During the spill, the USCDC (2014) established a 1 mg/L health advisory standard for MCHM.

Table 6.6 Physicochemical and sensory properties for *cis*- and *trans*-MMCHC and *cis*- and *trans*-4-MCHM.

Compound	Aq. Sol. (mg/L)	Henry's Law Constant Air/Water (–)		OTC (ppbv, air)	Odour Descriptors	Odour Activity Value (OAV)[e] in Air During Showering[a]
		22°C	40°C			
cis-MMCHC	173[a]	1.00×10^{-2}[a]	1.50×10^{-2}[a]	1.828[a]	Sweet, Fruity[a]	26.8
trans-MMCHC	173[a]	1.38×10^{-2}[a]	2.23×10^{-2}[a]	0.02[a]	Sweet, Fruity[a]	1546
cis-4-MCHM	1300[b]	2.00×10^{-4}[c]	6.38×10^{-4}[c]	120[d]	Fermented fruit[d]	0.8
trans-4-MCHM	1010[b]	4.85×10^{-4}[c]	1.52×10^{-3}[c]	0.06[d]	Liquorice[d]	6934

Notes: Aq.sol. = aqueous solubility; [a]Phetxumphou *et al.* (2016), [b]Dietrich *et al.* (2015), [c]Sain *et al.* (2015), [d]Gallagher *et al.* (2015), [e]OAV = chemical concentration/odour threshold concentration.

The AWDWTP used typical water treatment processes including permanganate oxidation (routine dose 0.6 mg/L), coagulation/flocculation, sedimentation, GAC, filtration, and chlorine disinfection (Weidhaas *et al.*, 2017). Soon after the contamination was detected at the AWDWTP, emergency measures included an increased permanganate dose of about 1.2 mg/L and addition of up to 19 mg/L PAC to the sedimentation basins (Weidhaas *et al.*, 2017). Although oxidant and PAC were added, consumers were still able to detect the odour of crude MCHM at their taps. Once monitoring began on January 10, 2014, a concentration of 3.4 mg/L 4-MCHM (measured as combined *cis*- and *trans*-4-MCHM isomers, which were the major components of the spill) was detected in river water and the research indicated that similar concentrations of MCHM were measured in the influent and effluent from the AWDWTP (Whelton *et al.*, 2015). As shown in Figure 6.26, a very slight reduction occurred during the first day of monitoring (January 10), with a reduction of approximately 0.5 mg/L. Subsequently, however, the effluent had a higher 4-MCHM concentration than the influent (January 11–12). Likely, desorption from the granular carbon led to these higher effluent values. Thus, conventional treatment, even with addition of PAC, failed to control the contamination and this was confirmed by consumer complaints of sweet, liquorice smelling tap water.

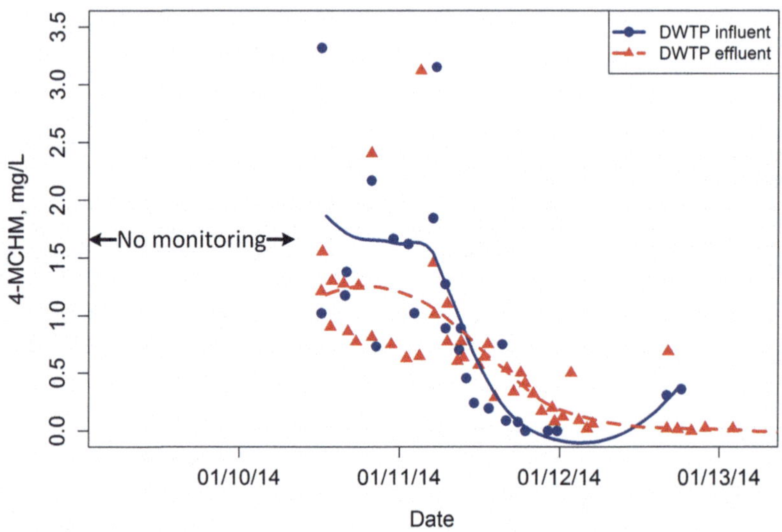

Figure 6.26 4-MCHM concentrations of influent and effluent at the water treatment plant following the spill. The spill occurred the morning of 9th January 2014. Curves are based on local polynomial regression smoothing.

Subsequent research confirmed the inability of conventional treatment to effectively remove *cis*- and *trans*-4-MCHM, the two major organic chemicals that comprised the crude MCHM (Ahart *et al.*, 2016; Weidhaas *et al.*, 2017). Oxidation by either permanganate or chlorine was demonstrated to be ineffective (Weidhaas *et al.*, 2017). A very high dose of 220 mg/L permanganate resulted in only a 15% reduction in *cis*- and *trans*-4-MCHM when the initial concentration was 12.1 mg/L crude MCHM. A dose of 2.6 mg/L free chlorine applied for 24 h resulted in no 4-MCHM removal. Activated carbon was demonstrated to preferentially remove the more odourous and less soluble *trans*-4-MCHM isomer but, to some extent, also sorb the *cis*-4-MCHM isomer (Ahart *et al.*, 2016;

Weidhaas *et al.*, 2017). For GAC, the *trans*-4-MCHM had a loading capacity of 26.5 ± 1.2 mg/g at a 1 mg/L aqueous equilibrium concentration while the *cis*-4-MCHM had a lower loading capacity of 12.8 ± 1.1 mg/g at a 1 mg/L aqueous equilibrium concentration (Ahart *et al.*, 2016). The 19 mg/L PAC added at the AWDWTP during this time was insufficient to prevent odourous levels from entering consumer homes. Ahart *et al.* (2016) and Weidhaas *et al.* (2017) demonstrated that 4-MCHM previously sorbed onto GAC/PAC could also desorb into water at levels that were readily detectable by the human sense of smell. The desorption route confirmed what was observed by residents of West Virginia who continued to smell sweet, liquorice odours of 4-MCHM for months after the spill. The carbon beds were eventually replaced in April 2014 and the odour complaints diminished. Since crude MCHM was poorly removed by engineered treatment and entered the distribution system, a concern was its sorption into polyethylene pipe materials. Research demonstrated that crude MCHM sorption into polyethylene was similar to that of polar 1-butanol. Crude MCHM sorbed slowly due to its low solubility of 0.003–0.008 g/cm^3 into polyethylene pipe; about 40% of the sorbed crude MCHM could be readily desorbed into water and thus created the possibility of contaminating drinking water with this odourous chemical mixture (Ahart *et al.*, 2016).

While *cis*- and *trans*-4-MCHM could be biodegraded, their rates of complete removal require more than a week and this is too slow to be effective during water treatment even if the appropriate microorganisms are present (Weidhaas *et al.*, 2016a; Yuan *et al.*, 2016a, b). A first order kinetic model fit the degradation data, and rate coefficients of 0.19–0.52 d^{-1} at room temperature were measured for *cis*- and *trans*-4-MCHM isomers under aerobic conditions and rates of 0.012–0.09 d^{-1} at room temperature under anaerobic conditions using organisms isolated from a regional river (Yuan *et al.*, 2016a). Degradation was faster when using microorganisms isolated from the Elk River that were presumably pre-exposed to crude MCHM (Weidhaas *et al.*, 2016a). Due to its slow biodegradation rate, crude MCHM travelled in the riverine system was still detectable nine days after the spill at a river site 632 km downstream of the spill (Bahadur & Samuels, 2015). Additionally, crude MCHM was revealed to be toxic to microorganisms present in microcosms constructed with water and sediment from the Elk River; this toxicity was demonstrated to delay the onset of biodegradation (Weidhaas *et al.*, 2016a).

In the aftermath of the spill, Freedom Industries declared bankruptcy and the owner and the president were fined and given one month jail sentences (Ward, 2016a, b). Legal settlements to reimburse the local businesses and consumers were agreed to be paid by Eastman Chemical Company, the manufacturer of crude MCHM and AWDWTP. Eastman Kodak agreed to pay USD 25,000,000 (Herald Media, 2016; Ward 2016c) and AWDWTP agreed to pay USD 126,000,000, with the stipulation that these costs could not be passed along to the utility customers (Herald Media, 2016; Ward 2016c).

This case study for crude MCHM contaminating a WTP and distribution system in West Virginia demonstrates the need for source water protection, vigilant attention to unusual conditions in the WTP, heeding consumer complaints, and understanding the limited role of conventional treatment to remove moderately soluble chemicals, the complexities of chemical mixtures especially complexities related to geometric isomers which can have drastically different physicochemical and biological properties (DeYonkers *et al.*, 2016), and the sensitivity of the human nose. Finally, the need to comprehensively assess individual and total chemicals in a mixture when assessing human exposure is urgent. The role of proactive research to prevent and contain spills cannot be understated (Weidhaas *et al.*, 2016b).

REFERENCES

Ahart M., Gallagher D. L., Scardina P. and Dietrich A. M. (2016). Industrial spills and water distribution: MCHM sorption and desorption in polymer pipes and linings. *Journal Environmental Engineering*, **142**, 04016045, doi: 10.1061/(ASCE)EE.1943-7870.0001116

Anonymous (1913). Unpleasant odour and taste in drinking water. *The Lancet*, **181**(4674), 912.

Bahadur R. and Samuels W. (2015). Modeling the fate and transport of a chemical spill in the Elk River, West Virginia. *Journal Environmental Engineering*, **141**, 05014007-1–05014007-9.

Bentley R. and Chasteen T. G. (2004). Environmental VOSCs – formation and degradation of dimethyl sulfide, methanethiol and related materials. *Chemosphere*, **55**, 291–317.

Bunmahotama W., Hung W. N. and Lin T. F. (2015). Predicting the adsorption of organic pollutants from water onto activated carbons based on the pore size distribution and molecular connectivity index. *Water Research*, **85**, 521–531.

Bunmahotama W., Hung W. N. and Lin T. F. (2017). Prediction of the adsorption capacities for four typical organic pollutants on activated carbons in natural waters. *Water Research*, **111**, 28–40.

Bunmahotama W., Hung W. N. and Lin T. F. (2018). Simulation of the adsorption capacity of polar organic compounds and dyes from water onto activated carbons: model development and validation. *Sustainable Environmental Research*, **28**, 57–64.

Chen J. J. and Yeh H. H. (2005). The mechanisms of potassium permanganate on algae removal. *Water Research*, **39**, 4420–4428.

Chiu Y. T., Yen H. K., Lin H. L. and Lin T. F. (2016). An alternative method to quantify 2-MIB producing cyanobacteria in drinking water reservoirs: method development and field applications. *Environmental Research*, **151**, 618–627.

Chow C. W. K., Drikas M., House J., Burch M. D. and Velzeboer R. M. A. (1999). The impact of conventional water treatment processes on cells of the cyanobacterium *Microcystis aeruginosa*. *Water Research*, **33**, 3253–3262.

Crittenden J. C., Trussell R. R., Hand D. W., Howe K. J. and Tchobanoglous G. (2012). MWH's Water Treatment: Principles and Design, 3rd edn. John Wiley, New York.

Daly R. I., Ho L. and Brookes J. D. (2007). Effect of chlorination on *Microcystis aeruginosa* cell integrity and subsequent microcystin release and degradation. *Environmental Science & Technology*, **41**, 4447–4453.

DeYonkers N., Charbonnet K. A. and Alexander W. A. (2016). Dipole moments of *trans*- and *cis*-(4-methylcyclohexyl)methanol (4-MCHM): obtaining the right conformer for the right reason. *Physical Chemistry Chemical Physics*, **18**, 17856–17867.

Dietrich A. M., Thomas A., Zhao Y., Smiley E., Shanaiah N., Ahart M., Charbonnet K. A., DeYonker N. J., Alexander W. A. and Gallagher D. L. (2015). Partitioning, aqueous solubility, and dipole moment data for *cis*- and *trans*-(4-methylcyclohexyl)methanol, principal contaminants of the West Virginia chemical spill. *Environmental Science & Technology Letters*, **2**, 23–127.

Dixon M. B., Falconet C., Ho L., Chow C. W. K., O'Neill B. K. and Newcombe G. (2011). Removal of cyanobacterial metabolites by nanofiltration from two treated waters. *Journal of Hazardous Materials*, **188**, 288–295.

Fan J. J., Rao L., Chiu Y. T. and Lin T. F. (2016). Impact of chlorine on the cell integrity and toxin release and degradation of colonial *Microcystis*. *Water Research*, **102**, 394–404.

Faruqi A., Henderson M., Henderson R. K., Stuetz R., Gladman B., McDowall B. and Zamyadi A. (2018). Removal of algal taste and odour compounds by granular and biological activated carbon in full-scale water treatment plants. *Water Sciene and Technology: Water Supply*, **18**, 1531–1544.

Foreman W. T., Rose D. L., Chambers D. B., Crain A. S., Murtagh L. K., Thakellapalli H. and Wang K. K. (2015). Determination of (4-methylcyclohexyl)methanol isomers by heated purge-and-trap GC/MS in water samples from the 2014 Elk River, West Virginia, chemical spill. *Chemosphere*, **131**, 217–224.

Gallagher D. L., Phetxumphou K., Smiley E. and Dietrich A. M. (2015). Tale of two isomers: complexities of human odour perception for *cis*- and *trans*-4-methylcyclohexane methanol from the chemical spill in West Virginia. *Environmental Science & Technology*, **49**, 1319–1327.

Gillogly T. E. T., Snoeyink V. L., Vogel J. C., Wilson C. M. and Royal E. P. (1999). Determining GAC bed life. *Journal American Water Works Association*, **91**, 98–110.

Greenwald M. J., Redding A. M. and Cannon F. S. (2015). A rapid kinetic dye test to predict the adsorption of 2-methylisoborneol onto granular activated carbons and to identify the influence of pore volume distributions. *Water Research*, **68**, 784–792.

Guo L. (2007). Doing battles with green monster of Taihu Lake. *Science*, **317**, 1166.

Henderson R., Parsons S. A. and Jefferson B. (2008). The impact of algal properties and pre-oxidation on solid–liquid separation of algae. *Water Research*, **42**, 1827–1845.

Herald Media (2016). The Latest: $151M settlement in chemical leak lawsuit. http://www.heraldmailmedia.com/news/nation/the-latest-m-settlement-in-chemical-leak-lawsuit/article_cc970d52-f976-503a-a5a5-192b61561998.html

Ho L. and Newcombe G. (2010). Granular activated carbon adsorption of 2-methylisoborneol (MIB): pilot- and laboratory-scale evaluations. *Journal of Environmental Engineering*, **136**, 965–974.

Ho L., Wijesundara S., Shaw G., O'Donohue M., Saint C. and Newcombe G. (2005). Biological filtration processes for the removal of algal metabolites. *Water Science and Technology: Water Supply*, **6**(2), 153–159.

Hsieh W. H., Chang D. W. and Lin T. F. (2014). Occurrence and removal of earthy-musty odorants in two waterworks in Kinmen Island, Taiwan. *Journal of Hazardous, Toxic, and Radioactive Waste*, **18**, 04014012.

Juttner F. (1984). Dynamics of the volatile organic substances associated with cyanobacteria and algae in a eutrophic shallow lake. *Applied and Environmental Microbiology*, **47**, 814–820.

Lalezary S., Pirbazari M. and McGuire M. J. (1986). Oxidation of five earthy-musty taste and odor compounds. *Journal American Water Works Association*, **78**, 62–69.

Li L., Wan N., Gan N. Q., Xia B. D. and Song L. R. (2007). Annual dynamics and origins of the odorous compounds in the pilot experimental area of Lake Dianchi, China. *Water Science Technology*, **55**, 43–50.

Li J., Zamyadi A. and Hofmann R. (2015). Effect of granular activated carbon type and age on quenching H_2O_2 residuals after UV/H_2O_2 drinking water treatment. *Journal of Water Supply: Research and Technology – AQUA*, **65**, 28–36.

Li X. (2016). Research on the Characteristics and Predicting Models of the Removal of Typical Odorants in Drinking Water. *PhD dissertation*, Tsinghua University, Beijing, P. R. China.

Li X., Wang J., Liu Y. Y., Shi J., Zhang X. J. and Chen C. (2015). Odorant screening and possible origin analysis of odor episodes in one reservoir in northern China. *Journal of Water Supply: Research and Technology-AQUA*, **64**, 847–856.

Li X., Lin P. F., Wang J., Liu Y. Y., Li Y., Zhang X. J. and Chen C. (2016). Treatment technologies and mechanisms for three odorants at trace level: IPMP, IBMP, and TCA. *Environmental Technology*, **37**, 308–315.

Li Y. (2009). Study on odorant component and removal technology in drinking water of china. PhD dissertation, Tsinghua University, Beijing, P. R. China.

Li Z. L., Yu J. W., Yang M., Zhang J., Burch M. D. and Han W. (2010). Cyanobacterial population and harmful metabolites dynamics during a bloom in Yanghe Reservoir, North China. *Harmful Algae*, **9**, 481–488.

Li Z. L., Hobson P., An W., Burch M. D., House J. and Yang M. (2012). Earthy odor compounds production and loss in three cyanobacterial cultures. *Water Research*, **46**, 5165–5173.

Lin T. F., Chang D. W., Lien S. K., Tseng Y. S., Chiu Y. T. and Wang Y. S. (2009). Effect of chlorination on the cell integrity of two noxious cyanobacteria and their release of odorants. *Journal of Water Supply: Research and Technology – AQUA*, **58**, 539–551.

Mackenzie J. A., Tennant M. F. and Mazyck D. W. (2005). Tailored GAC for the effective control of 2-methylisoborneol. *Journal American Water Works Association*, **97**, 76–87.

Matsui Y., Nakao S., Taniguchi T. and Matsushita T. (2013). Geosmin and 2-methylisoborneol removal using superfine powdered activated carbon: shell adsorption and branched-pore kinetic model analysis and optimal particle size. *Water Research*, **47**, 2873–2780.

Matsui Y., Nakao S., Sakamoto A., Taniguchi T., Pan L., Matsushita T. and Shirasaki N. (2015). Adsorption capacities of activated carbons for geosmin and 2-methylisoborneol vary with activated carbon particle size: effects of adsorbent and adsorbate characteristics. *Water Research*, **85**, 95–102.

McDowall B., Hoefel D., Newcombe G., Saint C. P. and Ho L. (2009). Enhancing the biofiltration of geosmin by seeding sand filter columns with a consortium of geosmin-degrading bacteria. *Water Research*, **43**, 433–440.

MOHC (Ministry of Health, China) (2006). Chinese Standard for Drinking Water Quality, GB5749-2006. Ministry of Health, China, Beijing, P. R. China.

Newcombe G., House J., Ho L., Baker P. and Burch M. (2010). Management Strategies for Cyanobacteria (blue-green algae): A Guide for Water Utilities, Research Report No 74, CRC for Water Quality and Treatment, Adelaide, SA, Australia.

Ömür-Özbek P. (2012). Global Taste and Odour Survey of Water Utilities: Final Report. American Water Works Association, Denver, CO, USA.

Phetxumphou K., Dietrich A. M., Shanaiah N., Smiley E. and Gallagher D. L. (2016). Subtleties of human exposure and response to chemical mixtures from spills. *Environmental Pollution*, **214**, 618–626.

Pikaar I., Sharma K. R., Hu S. H., Gernjak W., Keller J. and Yuan Z. G. (2014). Water engineering. Reducing sewer corrosion through integrated urban water management. *Science*, **345**(6198), 812–4.

Plummer J. D. and Edzwald J. K. (2002). Effects of chlorine and ozone on algal cell properties and removal by coagulation. *Journal Water Supply: Research Technology –AQUA*, **51**, 307–318.

Sain A. E., Dietrich A. M., Smiley E. and Gallagher D. L. (2015). Assessing human exposure and odor detection during showering with crude 4-(methylcyclohexyl)methanol (MCHM) contaminated drinking water. *Science of the Total Environment*, **538**, 298–305.

Schade C. P., Wright N., Gupta R., Latif D. A., Jha A. and Robinson J. (2015). Self-reported household impacts of large-scale chemical contamination of the public water supply, Charleston, West Virginia, USA. *PLoS One*, **10**, e0126744.

Schideman L., Strathmann T., Metz D., Isabel R. S. and Cummings J. (2012). Evaluating GAC Filters for Control of DBP Precursors and Trace Organic Contaminants. Web Report #4101. Water Research Foundation, Denver, CO, USA.

Suffet I. H., Ho J., Chou D., Khiari D. and Mallevialle J. (1995). Taste-and-odor problems observed during drinking water treatment. In: Advances in Taste-and-Odor Treatment and Control, I. H. Suffet, J. Mallevialle and E. Kawczynski (eds), American Water Works Association Research Foundation, Denver, CO, USA.

Suffet I. H., Corado A., Chou D., McGuire M. J. and Butterworth S. (1996). AWWA taste and odor survey. *Journal American Water Works Association*, **88**, 168–180.

Suffet I. H., Khiari D. and Bruchet A. (1999). The drinking water taste and odor wheel for the millennium: beyond geosmin and 2-methylisoborneol. *Water Science Technology*, **40**(6), 1–13.

Summers R. S., Kim S. M., Shimabuku K., Chae S.-H. and Corwin C. J. (2013). Granular activated carbon adsorption of MIB in the presence of dissolved organic matter. *Water Research*, **47**, 3507–3513.

Tung S. C., Lin T. F., Yang F. C. and Liu C. L. (2008). Seasonal change and correlation with environmental parameters for 2-MIB in Feng-Shen Reservoir, Taiwan. *Environmental Monitoring and Assessment*, **145**, 407–416.

USCDC (Centers for Disease Control and Prevention) (2014). Summary Report of Short-Term Screening Level Calculation and Analysis of Available Animal Studies for MCHM. Centers for Disease Control and Prevention, Atlanta, GA. http://www.bt.cdc.gov/chemical/MCHM/westvirginia2014/pdf/MCHM-Summary-Report.pdf

Ward K. (2016a). Freedom's Farrell gets 30 days in jail for chemical spill. Charleston Gazette-Mail, Feb 11, 2016. http://www.wvgazettemail.com/news/20160211/freedoms-farrell-gets-30-days-in-jail-for-chemical-spill

Ward K. (2016b). Ex-Freedom official Southern sentenced to 30 days, $20 K fine. Charleston Gazette-Mail, Feb 17, 2016. http://www.wvgazettemail.com/news/20160217/ex-freedom-official-southern-sentenced-to-30-days-20k-fine

Ward K. (2016c). WV water crisis settlements provide community up to $151M. Charleston Gazette-Mail, Oct. 31, 2016. http://www.wvgazettemail.com/news-cops-and-courts/20161031/wv-water-crisis-settlements-provide-community-up-to-151m

Watson S. (2003). Cyanobacteria and eukaryotic algal odour compounds: signals or by-products? A review of their biological activity. *Phycologia*, **42**, 332–350.

Weidhaas J., Lin L.-S., Buzby K. and Li X. (2016a). Biodegradation of MCHM and PPH in river microcosms and activated sludge. *Journal Environmental Engineering*, **142**, 04016056-1–04016056-9.

Weidhaas J., Dietrich A. M., DeYonker N., Dupont R. R., Foreman W., Gallagher D., Gallagher J., Whelton A. and Alexander W. (2016b). Enabling science support during incident response for better decision making. *Journal of Environmental Quality*, **45**, 1490–1500.

Weidhaas L., Lin L.-S. and Buzby K. (2017). A case study for orphaned chemicals: 4-methylcyclohexanemethanol (MCHM) and propylene glycol phenyl ether (PPH) in riverine sediment and water treatment processes. *Science of the Total Environment*, **574**, 1396–1404.

Whelton A. J., McMillan L., Connell M., Kelley K. M., Gill J. P., White K. D., Gupta R., Dey R. and Novy C. (2015). Residential tap water contamination following the Freedom Industries chemical spill: perceptions, water quality, and health impacts. *Environmental Science & Technology*, **49**, 813–823.

Yang M., Yu J. W., Li Z. L., Guo Z. H., Burch M. and Lin T. F. (2008). Taihu Lake not to blame for Wuxi's woe. *Science*, **319**, 158.

Yuan L., Zhi W., Liu Y., Smiley E., Gallagher D. L., Chen X., Dietrich A. M. and Zhang H. (2016a). Aerobic and anaerobic microbial degradation of crude (4-methylcyclohexyl) methanol in river sediments. *Science of the Total Environment*, **547**, 78–86.

Yuan L., Zhi W., Liu Y., Smiley E., Gallagher D. L., Chen X., Dietrich A. M. and Zhang H. (2016b). Degradation of cis- and trans-(4-methylcyclohexyl)methanol in activated sludge. *Jouuurnal of Hazardous Materials*, **306**, 247–256.

Zamyadi A., Henderson R. K., Stuetz R., Hofmann R., Ho L. and Newcombe G. (2015a). Fate of geosmin and 2-methylisoborneol in full-scale water treatment plants. *Water Research*, **83**, 171–183.

Zamyadi A., Sawade E., Ho L., Newcombe G. and Hofmann R. (2015b). Impact of UV–H_2O_2 advanced oxidation and aging processes on GAC capacity for the removal of cyanobacterial taste and odour compounds. *Environmental Health Insights*, **9** (Supplement 3), 1–10.

Zhang X. J. and Chen C. (2009). Emergency drinking water treatment in source water pollution incident: technology and practice in China. *Frontier of Environmental Science and Engineering in China*, **3**, 364–368.

Zhang X. J., Chen C., Ding J. Q., Hou A. X., Li Y., Niu Z. B., Su X. Y., Xu Y. J. and Laws E. (2010). The 2007 water crisis in Wuxi, China: analysis of the origin. *Journal of Hazardous Material*, **182**, 130–135.

Zhang X. J., Chen C., Lin P. F., Hou A. X., Niu Z. B. and Wang J. (2011). Emergency drinking water treatment during source water pollution accidents in China: origin analysis, framework and technologies. *Environmental Science & Technology*, **45**, 161–167.

Chapter 7

Management of T&O in source water

Zhiquan Liu[1], Tsair-Fuh Lin[2] and Michael Burch[3]

[1]Institute of Environmental Research at Greater Bay, Guangzhou University, Guangzhou, Guangdong, China
[2]Department of Environmental Engineering, National Cheng Kung University, Tainan, Taiwan
[3]Department of Ecology and Evolutionary Biology, University of Adelaide, South Australia, Australia

7.1 INTRODUCTION

Taste and odour (T&O) in drinking water is a significant problem for water utilities, and is often a cause of consumer complaints with the negative perception that the odourous water is associated with unsafe water (Bruchet & Lan, 2005). T&O compounds can originate from anthropogenic (e.g. chlorination of T&O precursors) and/or natural (e.g. geosmin and 2-methylisoborneol (2-MIB)) sources. The T&O compounds from anthropogenic sources can generally be controlled by optimizing disinfection processes, however, natural compounds are harder to treat and attract more research attention (Peter *et al.*, 2009; Proulx *et al.*, 2012). Geosmin and 2-MIB are the most common T&O compounds in source water and tap water, and are responsible for earthy/musty odour (Lin *et al.*, 2002). These T&O compounds can be produced by algae, cyanobacteria, fungi, bacteria and actinobacteria (actinomycetes) (Lloyd *et al.*, 1998).

Since conventional water treatment processes are largely ineffective for the removal of T&O compounds (Lin *et al.*, 2002; McGuire, 1986), a number of

Editor(s): Tsair-Fuh Lin, Susan Watson, Andrea M. Dietrich and I. H. (Mel) Suffet
doi: 10.2166/9781780406664_0211

advanced treatment processes have been applied to handle this problem, including powdered activated carbon (PAC) adsorption, advanced oxidation and biodegradation (Cook *et al.*, 2001; Ho *et al.*, 2007; Jo *et al.*, 2011). However, when there are high initial concentrations of T&O compounds in raw water, a high removal rate and efficiency of removal in these processes is required to meet drinking water demands due to the low tolerance threshold for these odour compounds (at low ng/L level, Table 7.1) (Krasner *et al.*, 1983). Furthermore, these processes are sometimes either too costly or impractical to be applied in practice (Lin *et al.*, 2002). For example, the residual concentration of 2-MIB may still be higher than the threshold value after adsorption by 20 mg/L PAC for one hour at an initial 2-MIB concentration of 100 ng/L (Liang *et al.*, 2005). Also, it may not be possible to obtain a higher removal rate by increasing PAC dose in the water treatment plants (WTP) because high doses of PAC can significantly reduce the filter run time and cause breakthrough of PAC (Najm *et al.*, 1992). Thus, the management of T&O in source water is important and necessary to mitigate the treatment burden of T&O to water works.

Table 7.1 Characteristics of earthy odourants.

Compounds	Chemical Name	Threshold Odour Conc. (ng/L)	Odour Description
Geosmin	*trans*-1,10-Dimethyl-*trans*-9-decalol	10	Earthy/musty
2-MIB	2-Methylisoborneol	29	Earthy/musty/camphorous
IPMP	2-Isopropyl-methoxypyrazine	2	Earthy/musty/potato-bin
IBMP	2-Isobutyl-methoxypyrazine	2	Earthy/musty/bell pepper
TCA	2,3,6-Trichloroanisole	7	Musty

Source: Krasner *et al.* (1983).

In this chapter, the fate of T&O compounds in lakes and reservoirs is analyzed, including the production and loss processes, and introduce some appropriate treatment methods for the management of T&O which can either inhibit or control the growth of T&O producers. The efficacy and cost of different management methods and the safety of these methods are also discussed.

7.2 THE FATE OF T&O COMPOUNDS IN LAKES AND RESERVOIRS

Understanding the fate of T&O compounds in lakes and reservoirs can provide insights into how to manage these compounds. Odourants are produced by microorganisms in the aquatic environment and eliminated by physical, chemical and biological processes. The influent (including storm runoff) and effluent of these water bodies can also affect the concentrations of T&O compounds in lakes and reservoirs. The mass balance of off-flavour compounds in lakes and reservoirs are shown in Figure 7.1 and can be expressed as Equation 7.1 (Hobson *et al.*, 2010):

$$\frac{\partial \sum_i M_i}{\partial t} = \sum_i M_{\text{in},i} - \sum_i M_{\text{out},i} + \sum_i S_i - \sum_i L_i \tag{7.1}$$

where, M is the mass of the odourant in the reservoir, M_{in} is the odourant entering the reservoir with inflow water, M_{out} is the odourant leaving the reservoir with outflow water, S represents the production of odourant from different sources, and L is the combination of sink terms for the odourants, which may include biodegradation, volatilization, adsorption and photolysis.

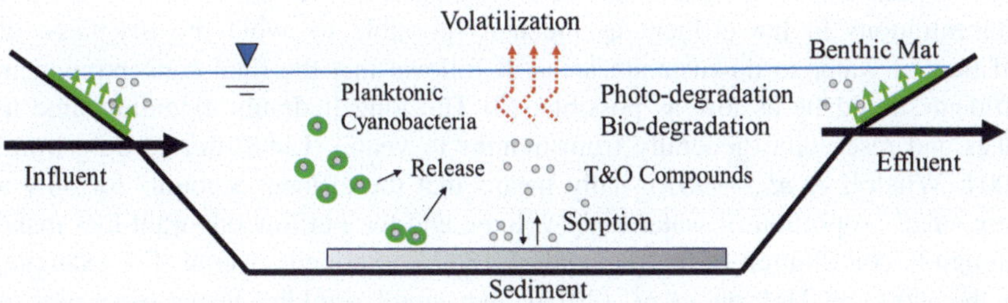

Figure 7.1 Schematic representation for the mass balance of T&O compounds in aquatic environment. (Modified from Hobson *et al.*, 2010).

On the right-hand side, the first and the second terms represent the mass entering and leaving with water flow as different phases (dissolved, cell-bound, and/or adsorbed on suspended particles); the third term represents the mass produced by microorganisms, including actinomycetes, planktonic and benthic cyanobacteria or algae; and the last term represents the elimination processes for odourants, including biodegradation, volatilization, adsorption and photolysis. Equation (7.1) does not consider the transformation of odourants from cell-bound or adsorbed form into dissolved form which may be important for the management of T&O, given that a well operated WTP can remove cells and particles as well as the odourants bound with them efficiently, but cannot perform a satisfactory removal of dissolved compounds (Jüttner & Watson, 2007). Thus, the successful management of odourants should focus on limiting the odourant influx, reducing the production by microorganisms, and enhancing the elimination or loss process, especially for the dissolved form.

7.2.1 T&O compounds in the influent and effluent

T&O compounds in the influent of lakes/reservoirs are produced by microorganisms in upstream areas and constitute a very small fraction of the total. Usually this proportion is negligible because of the dilution effect by large volume of lake water, but sometimes it may be important. For example, a strong spring run-off or storm flow can transfer off-flavour compounds bound to soil as well as their producers into surface waters and cause a T&O incident in lakes (Jensen *et al.*, 1994; Zaitlin & Watson, 2006). The impact of inflow water quality on source water is especially important for river types of sources. Sun *et al.* (2013) studied the musty odour problems of the major water source in Shanghai, China, the Huangpu River. In the study, they found that Tai Lake and a few tributary creeks were responsible for the high concentrations of 2-MIB, up to 28.6 and 71.0 ng/L, at the drinking water intake, and also contributed the 2-MIB producing cyanobacteria, *Phormidium* spp.

T&O compounds in the effluent play an insignificant role on the control of T&O for two reasons: (1) Successful T&O management should reduce the T&O concentrations in the effluent as much as possible to avoid the discharge of off-flavour water to downstream areas. It follows that the final concentration in effluent should be as low as possible; (2) The long hydraulic retention time in lakes and reservoirs (generally from months to years) (Lindström & Bergström, 2004; Windolf *et al.*, 1996) usually means that the effluent accounts for only a very small proportion of water body volume and the outflow odourant-loss make negligible contribution to the removal of T&O compounds overall. For example, in the work by Hobson *et al.* (2010), two small drinking water reservoirs in Taiwan and Australia were studied for geosmin and 2-MIB issues. The average residence times were 40 and 200 days for the Taiwanese and Australian reservoirs, respectively. The losses of the two T&O compounds associated with outflow were much less than those caused by biodegradation and also much smaller than from volatilization (Hobson *et al.*, 2010). In fact, although Westerhoff *et al.* (2005) indicate that inflow and outflow mass values are important to the variation of odourants in three reservoirs that they studied, the maximum daily inflow or outflow odourant mass only accounted for less than 9% of the total amounts in these reservoirs.

7.2.2 Production of T&O compounds in source water

Most of T&O compounds in source water originate from *in-situ* biosynthesis, which depends upon two factors: the abundance and the associated productivity of odourant producers in this aquatic environment. It is reported that odourants can be produced by a number of algae, cyanobacteria, fungi, actinomycetes, liverworts and amoebae (Jüttner & Watson, 2007). Among these microorganisms, cyanobacteria (including planktonic and benthic species) and actinomycetes are the most likely and significant producers in aquatic environments. It is well known that the density of planktonic cyanobacteria can be high in eutrophic lakes in summer (often $>10^4$ cells/mL) (Anderson *et al.*, 2002; Jin *et al.*, 2008), but there is little quantitative information regarding the abundance of benthic cyanobacteria. Benthic cyanobacteria can grow on the bottom sediments of water bodies where there is sufficient illumination and have been known to cover up to 70% of the total surface area of sediments (Mur *et al.*, 1999; Paul *et al.*, 2005). The densities of organisms reported in the few rigorous quantitative studies are in ranges from 4.3×10^6 to 1.1×10^8 cells/cm^2 in different seasons in a Spanish river (Sabater *et al.*, 2003); $1.1–3.4 \times 10^5$ cells/cm^2 in a South Australian reservoir and $1.1–3.4 \times 10^5$ cells/cm^2 in a slow sand filter in Taiwan (Chen *et al.*, 2010). Actinomycetes are also common bacteria in the aquatic environment, sometimes constituting a large percentage of microbial community (Warnecke *et al.*, 2005). Klausen *et al.* (2005) investigated the abundance of actinomycetes in oligotrophic and eutrophic lakes by using catalyzed reporter deposition-fluorescence *in-situ* hybridization method and found values ranging from 1×10^4 to 7×10^4 cells/mL, which are of the same order of magnitude as the abundance of cyanobacteria in eutrophic water bodies. Hobson *et al.* (2010) characterized actinomycetes from sediments of 17 locations in Hope Valley Reservoir and Happy Valley Reservoir in South Australia. Among the 92 isolates, all the isolates were from the genus *Streptomyces*, ~40% produced both geosmin and 2-MIB, and 60% produced geosmin only. Similarly, Tung *et al.* (2006) also isolated three *Streptomyces* from Feng-Shen Reservoir in Taiwan. Based on the analysis of 16S rDNA sequencing, they identified the strains to be *Streptomyces caelestis*, *Streptomyces roseoflavus* and *Streptomyces malaysiensis*. The strain *S. malaysiensis* produced both geosmin and 2-MIB, *S. roseoflavus* produced only geosmin, and *S. caelestis* produced only 2-MIB.

The odourant productivity by some common microorganisms in axenic laboratory culture are shown in Table 7.2. It is difficult to compare the productivities of odourants between studies due to the lack of standardized units of measurement. However, the data available show that the productivity can vary significantly among diverse microorganisms (as large as 200-fold), and for a specific microorganism, the value can change to 10-fold under different culture conditions, which includes the variables of temperature, pH, nutrient concentration, illumination, dissolved oxygen, amongst others (Dionigi & Ingram, 1994; Tung *et al.*, 2008; Walsh *et al.*, 1998; Westerhoff *et al.*, 2005). These lab-scale experiments separate and identify off-flavour producers as well as their production of odourants, but it is controversial whether the results are particularly representative of the actual values in the aquatic environment because of the dynamic range of variables and other influences in the environment. However, some field investigations have shown results similar to these lab

Table 7.2 Odourant producers and their productivities in axenic laboratory culture.

Taxon	Group	Odourant	Productivity	References
Penicillium expansum	Fungi	Geosmin	1.2–11.5 μg/g*	Dionigi and Ingram (1994)
Penicillium commune	Fungi	Geosmin, alcohols	10–100 ng/h for geosmin*; 10–1000 ng/h for alcohols*	Sunesson *et al.* (1995)
Streptomyces tendae	Actinomycetes	Geosmin	4–19 μg/g*	Dionigi and Ingram (1994)
S. malaysiensis	Actinomycetes	2-MIB, geosmin	2.4 μg/g for geosmin 4.5 μg/g for 2-MIB	Tung *et al.* (2008)
Streptomyces sp.	Actinomycetes	2-MIB, geosmin	7.84 ng/g for 2-MIB*; 18.84 μg/g for geosmin*	Xie *et al.* (2007)
Streptomyces sp.	Actinomycetes	2-MIB, geosmin	0.14–34.85 ag/cell/h for 2-MIB 0.16–15.34 ag/g for geosmin	Klausen *et al.* (2005)
Rothia-like sp.	Actinomycetes	2-MIB, geosmin	0.45 ag/cell/h for 2-MIB 0.83 ag/g for geosmin	Klausen *et al.* (2005)
Oscillatoria tenuis	Benthic cyanobacteria	2-MIB, geosmin	165.2 μg/g Chl-a for 2-MIB; 441.6 μg/g Chl-a for geosmin	Wu and Jüttner (1988)
Oscillatoria brevis	Benthic cyanobacteria	Geosmin	160 μg/g	Naes *et al.* (1989)
Oscillatoria sp.	Planktonic cyanobacteria	2-MIB	74.7–86.9 μg/L	Martin *et al.* (1991)
Anabaena circinalis	Planktonic cyanobacteria	Geosmin	2.5–8 fg/cell*	Rosen *et al.* (1992)
Anabaena PCC7120	Planktonic cyanobacteria	2-MIB	0.82 μg/L for 2-MIB at a OD_{750} of 0.638	Xie *et al.* (2007)
Microcystis NRC-1	Planktonic cyanobacteria	β-Cyclocitral	7–12 fg/cell*	Jüttner *et al.* (2010)
Microcystis aeruginosa	Planktonic cyanobacteria	β-Cyclocitral	1.92–11.20 μg/g geosmin equivalents	Walsh *et al.* (1998)
Symploca muscorum	Planktonic cyanobacteria	Geosmin	0.6 mg/L	Safferman *et al.* (1967)
Pseudanabaena sp.	Planktonic cyanobacteria	2-MIB	180–200 μg/g	Izaguirre and Taylor (1998)
Pseudanabaena galeata	Planktonic cyanobacteria	2-MIB	5.96–51.1 fg/cell	Chiu *et al.* (2016)

*Data were obtained either from figures in the references or by unit conversions from the published data.

studies, which indicates that the results from laboratory experiments are valuable to estimate the contributions of odourant producers to T&O problems in source water. Jones and Korth (1995) investigated *in-situ* production of odourants in one Australian river and one dam, and found that geosmin production by *Anabaena* sp. and β-cyclocitral production by *Microcystis* sp. were 10 and 2–10 fg/cell, respectively, while Hobson *et al.* (2010) found geosmin production by *Anabaena* sp. in three different shallow lakes ranged from 20 to 50 fg/cell, both of which correspond with the lab-scale results (Jüttner *et al.*, 2010; Rosen *et al.*, 1992). Thus, it can be inferred that there is much higher odour production potential from cyanobacteria than from fungi and actinomycetes in the aquatic environment (two orders of magnitude higher) according to lab-scale results (Table 7.2). Indeed, the actual productivity of odourants by cyanobacteria could be even higher. In a more recent study, Tsao *et al.* (2014) reported a high geosmin production of 15–838 fg/cell by *Anabaena circinalis* in Myponga Reservoir, South Australia.

To allow consistent comparison of results, the actual odourant production should be expressed as the mass of odourants produced by cells (or other units which can represent biomass, e.g. chlorophyll, dry weight) during a certain period (e.g. ng/cell/h) rather than the average concentration of mass produced by cells (e.g. ng/cell) as has been done in most previous studies. This actual odourant productivity can give an insight into the dynamic odourant production in source water. For example, Chen *et al.* (2010) have developed a novel method to measure the *in situ* odourant production by benthic cyanobacteria and indicate that the productivity of 2-MIB and gesomin by benthic cyanobacteria are $1.0 \times 10^{-6} - 5.5 \times 10^{-6}$ and $5.3 \times 10^{-5} - 1.8 \times 10^{-3}$ ng/cell/h, respectively, which are almost 3–5 orders of magnitude higher than actinomycetes reported by Klausen *et al.* (2005). On the basis of the data of Chen *et al.* (2010), it can be inferred that benthic cyanobacteria have sediment areal production rates of 42–44 μg/h/m^2 for 2-MIB and 180–1900 μg/h/m^2 for geosmin, respectively.

Considering both the abundance and the productivity of microorganisms in aquatic environment, it is proposed that planktonic and benthic cyanobacteria are the major source of odourants in source water (Jüttner & Watson, 2007; Peter *et al.*, 2009; Wnorowski, 1992). This is supported in that the concentrations of odourant correlate closely to cyanobacterial abundance but do not correlate to actinomycete numbers (Jüttner & Watson, 2007; Zaitlin & Watson, 2006). Yen *et al.* (2007) further found that there is a significant linear correlation between 2-MIB and water temperature, which is one of the most important environmental parameters affecting cyanobacterial growth in reservoirs. Furthermore, the presence of actinomycetes in the aquatic environment does not mean that they can grow actively there, and it is still some doubt whether odourant-producing actinomycetes in the aquatic environment are autochthonous inhabitants or are transported from terrestrial environments, which may significantly reduce their production of odourants (Zaitlin & Watson, 2006). In any case the control cyanobacteria in particular becomes one of the most important issues for the management of T&O problems in source water.

However, it should be noted that the production of T&O compounds is a complex process, and while cyanobacteria are the primary source they are not the only source of odourants in source water and variations can depend upon the differing seasonal contributions of the range of microorganisms. This is why for example odourants can occur in winter (when few cyanobacteria exist) and why sometimes there is a negative correlation between cyanobacterial density and odourant concentration (Dzialowski *et al.*, 2009). In other words, the range of odourant producers (e.g. actinomycetes) should also be considered in addition to the management and treatment of cyanobacteria.

7.2.3 Elimination of T&O compounds in source water

While odourants are being produced by various microorganisms in the aquatic environment, these compounds are also being eliminated at the same time by different mechanisms, including biodegradation, volatilization, adsorption and photolysis.

7.2.3.1 Biodegradation

Biodegradation is considered to be the primary elimination mechanism for odourants in the aquatic environment. The different species of bacteria that can degrade and utilize odourants (including Gram-negative and Gram-positive species) are shown in Table 7.3. According to the limited data available, most of these bacteria can be found occurring naturally in the aquatic environment, and *Bacillus* sp., *Pseudomonas* sp. and *Alphaproteobacteria* sp. are quite abundant in freshwater (Dang *et al.*, 2008; de Menezes *et al.*, 2012; Grüntzig *et al.*, 2001; Holmfeldt *et al.*, 2009; Janakiraman & Leff, 1999; Yang *et al.*, 2013). It follows that biodegradation of odourants may occur in lakes and reservoirs. Researchers have also investigated biodegradation kinetics, and most results indicate that biodegradation follows a pseudo-first-order reaction with a reaction constant of 0.074–0.70 1/d (equivalent to a half-life period of 0.99–9.37 d), depending upon different incubation conditions (Table 7.4). Field-scale studies can give us more insights into the biodegradation process. Westerhoff *et al.* (2005) indicated that the biodegradation elimination rates of 2-MIB are 0.14–1.1 mg/L/d in three lakes with different eutrophication levels, and the elimination rates were similar to their lab-scale degradation experiment results. Izaguirre *et al.* (1988) and Hobson *et al.* (2010) found that more than 99% of the total odourants can be degraded in less than 7 days which is much higher than that reported by Westerhoff *et al.* (2005). Clearly, biodegradation plays a very important role on the elimination of odourants in source water.

Table 7.3 Odourant decomposers reported in previous publications.

Taxon	Affiliation	Odourant	References
Enterobacter sp.	Gram-negative	2-MIB	Tanaka *et al.* (1996) Izaguirre *et al.* (1988)
Pseudomonas sp.	Gram-negative	2-MIB	Tanaka *et al.* (1996) Oikawa *et al.* (1995)
Rhodococcus sp.	Gram-positive	2-MIB	Eaton and Sandusky (2009)
Bacillus subtilis	Gram-positive	2-MIB	Yagi *et al.* (1988)
Bacillus fusiformis *Bacillus sphaericus*	Gram-positive	2-MIB	Lauderdale *et al.* (2004)
Chryseobacterium sp.	Gram-negative	Geosmin	Zhou *et al.* (2011)
Sinorhizobium sp.	Gram-negative nitrogen-fixing	Geosmin	Zhou *et al.* (2011)
Stenotrophomonas sp.	Gram-negative	Geosmin	Zhou *et al.* (2011)
Alphaproteobacteria sp.	Gram-negative	Geosmin	Hoefel *et al.* (2009)
Sphingopyxis alaskensis *Novosphingobium stygiae* *Pseudomonas veronii*	Gram-negative	Geosmin	Hoefel *et al.* (2006)
Arthrobacter sp.	Gram-positive	Geosmin	Saadoun and El-Migdadi (1998)
Chlorophenolicus N-1053	Gram-positive	Geosmin	Saadoun and El-Migdadi (1998)
Rhodococcus maris	Gram-positive	Geosmin	Saadoun and El-Migdadi (1998)
B. subtilis	Gram-positive	Geosmin	Yagi *et al.* (1988)
B. subtilis *Bacillus cereus*	Gram-positive	Geosmin	Narayan and Nunez (1974)

Table 7.4 Summary of biodegradation kinetic constants of geosmin and 2-MIB.

Odourants	Reaction Constant	Rate Law	Conditions	References
Geosmin	0.25–0.67 1/d	First-order	22°C, BH medium	Hoefel *et al.* (2009)
Geosmin	0.12–0.58 1/d	First-order	20°C, natural water	Ho *et al.* (2007)
Geosmin	0.097 1/d 0.089 1/d 0.074 1/d	First-order	30°C, MSM medium, for *Chryseobacterium* sp., *Sinorhizobium* sp., *Stenotrophomonas* sp.	Zhou *et al.* (2011)
Geosmin	0.20–0.70 1/d	First-order	10–25°C, natural water	Hobson *et al.* (2010)
Geosmin	0.8–0.9 ng/L/d	Zero-order	22°C, natural water	Westerhoff *et al.* (2005)
2-MIB	0.10–0.36 1/d	First-order	20°C, natural water	Ho *et al.* (2007)
2-MIB	0.8–1.2 ng/L/d	Zero-order	22°C, natural water	Westerhoff *et al.* (2005)

7.2.3.2 Volatilization

Volatilization is another important loss mechanism for odourants in the aquatic environment. It usually follows a pseudo-first-order decay with respect to the concentration of odourant. The loss rate (R) is proportional to the air-water interface area (A) and mass transfer coefficient (K) and is inversely proportional to the volume of water (V), i.e. $R = KA/V$. The K can be estimated from two-film theory as Chen *et al.* (2010) described, which depends on the Henry's constant of the odourant, water temperature and wind speed. Thus, the depth of water and the wind speed in a specific lake or reservoir determine the contribution of volatilization to total mass loss. This is why a high volatilization constant (0.041–0.062 1/d) can be found in some shallow lakes with an average wind speed of 2.9–4.0 m/s (Hobson *et al.*, 2010), which is slightly less than the biodegradation rate shown in Table 7.4, while volatilization is negligible in some thermally stratified deep lakes (average depth $>$35 m) with a low average wind speed of 0.5 m/s (Westerhoff *et al.*, 2005).

7.2.3.3 Adsorption

Suspended solids (SS) may adsorb odourants onto their surface, but this effect as a loss mechanism can be negligible. The adsorption of odourants by SS has been rarely studied, but the value can be estimated from empirical calculation. The adsorption of odourants onto SS depends upon the partition coefficient for organic compounds between organic carbon and water (K_{OC}), the organic carbon content of SS (f_{OC}), the concentration of odourants in water (C) and the SS concentration in water (C_{SS}). K_{OC} is in proportion to the octanol/water partition coefficient, K_{ow} ($K_{OC} = 0.63\ K_{ow}$), according to the study by Karickhoff *et al.* (1979). It can be inferred that K_{OC} values of geosmin and 2-MIB are 3090 and 832, respectively, calculated from the log K_{ow} value of geosmin and 2-MIB (3.7 for geosmin and 3.1 for 2-MIB: Choi *et al.*, 2010; Westerhoff *et al.*, 2005). Assuming that f_{OC} is 0.02 (Kile *et al.*, 1995) and C_{SS} is 2000 mg/L, the adsorbed odourants only account for less than 6% (geosmin) and 2% (2-MIB) of the total, inferring that adsorption of the odourants onto SS in a lake or reservoir is not important under most conditions.

7.2.3.4 Photolysis

Photolysis is also insignificant to the loss of odourants in the aquatic environment. Although it is reported that 2-MIB and geosmin can be decomposed by UV-irradiation, only a small fraction of odourants can be removed by photolysis with low-pressure UV-irradiation (less than 5,000 J/m^2) or long-wave UV (Jo *et al.*, 2011; Kutschera *et al.*, 2009). In addition, the attenuation of solar UV radiation in lakes can be enhanced by the high concentration of dissolved organic carbon (DOC), especially for short-wave UV (Morris *et al.*, 1995). High concentrations of natural odourant often occur in eutrophic lakes and reservoirs, which are also characterized as having high concentrations of DOC and low transparency. Thus, photolysis of odourants is unlikely to occur under these conditions. In the study by Hobson *et al.* (2010) they investigated the photolysis of geosmin by solar UV-irradiation in ultra-pure water, and found no differences between the samples with and without illumination during a 14 d exposure.

In conclusion, biodegradation and volatilization are considered the primary loss mechanisms for odourants in source water, while adsorption and photolysis are negligible in most cases. In typical lakes and reservoirs with odourant-producing microorganisms, odourous chemicals are produced and eliminated at the same time, which will maintain the overall total for odourants in source water at a dynamic balance at mostly low concentrations.

7.3 CONTROL OF ODOURANT PRODUCERS

As discussed in Section 7.2, cyanobacteria are the primary odourant producers in source water, and their metabolites account for a large proportion of the total T&O compounds in the aquatic environment. Thus, it is very important for the effective management of T&O in source water to try to limit the growth of odour-producing cyanobacteria.

7.3.1 Management of nutrients

Nutrient enrichment in lakes and reservoirs is by definition the cause of eutrophication, and results in excessive plant production, blooms of harmful algae, increased frequency of anoxic events, and fish kills, and has become a global problem in many aquatic environments (Carpenter, 2005). Management of nutrients, including in watersheds and in reservoirs (or in lakes), is considered to be essential for the successful mitigation of eutrophication (Søndergaard *et al.*, 2000b) and is also important for T&O management due to the link with the production of odourants by excessive growth of cyanobacteria.

7.3.1.1 Watershed management

The most important prerequisite for high biomass density of any phytoplankton, including nuisance cyanobacteria, is an elevated concentration of nutrients. Due to the ability of cyanobacteria to fix atmospheric nitrogen in most cases the limiting nutrient is phosphorus, and the concentration of total phosphorus (TP) can be used to estimate the potential of bloom development in a water body. There is still however some debate on which nutrients (nitrogen or phosphorus) should be controlled as a priority to reverse the adverse effects of eutrophication.

Some researchers (Carpenter, 2008; Schindler *et al.*, 2008) argued that phosphorus (P) is the primary limiting factor to mitigate eutrophication in contrast to nitrogen (N). Many bloom-forming cyanobacteria can fix atmospheric nitrogen (from N_2 to reactive N, such as nitrate, ammonia, or organic N compounds), which can then significantly alleviate nitrogen shortages and maintain phosphorus-limited conditions in lakes. Schindler *et al.* (2008) also indicated that reducing nitrogen inputs will cause N-fixing cyanobacteria to be the dominant species. Thus, control of nitrogen alone cannot mitigate eutrophication in lakes. Other researchers (Conley *et al.*, 2009) argue that both nitrogen and phosphorus should be controlled for mitigating eutrophication. N_2 fixation depends on the abundance of N-fixing cyanobacteria in aquatic ecosystems and is rarely observed in high salinity water bodies. In estuaries and coastal regions, primary production of phytoplankton is largely limited by nitrogen (Paerl, 2006). Thus, comprehensive catchment management strategies with overall nutrient reduction activities are generally important as the first and central priority in control of eutrophication and cyanobacterial blooms.

7.3.1.2 In-reservoir management

Both the external and internal nutrient load can affect the available nutrient pool in source water which supports the excessive growth of cyanobacteria. In some cases, external load reduction may be effective, but the internal phosphorus load, released from sediments, can still be sufficiently high to sustain cyanobacterial biomass above target levels. This is particularly so in lakes with very low flushing rates where the internal phosphorus supply in the lake may remain high for many years, even after substantial external load reduction. For example, it has been reported that an acceleration of internal loading can counteract decreases in external loading and promote the growth of cyanobacteria in the Baltic Sea (Pitkänen *et al.*, 2001). It follows that it is important to undertake in-reservoir management to control the internal nutrient cycle as part of reducing the overall nutrient loading to a lake or reservoir.

In eutrophic or hyper-eutrophic conditions, it is possible that in-lake interventions may give the ecosystem a 'jolt' or interruption to the prevailing feedback mechanisms that stabilize the eutrophic situation. Internal measures can target changes in chemistry to reduce phosphorus release rates from the sediments or they can target conditions other than nutrient concentrations to make them less favorable for cyanobacteria. The latter non-nutrient based approaches can target blooms in different ways, for example: by suppressing the dominance of cyanobacteria in favour of other phytoplankton (non-toxic, non-scum forming) algae and by increasing loss rates of the cyanobacterial population.

Usually in-reservoir management of nutrients focuses on the control of phosphorus, because most of lakes and reservoirs are P-limited and the N-limiting conditions can be alleviated by the growth of N-fixing cyanobacteria as described above. The internal cycle of phosphorus in lakes and reservoirs is dependent upon the prevailing redox conditions in the aquatic environment. When phosphorus flows into a lake, only a small proportion is immediately assimilated by phytoplankton, while most of the external phosphorus load binds with ferric iron (Fe) and deposits in sediments. The Fe-bound phosphorus is stable and is not available to cyanobacteria under oxic conditions. However, under anoxic conditions, which may form via microbially-mediated decomposition of organic compounds, the Fe-bound phosphorus can be released into the water due to the competition of sulfide (to form insoluble FeS_x), which is reduced from sulfate or released from the decomposition of organics, and what follows is an increase in bio-available phosphorus. This process of internal supply can then contribute significantly to eutrophication (Mun, 2003; Smolders *et al.*, 2006). It has been reported that the release rate of phosphorus can be up to 6.49 $mg/m^2/d$ under anoxic conditions, and this internal release of P is sufficient to maintain significant growth of cyanobacteria in shallow lakes (Welch & Cooke, 1995).

Nutrient inactivation, sediment capping and dredging are the most common methods to control internal phosphorus loading for in-reservoir management (Jin *et al.*, 2008). As discussed above, only dissolved phosphorus can be assimilated by cyanobacteria and algae, while phosphorus adsorbed to metal oxides and hydroxides is generally not biologically available (Reynolds & Davies, 2001). This available phosphorus can be inactivated by aluminum or ferric salts, which can significantly reduce phosphorus supply and mitigate eutrophication in lakes and reservoirs (Deppe & Benndorf, 2002; Kennedy & Cook, 1982; Rydin *et al.*, 2000). Another option for phosphorus inactivation is to maintain oxic conditions in the hypolimnion of lakes and reservoirs by aeration or nitrate treatment, which can prevent the reduction of sulfate and keep phosphorus in a non-available form (Singleton & Little, 2006; Søndergaard *et al.*, 2000a; Wauer *et al.*, 2005). Nutrient exchange between sediment and water column can also be prevented by capping sediments with chemical agents, e.g. lanthanum-activated modified clay (LAMC) and calcium carbonate (Haghseresht *et al.*, 2009; Meis *et al.*, 2012; Otsuki & Wetzel, 1972). LAMC has been shown to strip dissolved phosphorus from the water column and increase the sediment P-sorption capacity (removal of available $P > 80\%$), and the phosphorus released under anoxic conditions is found to be negligible (Ross *et al.*, 2008). However, the potential environmental impact and safety of inactivation treatments and sediment capping should be considered with the use of these agents, as they may be released back into water column and lead to elevated concentrations of associated metals (Hickey & Gibbs, 2009). Dredging can remove nutrient-rich sediments and reduce the total phosphorus concentration, and this practice has been widely performed for lake restoration for many years (Ryding, 1982; Van der Does *et al.*, 1992; Zhong *et al.*, 2008). However, dredging can also affect benthic and zooplankton communities in lakes and reservoirs (de Grave & Whitaker, 1999; Zhang *et al.*, 2010), and the cost of dredging is high. It is important to perform a thorough overall evaluation of the feasibility, cost-benefit and impact of dredging before adopting this technique.

The above-mentioned techniques are important for nutrient management but cannot be performed as stand-alone solutions without addressing watershed management. The impact of the reduction in internal available phosphorus by careful application of a range of techniques (inactivation, sediment capping and dredging) may not be sufficient where there is a significant on-going external loading. On-going run-off will lead to the development of a new sediment-water interface within a short time period without catchment intervention to reduce the external load. These treatments may therefore provide only temporary improvements followed by a slow return to previous conditions (Kleeberg & Kohl, 1999; Søndergaard *et al.*, 2000b).

7.3.2 Control of planktonic cyanobacteria

Nutrient management is important for the control of T&O problems but may not bring about the required short-term reductions in densities of cyanobacteria to meet T&O objectives and targets. In this case there are range of other approaches, including chemical, physical and biological methods, that can be applied to achieve immediate effects, albeit with a range of advantages and disadvantages.

7.3.2.1 Algaecides

Algaecide application is often suggested as being more cost-effective than T&O removal by water treatment, as persistent blooms may require extended periods of additional enhanced water treatment for removal of odourants. It must be balanced against environmental concerns which are regularly raised because the most commonly used algaecide, copper sulfate, has broad ecological impact, and the copper may accumulate in the lake sediments (Prepas & Murphy, 1988). These concerns tend to limit algaecide treatment to dedicated water supply reservoirs and algaecide treatment is often viewed as unsatisfactory as a long-term solution, and where possible, other management measures that reduce the growth of cyanobacteria are preferable. In addition, in many countries, environmental regulations prohibit or limit the use of algaecides due to their adverse environmental impact. These requirements therefore need to be clarified prior to considering the use of algaecides.

The most common algaecide is copper sulfate and this has been used for well over 100 years in both Europe (Sawyer & Hazzard, 1962) and the USA (Moore & Kellerman, 1905). Copper sulfate is used widely as it is regarded as economical, effective, relatively safe and easy to apply. It is also considered to be of limited significance to human health at the doses commonly employed (WHO, 2011) and has been considered to not cause extensive environmental damage (Elder & Horne, 1978; McKnight *et al.*, 1983). This is however controversial as copper tends to accumulate in lake sediments (Hanson & Stefan, 1984; Qian *et al.*, 2010; Sanchez & Lee, 1978). In some cases, it appears not to be remobilised and is permanently bound to sediments (Elder & Horne, 1978; Sanchez & Lee, 1978). However, other studies show that it can be released back into the water column under low dissolved oxygen conditions in the hypolimnion (Prepas & Murphy, 1988). The effect of copper as an algaecide can generally be short-lived due to rapid loss via precipitation as insoluble salts and hydroxylates depending upon the water chemistry of the receiving water (Cooke *et al.*, 1993; Fan *et al.*, 2013; Qian *et al.*, 2010; Zhou *et al.*, 2013). Copper and other heavy metals differ from some other toxic contaminants in that they are not biodegradable, and in the environment their potential toxicity is controlled largely by their speciation or physicochemical form (Florence, 1982; Mastin & Rodgers Jr, 2000). Copper sulfate treatment has also been shown to cause short-term changes in both phytoplankton abundance and species succession (Effler *et al.*, 1980; McKnight, 1981). In addition, the on-going application of copper algaecides may encourage copper-resistance to develop and cause shifts in the community composition to copper-resistant green algae (Qian *et al.*, 2010; Rouco *et al.*, 2014). Fish kills may also occur following copper sulfate treatment, although it is not clear whether this is consequence of direct copper toxicity or oxygen depletion caused by the bloom decay (Hanson & Stefan, 1984).

Recently, some novel and selective algaecides, including chelated copper, oxidants (e.g. stabilized hydrogen peroxide) and organic agents, have been developed to minimize the adverse effects of copper sulfate described above (Chang *et al.*, 2018; Fan *et al.*, 2013; Huo *et al.*, 2015; Masuda & Boyd 1993; Straus, 1993; Tsai, 2016; Yang *et al.*, 2012). Among these compounds, hydrogen peroxide-based oxidants appear to be a promising option, as it can kill cyanobacteria and also simultaneously decompose their metabolites with no long-term residual in the aquatic environment.

Hydrogen peroxide has been recently applied in the field, including in lakes and streams in the Netherlands (Drábková *et al.*, 2007); in a wastewater pond (Barrington & Ghadouani, 2008); in China (Pflaumer, 2016; Wang *et al.*, 2017; Zhou *et al.*, 2018); and in Argentina (Bauzá *et al.*, 2014), demonstrating a growing interest in the technique. However, the parameters that determine the success of application, such as the dose rate, incident irradiance intensity (which facilitates radical formation), mixing of hydrogen peroxide, species sensitivity of cyanobacteria, and the effects upon their metabolites, need to be further studied (Huo *et al.*, 2015). Huo *et al.* (2015) applied a sequential reaction model to describe the kinetics of cell integrity of *M. aeruginosa* and the degradation of the cyanotoxins released from cyanobacteria exposed to UV and hydrogen peroxide. The models may have the potential to be used to characterize use in systems with other cyanobacteria with different metabolites. In addition, there is also potential to combine this for example with a reservoir hydrodynamic model to simulate spatial effects for field applications.

In terms of the mechanism of effect of this oxidant, radicals that are generated from natural organic matter (NOM) or nitrite and nitrate photolysis by incident irradiance in the environment (Mopper & Zhou, 1990) interact with these oxidants which may catalyse and further enhance their oxidative potential. Oxidants impair cell integrity and cause DNA strand breaks, leading to cell death and loss (Fan *et al.*, 2013; He & Häder, 2002). It is important to note that NOM is not only a source of radicals but is also a radical scavenger, and can significantly reduce the reaction rate and increase the oxidant dosage required (Westerhoff *et al.*, 1999). Therefore, the actual effect of oxidants in natural situations is highly dependent upon environmental factors, especially the concentration and the properties of NOM, which supply an 'oxidant demand'. Similarly, some caution should also be applied when interpreting the likely effects of other novel algaecides, which may have only been evaluated based upon their chemistry or upon limited laboratory experiments. The ecological impacts and the environmental safety of new chemicals should always be carefully evaluated before any field application.

Determining the best time to use algaecides is also important. It is preferable to apply algaecides at the early stages of bloom development before cyanobacteria become well established (Lüderitz *et al.*, 1989). Lower population cell densities will allow the algaecide to be more effective and may also reduce the dosage required and mitigate potential adverse effects of chemicals upon the ecosystem and human health (Liu *et al.*, 2013; Stratton & Giles, 1990). Conversely, when dosing a well established bloom, the death and decay of high cell loads may deplete dissolved oxygen (Paerl & Tucker, 1995) and release high concentrations of intracellular toxins and odourants into the dissolved state, which may worsen T&O problems and cause environmental effects. Therefore, early stage treatment is advisable, however this requires an active and careful monitoring program as part of an algaecide management plan.

7.3.2.2 Phytoremediation and biomanipulation

Biological methods which use approaches to avoid the addition of chemicals to the aquatic environment are often seen as attractive when compared with algaecide treatment. The most common biological methods promoted for the control of phytoplankton are phytoremediation (bioremediation with plants) and biomanipulation (food-web manipulation). Phytoremediation utilizes aquatic macrophytes, including submerged, emergent and floating plants, to assimilate and remove nutrients and contaminants from water (Li *et al.*, 2009; Srivastava *et al.*, 2008; Zhao *et al.*, 2012). These plants compete with cyanobacteria and algae for nutrients and reduce turbulence by slowing and dampening water currents, which encourages particle sedimentation and assists in preventing sediment re-suspension (Srivastava *et al.*, 2008). Plants may also secrete allelochemicals, which can inhibit the growth of cyanobacteria and algae (Comín *et al.*, 1990; Gross, 2003). Macrophytes also provide refuges for zooplankton, which are predators of cyanobacteria and algae, allowing them to avoid fish predation, which can then be beneficial for reducing phytoplankton biomass (Schriver *et al.*, 1995). Case studies have demonstrated positive examples where phytoremediation has been effective for the restoration of eutrophic lakes and reservoirs (Gulati & van Donk, 2002; Qiu *et al.*, 2001). The process can provide positive outcomes by significantly reducing the concentrations of COD, suspended solids, nitrogen, phosphorus, chlorophyll and algal densities and can lead to an increase in water transparency.

Biomanipulation often refers to the process which aims to artificially increase the density of the zooplankton predators of cyanobacteria and algae in aquatic ecosystems. This aims to facilitate the reduction in biomass of nuisance phytoplankton and thereby assist lakes and reservoirs change from high turbidity status to clear water with low turbidity and low algal density. Traditional biomanipulation, is based upon the 'top-down' and 'bottom-up' theory (McQueen *et al.*, 1986), which aims to reduce the density of planktivores (e.g. roach and bream) in lakes and reservoirs by fishing and/or stocking piscivorous fish (e.g. pike and perch), which then allows for an increase in the density of zooplankton (e.g. *Daphnia* spp.) and indirectly affects the phytoplankton community (Lammens, 2001). Another approach to biomanipulation is to stock planktivores (e.g. silver carp) in lakes and reservoirs, which directly graze upon phytoplankton (Ma *et al.*, 2012).

Both of these methods have been applied to control eutrophication, but the results appear inconsistent (some succeed while the others fail) (Hansson *et al.*, 1998; Jackson *et al.*, 2010; Leventer & Teltsch, 1990; Søndergaard *et al.*, 2000b; Spataru & Gophen, 1985; Starling, 1993; Starling & Rocha, 1990). Failure of biomanipulation can have several explanations. The first is that continued high nutrient loading (including both the external and the internal loading) enhances the growth of cyanobacteria which effectively neutralizes the population reduction by biomanipulation (Mehner *et al.*, 2002; Peretyatko *et al.*, 2012; Vašek *et al.*, 2013). Another explanation is ascribed to having insufficient intensity of the biomanipulation intervention. It is reported that the reduction rate required from planktivores should be more than 75% over a short period of time (less than one year) to be effective (Hansson *et al.*, 1998; Søndergaard *et al.*, 2000b). In addition to these two theories, the structure of the phytoplankton community will also affect the performance of biomanipulation, as both zooplankton and planktivorous fish can only feed on particular components of the cyanobacterial community. It is reported that *Daphnia* spp. cannot successfully feed on phytoplankton species larger than 50 μm (McCauley *et al.*, 1988), while silver carp cannot take up algae smaller than 10 μm (Vörös *et al.*, 1997). A case of selective predation in this case can reduce the biomass of a particular size class of edible cyanobacteria in a short period, but the total biomass of cyanobacteria may recover over time with the rapid growth and subsequent dominance of less edible cyanobacteria (McQueen, 1998), i.e. increasing zooplankton biomass may encourage dominance of large colonial phytoplankton species (e.g. *Microcystis* sp.) while increasing planktivore density may lead to the dominance of small unicellular phytoplankton species (e.g. *Chlorella* sp.).

In addition, a further confounding factor is that planktivorous fish can also consume zooplankton as well as phytoplankton, which leads to a competitive loss of zooplankton pressure. Separating planktivores and zooplankton to avoid the predation of zooplankton has been proposed as an approach to address this problem. Ma *et al.* (2009) conducted a pilot-scale experiment to investigate the performance of zooplankton coupling with silver carp on the control of phytoplankton. The results showed that 86.8% of the total phytoplankton and 59.4% of chlorophyll a was removed by this system. However, the performance of this approach has not been validated in a field-scale application.

In practice, phytoremediation and biomanipulation can be operated simultaneously to achieve a potentially more successful outcome. Phytoplankton, macrophytes, zooplankton, planktivores and piscivores will coexist with complex interdependent interactions in aquatic ecosystems (see Figure 7.2). Biomanipulation aims to reduce the biomass of phytoplankton in a short period and increase water transparency, which promotes macrophyte growth, thereby providing competition with phytoplankton for nutrients. The growth of macrophytes, in addition to reducing available nutrients, provides refuges to protect zooplankton from predation by fish. In practice a reduction in phytoplankton density by grazing is a desirable first step, however a corresponding recovery of macrophytes can determine whether the treatment can be successfully sustained (Hansson *et al.*, 1998). The complementary synergistic benefits of encouraging grazing and macrophyte recovery is considered to be a reason that biomanipulation can be more effective in shallow lakes and reservoirs than that in deeper stratified ones (McQueen, 1998; Vašek *et al.*, 2013). It is reported that phytoplankton are more competitive for both light and nutrients than macrophytes (Sand-Jensen & Borum, 1991). Thus, biomanipulation to significantly reduce cyanobacterial density can enhance the establishment and survival of macrophytes in shallower eutrophic lakes and reservoirs.

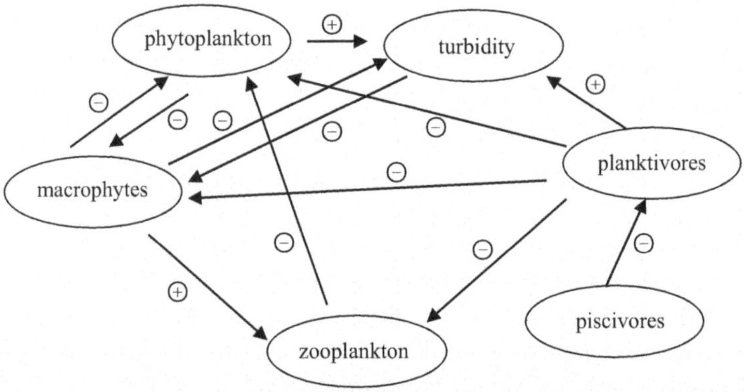

Figure 7.2 Interactions between phytoplankton, macrophytes, zooplankton, planktivores and piscivores and their effects on turbidity. ⊕ positive effects; ⊖ negative effects (modified from Lammens, 2001).

Eutrophication is regarded as shifting the ecological balance to cause excessive growth of cyanobacteria and the loss of biodiversity. Phytoremediation and biomanipulation are interventions which aim to artificially prevent this environmental deterioration by biological methods and assist restoration of lakes and reservoirs from turbid water to clear water. These biological methods are not only an approach for environmental remediation but can also be regarded as working to achieve an ultimate goal of overall improvements in water quality.

7.3.2.3 Physical methods

A range of approaches for algal and phytoplankton control are included under the category of physical methods and include destratification and mixing, application of ultrasound, light-shading, manual harvesting and removal of biomass.

Destratification with aeration or mechanical mixers can enhance the vertical mixing of the water column and change the light climate experienced by the phytoplankton community, and can also reduce internal phosphorus loading by providing oxygen to the sediment water interface (Antenucci *et al.*, 2005). The most common principle involved in the application destratification is to reduce and prevent the rapid development of surface scums by preventing the vertical migration of cyanobacterial colonies by mixing an otherwise stably-stratified lake or reservoir. Mixing can also reduce the growth and dominance of cyanobacteria in the phytoplankton if it is strong enough to push buoyant colonies into deeper layers where light is limiting (Visser *et al.*, 1996). Mixing may also enhance the growth rate of other phytoplankton such as diatoms and non-motile green algae that grow better under mixed conditions. However, the results of field applications for successful algal and cyanobacterial control are inconsistent. The application of artificial destratification has been clearly successful in a number of cases (Becker *et al.*, 2006; Hawkins & Griffiths, 1993; Heo & Kim, 2004; Lewis, 2004; Reynolds *et al.*, 1983). There are examples where it reduces the biomass of phytoplankton and/or changes the community structure (Simmons, 1998; Steinberg, 1983), while in other cases, it plays a negligible role in the control phytoplankton (Costa-Pierce & Laws, 1985; Lawson & Anderson, 2007).

The most common and effective destratification devices are bubble-plume aerators. They work by pumping air to a diffuser line placed at depth in the reservoir. This generates upward flow and generates a plume which induces circulation currents to disrupt density gradients and generate basin-scale circulation (Schladow, 1993; Whittington *et al.*, 2000). A number of other techniques have also been proposed to directly disrupt the thermal stratification of a water body. Surface-mounted mechanical mixers have also been evaluated in limited trials, however these have not yet proved to be more successful than bubble-plume aerators (Lewis, 2004). In many cases solar powered up-flow water circulators have been installed to control cyanobacteria. Upadhyay *et al.* (2013) show that these mixers tend to circulate only the surface or epilimnetic water and stratification is still maintained. These mixers therefore have a limited zone of influence and are unlikely to adequately mix entire lakes to effectively control cyanobacteria.

In general, for mixing to be successful in suppressing cyanobacteria, a large part of the water body must be sufficiently deep. Waters with extensive shallow areas are difficult to mix with aerators as entrainment of water into the bubble plume is not efficient. In addition, even a thermally stratified water body needs to be sufficiently deep so that mixing can reduce the light-controlled growth sufficiently to prevent cyanobacterial populations from developing. It is important to note this cannot be readily generalised, as the mixing depth that is needed to reduce cyanobacterial growth also depends upon the depth of light penetration.

In many cases some stratification will still occur outside of the immediate area of influence of the aerator (Sherman *et al.*, 2000) and while this stratification may be weaker than without mixing, cyanobacterial cells near the surface may not be entrained (Visser *et al.*, 1996) and exploit this to continue to grow successfully (Sherman *et al.*, 2000).

Many mixing and circulating systems are available commercially and good engineering expertise is important to design systems that can meet the mixing requirements. Aerators that are poorly designed may transport nutrients to the epilimnion without reducing stratification sufficiently to meet targets, and this will favour the growth of cyanobacteria (Tsujimura, 2004).

The use of ultrasound for the control of nuisance cyanobacteria has been seen as attractive as a non-chemical technique. Its use has become widespread over the last 20 years due to the availability of relatively low-cost commercial ultrasound systems for cyanobacterial control. Research to support the effectiveness and mechanism of action of ultrasound in aquatic systems has been building in the literature, however there are still not clear and objective scientific assessments and comparing studies is not straightforward. Most studies have used experimental equipment with a range of frequencies and power intensity which may be different to the commercial units. In addition, the studies and reviews of the mechanisms and effectiveness of ultrasound for control of cyanobacterial provide a wide range of test conditions that have been applied in both laboratory and field studies and with both cultures or natural material (Lürling *et al.*, 2016; Lürling & Tolman, 2014; Park *et al.*, 2017; Rajasekhar *et al.*, 2012).

Ultrasound has been reported to reduce cyanobacterial growth through a range of mechanisms that affect the structural integrity of cells (membrane damage or gas vesicle collapse and destruction) or impact cell physiology (photosynthetic activity and growth reduction (Kotopoulis *et al.*, 2009; Mason *et al.*, 2003; Park *et al.*, 2017; Rajasekhar *et al.*, 2012; Wu *et al.*, 2011). The key operational parameters that are important in the application of ultrasound for algal control are frequency, intensity (or power) and duration of exposure (Park *et al.*, 2017).

The studies that have shown clear effects upon cyanobacteria have used high-power experimental devices rather than commercial units sold for algal control, which are relatively low power (Lürling *et al.*, 2016). The low-energy ultrasound devices appear to be similar in design and recommended means of application and they use a range of different ultrasound frequencies in the lower range that are not always disclosed. The conflicting reports from the literature on the effectiveness of ultrasound in natural water bodies with cyanobacteria is most likely due both to the design and configuration of the sonication equipment when deployed in different size water bodies. In addition, there is considerable difficulty in designing studies to account for the varying ecological effects that may influence growth.

Some examples from the published studies highlight the conflicting results: Field trials with ultrasound devices conducted in the Netherlands in 2007 concluded ultrasound was not effective in reducing cyanobacteria (Kardinaal *et al.*, 2008). Similarly, in the UK, Purcell *et al.* (Purcell *et al.*, 2013) tested two models of commercial ultrasound devices with inconsistent results. By contrast, an evaluation of commercial ultrasound devices in reservoirs in New Jersey (USA) found positive results (Schneider *et al.*, 2015). A further comprehensive study in the Netherlands that was based both upon experiments and a literature review, concluded that the commercially available ultrasound transducers tested would not control cyanobacteria in natural water bodies (Lurling & Tolman, 2014).

To sum up, the use of ultrasound for control of cyanobacterial remains inconclusive because of both the limited number of carefully validated field trials and the lack of information on the practicality of applying both commercial and experimental devices in larger water bodies.

7.3.3 Control of benthic cyanobacteria

In comparison with the control of planktonic cyanobacteria, there are few reports on the management of benthic cyanobacteria, even though it is increasingly recognized that benthic cyanobacteria are responsible for significant odour events (Baker *et al.*, 2001; Sabater *et al.*, 2003). Based upon anecdotal industry experience, it is known that benthic cyanobacteria are more difficult to control than planktonic cyanobacteria. In ecological terms benthic growing types may be less sensitive to nutrient variation than phytoplankton (Bonilla *et al.*, 2005), which implies that nutrient management is likely to be less effective to limit their growth.

Chemical treatment, such as dosing with copper sulfate, may be effective, but the potential environmental impact should be carefully considered. Given that benthic cyanobacteria are often very localized and/or distributed around the margins of lakes and reservoirs, it is important to undertake a detailed survey to target the application of algaecides and achieve the effective minimum treatment doses (Hobson *et al.*, 2010).

Physical removal by disturbance treatments, such as towing drag lines across the sediment surface, is another method that has been used for the control of nuisance benthos. This can scour the surface of sediments and dislodge the benthic communities. However, this technique will also most likely re-suspend material and may release hazardous or odourous compounds previously adsorbed to the sediments. These potential negative effects may be important when the target area is nearby the intake of a water treatment plant.

Reducing the water level to dry out or desiccate benthic cyanobacteria have also been used for control. Desiccation both reduces the growth and will kill a proportion of the benthic cyanobacterial community (Coops & Hosper, 2002). It is important to note that many cyanobacteria are very resistant to desiccation, and the period required to achieve a high loss rate and reduce recovery can be substantial (Hawes *et al.*, 1992; Hobson *et al.*, 2010). Furthermore, it is must be recognized that lowering the water level for a long period may allow cyanobacteria to colonize a new band of sediments which was previously too deep for their growth at the normal operating level. The wider potential water quality effects and ecological impacts of water level management have not been well studied to date, and further work is required.

7.4 CONCLUSIONS

In this chapter, we discussed the production and the elimination of odourants in source water as well as some potential control methods for T&O compounds. The key points are summarised below.

There is a dynamic balance of production and loss of odourants in source water. Odourants are expected to be simultaneously produced and eliminated at quite high concentrations, and the resulting net concentration of T&O maintained in lakes and reservoirs is generally low and is the result of this balance.

Cyanobacteria are the primary odour-producing organisms in source waters, while actinomycetes may also play a role, but the actual contribution of these microorganisms has not been confirmed.

Biodegradation is the primary mechanism of odourant elimination in source water. Volatilization is likely to be equally as important in shallow water, but is negligible in deep water.

Nutrient management is essential for the remediation of eutrophic lakes and reservoirs and includes watershed management and in-reservoir (or in-lake) management. Watershed management should aim to limit external nutrients inputs including point sources and non-point sources to source water and prevent soil erosion. In-reservoir management of nutrients usually focuses on the control of phosphorus with nutrient inactivation, sediment capping and dredging being the most common control methods.

Planktonic cyanobacteria can be controlled by chemical, physical and biological methods. There is no clear recommended method and all techniques have advantages and disadvantages. Algaecides are effective in the short-term, but require local regulatory approval and have adverse environmental impacts. Techniques such destratification and mixing are site-specific and best for relatively deep reservoirs and require careful engineering design. While lake rehabilitation with macrophytes and biomanipulation are attractive they require significant local ecological knowledge, skill and monitoring to be effectively implemented.

To date, there is no satisfactory method to either monitor, evaluate or control benthic cyanobacteria, which is important as their contribution to T&O is becoming increasingly recognized as significant in some cases. This is a priority area for further study in this field.

REFERENCES

Anderson D. M., Glibert P. M. and Burkholder J. M. (2002). Harmful algal blooms and eutrophication: nutrient sources, composition, and consequences. *Estuaries*, **25**, 704–726.

Antenucci J. P., Ghadouani A., Burford M. A. and Romero J. R. (2005). The long-term effect of artificial destratification on phytoplankton species composition in a subtropical reservoir. *Freshwater Biology*, **50**, 1081–1093.

Baker P. D., Steffensen D. A., Humpage A. R., Nicholson B. C., Falconer I. R., Lanthois B., Fergusson K. M. and Saint C. P. (2001). Preliminary evidence of toxicity associated with the benthic cyanobacterium *Phormidium* in South Australia. *Environmental Toxicology*, **16**, 506–511.

Barrington D. J. and Ghadouani A. (2008). Application of hydrogen peroxide for the removal of toxic cyanobacteria and other phytoplankton from wastewater. *Environmental Science & Technology*, **42**, 8916–8921.

Bauzá L., Aguilera A., Echenique R., Andrinolo D. and Giannuzzi L. (2014). Application of Hydrogen Peroxide to the control of eutrophic lake systems in laboratory assays. *Toxins*, **6**, 2657–2675.

Becker A., Herschel A. and Wilhelm C. (2006). Biological effects of incomplete destratification of hypertrophic freshwater reservoir. *Hydrobiologia*, **559**, 85–100.

Bonilla S., Villeneuve V. and Vincent W. F. (2005). Benthic and planktonic algal communities in a high arctic lake: Pigment structure and contrasting responses to nutrient enrichment. *Journal of Phycology*, **41**, 1120–1130.

Bruchet A. and Lan J. (2005). Efficiency of membrane processes for taste and odor removal. *Water Science & Technology*, **51**(6–7), 257–265.

Carpenter S. R. (2005). Eutrophication of aquatic ecosystems: bistability and soil phosphorus. *Proceedings of the National Academy of Sciences of the United States of America*, **102**, 10002–10005.

Carpenter S. R. (2008). Phosphorus control is critical to mitigating eutrophication. *Proceedings of the National Academy of Sciences of the United States of America*, **105**, 11039–11040.

Chang C. W., Huo X. C. and Lin T. F. (2018). Exposure of *Microcystis aeruginosa* to hydrogen peroxide and titanium dioxide under visible light conditions: modeling the impact of hydrogen peroxide and dydroxyl radical on cell rupture and microcystin degradation. *Water Research*, **141**, 217–226.

Chen Y.-M., Hobson P., Burch M. D. and Lin T.-F. (2010). In situ measurement of odor compound production by benthic cyanobacteria. *Journal of Environmental Monitoring*, **12**, 769–775.

Chiu Y. T., Yen H. K., Lin H. L. and Lin T.-F. (2016). An alternative method to quantify 2-MIB producing cyanobacteria in drinking water reservoirs: method development and field applications. *Environmental Research*, **151**, 618–627.

Choi Y., Oh H., Lee S., Choi Y., Hwang T.-M., Jeon J.-C. and Choung Y.-K. (2010). Removal of taste and odor model compounds (2-MIB and geosmin) with the NF membrane. *Desalination and Water Treatment*, **15**, 141–148.

Comín F. A., Menéndez M. and Lucena J. R. (1990). Proposals for macrophyte restoration in eutrophic coastal lagoons. In: Biomanipulation Tool for Water Management, R. D. Gulati, E. H. R. R. Lammens, M. L. Meijer and E. van Donk (eds), Springer, Dordrecht, the Netherlands, Vol. **61**.

Conley D. J., Paerl H. W., Howarth R. W., Boesch D. F., Seitzinger S. P., Havens K. E., Lancelot C. and Likens G. E. (2009). Controlling eutrophication: nitrogen and phosphorus. *Science*, **323**(5917), 1014–1015.

Cook D., Newcombe G. and Sztajnbok P. (2001). The application of powdered activated carbon for MIB and geosmin removal: predicting PAC doses in four raw waters. *Water Research*, **35**, 1325–1333.

Cooke G. D., Welch E. B., Peterson S. A., Newroth P. R. *et al.* (1993). Artificial circulation. In: Restoration and Management of Lakes and Reservoirs, G. D. Cooke, E. B. Welch and S. A. Peterson (eds), Lewis Publishers, Boca Raton, USA, pp. 419–449.

Coops H. and Hosper S. H. (2002). Water-level management as a tool for the restoration of shallow lakes in the Netherlands. *Lake and Reservoir Management*, **18**, 293–298.

Costa-Pierce B. A. and Laws E. A. (1985). Effects of destratification on autotrophic and heterotrophic microplankton productivity in eutrophic aquaculture ponds. *Aquaculture*, **50**, 141–151.

Dang H., Ren J., Song L., Sun S. and An L. (2008). Diverse tetracycline resistant bacteria and resistance genes from coastal waters of Jiaozhou Bay. *Microbial Ecology*, **55**, 237–246.

de Grave S. and Whitaker A. (1999). Benthic community re-adjustment following dredging of a muddy-maerl matrix. *Marine Pollution Bulletin*, **38**, 102–108.

de Menezes A. B., McDonald J. E., Allison H. E. and McCarthy A. J. (2012). Importance of *M. icromonospora* spp. as colonizers of cellulose in freshwater lakes as demonstrated by Quantitative Reverse Transcriptase PCR of 16S rRNA. *Applied and Environmental Microbiology*, **78**, 3495–3499.

Deppe T. and Benndorf J. (2002). Phosphorus reduction in a shallow hypereutrophic reservoir by in-lake dosage of ferrous iron. *Water Research*, **36**, 4525–4534.

Dionigi C. P. and Ingram D. A. (1994). Effects of temperature and oxygen concentration on geosmin production by *Streptomyces tendae* and *Penicillium expansum*. *Journal of Agricultural and Food Chemistry*, **42**, 143–145.

Drábková M., Admiraal W. and Maršálek B. (2007). Combined exposure to hydrogen peroxide and light – selective effects on cyanobacteria, green algae and diatoms. *Environmental Science and Technology*, **41**, 309–314.

Dzialowski A. R., Smith V. H., Huggins D. G., de Noyelles F., Lim N.-C., Baker D. S. and Beury J. H. (2009). Development of predictive models for geosmin-related taste and odor in Kansas, USA, drinking water reservoirs. *Water Research*, **43**, 2829–2840.

Eaton R. W. and Sandusky P. (2009). Biotransformations of 2-methylisoborneol by camphor-degrading bacteria. *Applied and Environmental Microbiology*, **75**, 583–588.

Effler S., Litten S., Field S., Tong-Ngork T., Hale F., Meyer M. and Quirk M. (1980). Whole lake responses to low level copper sulfate treatment. *Water Research*, **14**, 1489–1499.

Elder J. F. and Horne A. J. (1978). Copper cycles and CuSO4 algicidal capacity in two California lakes. *Environmental Management*, **2**, 17–30.

Fan J., Ho L., Hobson P. and Brookes J. (2013). Evaluating the effectiveness of copper sulfate, chlorine, potassium permanganate, hydrogen peroxide and ozone on cyanobacterial cell integrity. *Water Research*, **47**, 5153–5164.

Florence T. (1982). The speciation of trace elements in waters. *Talanta*, **29**, 345–364.

Gross E. M. (2003). Allelopathy of aquatic autotrophs. *Critical Reviews in Plant Sciences*, **22**, 313–339.

Grüntzig V., Nold S. C., Zhou J. and Tiedje J. M. (2001). Pseudomonas stutzeri nitrite reductase gene abundance in environmental samples measured by real-time PCR. *Applied and Environmental Microbiology*, **67**, 760–768.

Gulati R. D. and van Donk E. (2002). Ecological Restoration of Aquatic and Semi-Aquatic Ecosystems in the Netherlands (NW Europe). Springer, Dordrecht, the Netherlands, pp. 73–106.

Haghseresht F., Wang S. and Do D. (2009). A novel lanthanum-modified bentonite, Phoslock, for phosphate removal from wastewaters. *Applied Clay Science*, **46**, 369–375.

Hanson M. J. and Stefan H. G. (1984). Side effects of 58 years of copper sulfate treatment of the Fairmont Lakes, Minnesota. *Water Resources Bulletin*, **20**, 889–900.

Hansson L.-A., Annadotter H., Bergman E., Hamrin S. F., Jeppesen E., Kairesalo T., Luokkanen E., Nilsson P.-Å., Søndergaard M. and Strand J. (1998). Biomanipulation as an application of food-chain theory: constraints, synthesis, and recommendations for temperate lakes. *Ecosystems*, **1**, 558–574.

Hawes I., Howard-Williams C. and Vincent W. F. (1992). Desiccation and recovery of Antarctic cyanobacterial mats. *Polar Biology*, **12**, 587–594.

Hawkins P. and Griffiths D. (1993). Artificial destratification of a small tropical reservoir: effects upon the phytoplankton. *Hydrobiologia*, **254**, 169–181.

He Y.-Y. and Häder D.-P. (2002). UV-B-induced formation of reactive oxygen species and oxidative damage of the cyanobacterium *Anabaena* sp.: protective effects of ascorbic acid and *N*-acetyl-l-cysteine. *Journal of Photochemistry and Photobiology B: Biology*, **66**, 115–124.

Heo W. M. and Kim B. (2004). The effect of artificial destratification on phytoplankton in a reservoir. *Hydrobiologia*, **524**, 229–239.

Hickey C. W. and Gibbs M. M. (2009). Lake sediment phosphorus release management – decision support and risk assessment framework. *New Zealand Journal of Marine and Freshwater Research*, **43**, 819–856.

Ho L., Hoefel D., Bock F., Saint C. P. and Newcombe G. (2007). Biodegradation rates of 2-methylisoborneol (MIB) and geosmin through sand filters and in bioreactors. *Chemosphere*, **66**, 2210–2218.

Hobson P., Fazekas C., House J., Daly R., Kildea T., Giglio S., Burch M., Lin T.-F. and Chen Y.-M. (2010). Tastes and Odours in Reservoirs. Water Quality Research Australia, Adelaide, Australia.

Hoefel D., Ho L., Aunkofer W., Monis P., Keegan A., Newcombe G. and Saint C. (2006). Cooperative biodegradation of geosmin by a consortium comprising three gram-negative bacteria isolated from the biofilm of a sand filter column. *Letters in Applied Microbiology*, **43**, 417–423.

Hoefel D., Ho L., Monis P. T., Newcombe G. and Saint C. P. (2009). Biodegradation of geosmin by a novel Gram-negative bacterium; isolation, phylogenetic characterisation and degradation rate determination. *Water Research*, **43**, 2927–2935.

Holmfeldt K., Dziallas C., Titelman J., Pohlmann K., Grossart H. P. and Riemann L. (2009). Diversity and abundance of freshwater Actinobacteria along environmental gradients in the brackish northern Baltic Sea. *Environmental Microbiology*, **11**, 2042–2054.

Huo C. H., Chang D. W., Tseng J. H., Burch M. D. and Lin T. F. (2015). Exposure of *Microcystis aeruginosa* to hydrogen peroxide under light: kinetic modeling of cell rupture and simultaneous microcystin degradation. *Environmental Science and Technology*, **49**, 5502–5510.

Izaguirre G. and Taylor W. D. (1998). A *Pseudanabaena species* from Castaic Lake, California, that produces 2-methylisoborneol. *Water Research*, **32**, 1673–1677.

Izaguirre G., Wolfe R. L. and Means E. G. (1988). Degradation of 2-methylisoborneol by aquatic bacteria. *Applied and Environmental Microbiology*, **54**, 2424–2431.

Jackson Z. J., Quist M. C., Downing J. A. and Larscheid J. G. (2010). Common carp (*Cyprinus carpio*), sport fishes, and water quality: ecological thresholds in agriculturally eutrophic lakes. *Lake and Reservoir Management*, **26**, 14–22.

Janakiraman A. and Leff L. G. (1999). Comparison of survival of different species of bacteria in freshwater microcosms. *Journal of Freshwater Ecology*, **14**, 233–240.

Jensen S. E., Anders C. L., Goatcher L. J., Perley T., Kenefick S. and Hrudey S. E. (1994). Actinomycetes as a factor in odour problems affecting drinking water from the North Saskatchewan River. *Water Research*, **28**, 1393–1401.

Jin X., Lu S., Hu X., Jiang X. and Wu F. (2008). Control concept and countermeasures for shallow lakes' eutrophication in China. *Frontiers of Environmental Science & Engineering in China*, **2**, 257–266.

Jo C. H., Dietrich A. M. and Tanko J. M. (2011). Simultaneous degradation of disinfection byproducts and earthy-musty odorants by the UV/H_2O_2 advanced oxidation process. *Water Research*, **45**, 2507–2516.

Jones G. J. and Korth W. (1995). In situ production of volatile odour compounds by river and reservoir phytoplankton populations in Australia. *Water Science & Technology*, **31**(11), 145–151.

Jüttner F. and Watson S. B. (2007). Biochemical and ecological control of geosmin and 2-methylisoborneol in source waters. *Applied and Environmental Microbiology*, **73**, 4395–4406.

Jüttner F., Watson S. B., von Elert E. and Köster O. (2010). β-Cyclocitral, a grazer defence signal unique to the cyanobacterium *Microcystis*. *Journal of Chemical Ecology*, **36**, 1387–1397.

Kardinaal E., De Haan M. and Ruiter H. (2008). Maatregelen ter voorkoming blauwalgen werken onvoldoende. *H_2O – Magazine for Water Supply and Wastewater Treatment*, **41**, 4–7 (In Dutch).

Karickhoff S. W., Brown D. S. and Scott T. A. (1979). Sorption of hydrophobic pollutants on natural sediments. *Water Research*, **13**, 241–248.

Kennedy R. H. and Cook G. D. (1982). Control of lake phosphorus with aluminum sulfate: dose determination and application techniques. *Journal of the American Water Resources Association*, **18**, 389–395.

Kile D. E., Chiou C. T., Zhou H., Li H. and Xu O. (1995). Partition of nonpolar organic pollutants from water to soil and sediment organic matters. *Environmental Science & Technology*, **29**(5), 1401–1406.

Klausen C., Nicolaisen M. H., Strobel B. W., Warnecke F., Nielsen J. L. and Jørgense N. O. (2005). Abundance of actinobacteria and production of geosmin and 2-methylisoborneol in Danish streams and fish ponds. *FEMS Microbiology Ecology*, **52**, 265–278.

Kleeberg A. and Kohl J.-G. (1999). Assessment of the long-term effectiveness of sediment dredging to reduce benthic phosphorus release in shallow Lake Müggelsee (Germany). *Hydrobiologia*, **394**, 153–161.

Kotopoulis S., Schommartz A. and Postema M. (2009). Sonic cracking of blue-green algae. *Applied Acoustics*, **70**, 1306–1312.

Krasner S., Hwang C. and McGuire M. (1983). A standard method for quantification of earthy-musty odorants in water, sediments, and algal cultures. *Water Science & Technology*, **15**(6–7), 127–138.

Kutschera K., Börnick H. and Worch E. (2009). Photoinitiated oxidation of geosmin and 2-methylisoborneol by irradiation with 254 nm and 185 nm UV light. *Water Research*, **43**, 2224–2232.

Lammens E. H. (2001). Consequences of Biomanipulation for Fish and Fisheries. FAO Fisheries Circular No. 952. Food and Agriculture Organization of the United Nations, Rome, Italy.

Lauderdale C. V., Aldrich H. C. and Lindner A. S. (2004). Isolation and characterization of a bacterium capable of removing taste-and odor-causing 2-methylisoborneol from water. *Water Research*, **38**, 4135–4142.

Lawson R. and Anderson M. A. (2007). Stratification and mixing in Lake Elsinore, California: an assessment of axial flow pumps for improving water quality in a shallow eutrophic lake. *Water Research*, **41**, 4457–4467.

Leventer H. and Teltsch B. (1990). The contribution of silver carp (*Hypophthalmichthys molitrix*) to the biological control of Netofa reservoirs. In: Trophic Relationships in Inland Waters, P. Biró and J. F. Talling (eds), Springer, Dordrecht, the Netherlands, Vol. **53**.

Lewis D. M. (2004). Surface Mixers for Destratification and Management of *Anabaena circinalis*. *Thesis*, University of Adelaide, Australia, 249 p.

Li M., Wu Y.-J., Yu Z.-L., Sheng G.-P. and Yu H.-Q. (2009). Enhanced nitrogen and phosphorus removal from eutrophic lake water by *Ipomoea aquatica* with low-energy ion implantation. *Water Research*, **43**, 1247–1256.

Liang C., Wang D., Yang M., Sun W. and Zhang S. (2005). Removal of earthy-musty odorants in drinking water by powdered activated carbon. *Journal of Environmental Science and Health, Part A*, **40**, 767–778.

Lin T.-F., Wong J.-Y. and Kao H.-P. (2002). Correlation of musty odor and 2-MIB in two drinking water treatment plants in South Taiwan. *Science of the Total Environment*, **289**, 225–235.

Lindström E. S. and Bergström A.-K. (2004). Influence of inlet bacteria on bacterioplankton assemblage composition in lakes of different hydraulic retention time. *Limnology and Oceanography*, **49**, 125–136.

Liu Z., Cui F., Ma H., Fan Z., Zhao Z., Hou Z., Liu D. and Jia X. (2013). The interaction between nitrobenzene and *Microcystis aeruginosa* and its potential to impact water quality. *Chemosphere*, **92**, 1201–1206.

Lloyd S. W., Lea J. M., Zimba P. V. and Grimm C. C. (1998). Rapid analysis of geosmin and 2-methylisoborneol in water using solid phase micro extraction procedures. *Water Research*, **32**, 2140–2146.

Lüderitz V., Nicklisch A. and Kohl J. G. (1989). Kupfer als algizid. *Acta Hydrochimica et Hydrobiologica*, **17**, 61–73, (In German).

Lurling M. and Tolman Y. (2014). Effects of commercially available ultrasound on the zooplankton grazer *Daphnia* and consequent water greening in laboratory experiments. *Water*, **6**, 3247–3263.

Lürling M. and Tolman Y. (2014). Beating the blues: Is there any music in fighting cyanobacteria with ultrasound? *Water Research*, **66**, 361–373.

Lürling M., Waajen G. and de Senerpont Domis L. N. (2016). Evaluation of several end-of-pipe measures proposed to control cyanobacteria. *Aquatic Ecology*, **50**, 499–519.

Ma H., Cui F., Liu Z. and Fan Z. (2009). Pilot study on control of phytoplankton by zooplankton coupling with filter-feeding fish in surface water. *Water Science Technology*, **60**(3), 737–743.

Ma H., Cui F., Fan Z., Liu Z. and Zhao Z. (2012). Efficient control of *Microcystis* blooms by promoting biological filter-feeding in raw water. *Ecological Engineering*, **47**, 71–75.

Martin J. F., Izaguirre G. and Waterstrat P. (1991). A planktonic *Oscillatoria* species from Mississippi catfish ponds that produces the off-flavor compound 2-methylisoborneol. *Water Research*, **25**, 1447–1451.

Mason T., Joyce E., Phull S. and Lorimer J. (2003). Potential uses of ultrasound in the biological decontamination of water. *Ultrasonics Sonochemistry*, **10**, 319–323.

Mastin B. and Rodgers Jr. J. (2000). Toxicity and bioavailability of copper herbicides (Clearigate, Cutrine-Plus, and copper sulfate) to freshwater animals. *Archives of Environmental Contamination and Toxicology*, **39**, 445–451.

Masuda K. and Boyd C. E. (1993). Comparative evaluation of the solubility and algal toxicity of copper sulfate and chelated copper. *Aquaculture*, **117**, 287–302.

McCauley E., Murdoch W. W. and Watson S. (1988). Simple models and variation in plankton densities among lakes. *American Naturalist*, **132**, 383–403.

McGuire M. J. (1986). Oxidation of five earthy musty taste and odor compounds. *Journal of the American Water Works Association*, **78**, 62–69.

McKnight D. (1981). Chemical and biological processes controlling the response of a freshwater ecosystem to copper stress: a field study of the $CuSO_4$ treatment of Mill Pond Reservoir, Burlington, Massachusetts. *Limnology and Oceanography*, **26**, 518–531.

McKnight D. M., Chisholm S. W. and Harleman D. R. (1983). $CuSO_4$ treatment of nuisance algal blooms in drinking water reservoirs. *Environmental Management*, **7**, 311–320.

McQueen D. (1998). Freshwater food web biomanipulation: a powerful tool for water quality improvement, but maintenance is required. *Lakes & Reservoirs: Research & Management*, **3**, 83–94.

McQueen D. J., Post J. R. and Mills E. L. (1986). Trophic relationships in freshwater pelagic ecosystems. *Canadian Journal of Fisheries and Aquatic Sciences*, **43**, 1571–1581.

Mehner T., Benndorf J., Kasprzak P. and Kosche R. (2002). Biomanipulation of lake ecosystems: successful applications and expanding complexity in the underlying science. *Freshwater Biology*, **47**(12), 2453–2465.

Meis S., Spears B. M., Maberly S. C., O'Malley M. B. and Perkins R. G. (2012). Sediment amendment with Phoslock in Clatto Reservoir (Dundee, UK): investigating changes in sediment elemental composition and phosphorus fractionation. *Journal of Environmental Management*, **93**, 185–193.

Moore G. T. and Kellerman K. F. (1905). Copper as an Algicide and Disinfectant in Water Supplies. US Government Printing Office, Washington, DC.

Mopper K. and Zhou X. (1990). Hydroxyl radical photoproduction in the sea and its potential impact on marine processes. *Science*, **250**, 661–664.

Morris D. P., Zagarese H., Williamson C. E., Balseiro E. G., Hargreaves B. R., Modenutti B., Moeller R. and Queimalinos C. (1995). The attenuation of solar UV radiation in lakes and the role of dissolved organic carbon. *Limnology and Oceanography*, **40**, 1381–1391.

Mun T. K. (2003). Effect of phototrophic bacteria on water blooming. *Water Resources*, **30**, 325–332.

Mur R., Skulberg O. M. and Utkilen H. (1999). Cyanobacteria in the environment. In: Toxic Cyanobacteria in Water: A Guide to their Public Health Consequences, Monitoring and Management, I. Chorus and J. Bartram (eds), Spon Press, London.

Naes H., Utkilen H. and Post A. (1989). Geosmin production in the cyanobacterium *Oscillatoria brevis*. *Archives of Microbiology*, **151**, 407–410.

Najm I., Snoeyink V., Lykins Jr. B. and Adams J. (1992). Using powdered activated carbon: a critical review. *Journal American Water Works Association*, **83**, 65–76.

Narayan L. V. and Nunez W. J. (1974). Biological control: isolation and bacterial oxidation of the taste-and-odor compound geosmin. *Journal American Water Works Associatoion*, **66**, 532–536.

Oikawa E., Shimizu A. and Ishibashi Y. (1995). 2-methylisoborneol degradation by the cam operon from Pseudomonas putida PpG1. *Water Science & Technology*, **31**(11), 79–86.

Otsuki A. and Wetzel R. G. (1972). Coprecipitation of phosphate with carbonates in a Marl lake. *Limnology and Oceanography*, **17**, 763–767.

Paerl H. W. (2006). Assessing and managing nutrient-enhanced eutrophication in estuarine and coastal waters: interactive effects of human and climatic perturbations. *Ecological Engineering*, **26**, 40–54.

Paerl H. W. and Tucker C. S. (1995). Ecology of Blue-Green algae in aquaculture ponds. *Journal of the World Aquaculture Society*, **26**, 109–131.

Park J., Church J., Son Y., Kim K.-T. and Lee W. H. (2017). Recent advances in ultrasonic treatment: challenges and field applications for controlling harmful algal blooms (HABs). *Ultrasonics Sonochemistry*, **38**, 326–334.

Paul V. J., Thacker R. W., Banks K. and Golubic S. (2005). Benthic cyanobacterial bloom impacts the reefs of South Florida (Broward County, USA). *Coral Reefs*, **24**, 693–697.

Peretyatko A., Teissier S., De Backer S. and Triest L. (2012). Biomanipulation of hypereutrophic ponds: when it works and why it fails. *Environmental Monititoring Assessment*, **184**, 1517–1531.

Peter A., Koster O., Schildknecht A. and von Gunten U. (2009). Occurrence of dissolved and particle-bound taste and odor compounds in Swiss lake waters. *Water Research*, **43**, 2191–2200.

Pflaumer A. L. (2016). Hydrogen Peroxide in Eutrophic Lake Taihu, China: Addition Effects on Phytoplankton and Diel Variability in Natural Concentrations. Masters thesis, University of North Carolina, Chapel Hill, NC, USA.

Pitkänen H., Lehtoranta J. and Räike A. (2001). Internal nutrient fluxes counteract decreases in external load: the case of the estuarial eastern Gulf of Finland, Baltic Sea. *AMBIO*, **30**, 195–201.

Prepas E. and Murphy T. (1988). Sediment–water interactions in farm dugouts previously treated with copper sulfate. *Lake and Reservoir Management*, **4**, 161–168.

Proulx F., Rodriguez M. J., Serodes J. B. and Bouchard C. (2012). Spatio-temporal variability of tastes and odors of drinking water within a distribution system. *Journal of Environmental Management*, **105**, 12–20.

Purcell D., Parsons S. A., Jefferson B., Holden S., Campbell A., Wallen A., Chipps M., Holden B. and Ellingham A. (2013). Experiences of algal bloom control using green solutions barley straw and ultrasound, an industry perspective. *Water and Environment Journal*, **27**, 148–156.

Qian H., Yu S., Sun Z., Xie X., Liu W. and Fu Z. (2010). Effects of copper sulfate, hydrogen peroxide and N-phenyl-2-naphthylamine on oxidative stress and the expression of genes involved photosynthesis and microcystin disposition in *Microcystis aeruginosa*. *Aquatic Toxicology*, **99**, 405–412.

Qiu D., Wu Z., Liu B., Deng J., Fu G. and He F. (2001). The restoration of aquatic macrophytes for improving water quality in a hypertrophic shallow lake in Hubei Province, China. *Ecological Engineering*, **18**, 147–156.

Rajasekhar P., Fan L., Nguyen T. and Roddick F. A. (2012). A review of the use of sonication to control cyanobacterial blooms. *Water Research*, **46**, 4319–4329.

Reynolds C. and Davies P. (2001). Sources and bioavailability of phosphorus fractions in freshwaters: a British perspective. *Biological Reviews*, **76**, 27–64.

Reynolds C., Wiseman S., Godfrey B. and Butterwick C. (1983). Some effects of artificial mixing on the dynamics of phytoplankton populations in large limnetic enclosures. *Journal of Plankton Research*, **5**, 203–234.

Rosen B., MacLeod B. and Simpson M. (1992). Accumulation and release of geosmin during the growth phases of *Anabaena circinalis* (Kutz.) Rabenhorst. *Water Science & Technology*, **25**(2), 185–190.

Ross G., Haghseresht F. and Cloete T. E. (2008). The effect of pH and anoxia on the performance of Phoslock®, a phosphorus binding clay. *Harmful Algae*, **7**, 545–550.

Rouco M., López-Rodas V., González R., Huertas I. E., García-Sánchez M. J., Flores-Moya A. and Costas E. (2014). The limit of the genetic adaptation to copper in freshwater phytoplankton. *Oecologia*, **175**, 1179–1788.

Rydin E., Huser B. and Welch E. B. (2000). Amount of phosphorus inactivated by alum treatments in Washington lakes. *Limnology and Oceanography*, **45**, 226–230.

Ryding S.-O. (1982). Trehörningen restoration project. Changes in water quality after sediment dredging. *Hydrobiologia*, **92**, 549–558.

Saadoun I. and El-Migdadi F. (1998). Degradation of geosmin-like compounds by selected species of Gram-positive bacteria. *Letters in Applied Microbiology*, **26**, 98–100.

Sabater S., Vilalta E., Gaudes A., Guasch H., Munoz I. and Romani A. (2003). Ecological implications of mass growth of benthic cyanobacteria in rivers. *Aquatic Microbial Ecology*, **32**, 175–184.

Safferman R., Rosen A. A., Mashni C. I. and Morris M. E. (1967). Earthy-smelling substance from a blue-green alga. *Environmental Science & Technology*, **1**, 429–430.

Sanchez I. and Lee G. F. (1978). Environmental chemistry of copper in Lake Monona, Wisconsin. *Water Research*, **12**, 899–903.

Sand-Jensen K. and Borum J. (1991). Interactions among phytoplankton, periphyton, and macrophytes in temperate freshwaters and estuaries. *Aquatic Botany*, **41**, 137–175.

Sawyer C. N. and Hazzard A. S. (1962). Causes, effects, and control of aquatic growths. *Journal of the Water Pollution Control Federation*, **34**, 279–290.

Schindler D. W., Hecky R., Findlay D., Stainton M., Parker B., Paterson M., Beaty K., Lyng M. and Kasian S. (2008). Eutrophication of lakes cannot be controlled by reducing nitrogen input: results of a 37-year whole-ecosystem experiment. *Proceedings of the National Academy of Sciences of the United States of America*, **105**, 11254–11258.

Schladow S. G. (1993). Lake destratification by bubble-plume systems: Design methodology. *Journal of Hydraulic Engineering*, **119**, 350–368.

Schneider O. D., Weinrich L. A. and Brezinski S. (2015). Ultrasonic treatment of algae in a New Jersey reservoir. *Journal of the American Water Works Association*, **107**, E533–E542.

Schriver P., Bøgestrand J., Jeppesen E. and Søndergaard M. (1995). Impact of submerged macrophytes on fish-zooplanl phytoplankton interactions: large-scale enclosure experiments in a shallow eutrophic lake. *Freshwater Biology*, **33**, 255–270.

Sherman B., Whittington J. and Oliver R. (2000). The impact of artificial destratification on water quality in Chaffey Reservoir. *Archiv fur Hydrobiologie Spec. Issues Advanc. Limnol.*, **55**, 15–29.

Simmons J. (1998). Algal control and destratification at Hanningfield Reservoir. *Water Science & Technology*, **37**(2), 309–316.

Singleton V. L. and Little J. C. (2006). Designing hypolimnetic aeration and oxygenation systems – a review. *Environmental Science & Technology*, **40**, 7512–7520.

Smolders A., Lamers L., Lucassen E., Van der Velde G. and Roelofs J. (2006). Internal eutrophication: how it works and what to do about it – a review. *Chemistry and Ecology*, **22**, 93–111.

Søndergaard M., Jeppesen E. and Jensen J. P. (2000a). Hypolimnetic nitrate treatment to reduce internal phosphorus loading in a stratified lake. *Lake and Reservoir Management*, **16**, 195–204.

Søndergaard M., Jeppesen E., Jensen J. P. and Lauridsen T. (2000b). Lake restoration in Denmark. *Lakes & Reservoirs: Research & Management*, **5**, 151–159.

Spataru P. and Gophen M. (1985). Feeding behaviour of silver carp *Hypophthalmichthys molitrix* Val. and its impact on the food web in Lake Kinneret, Israel. *Hydrobiologia*, **120**, 53–61.

Srivastava J., Gupta A. and Chandra H. (2008). Managing water quality with aquatic macrophytes. *Reviews in Environmental Science and Bio/Technology*, **7**, 255–266.

Starling F. L. and Rocha A. J. (1990). Experimental study of the impacts of planktivorous fishes on plankton community and eutrophication of a tropical Brazilian reservoir. *Hydrobiologia*, **200**, 581–591.

Starling F. L. R. M. (1993). Control of eutrophication by silver carp (Hypophthalmichthys molitrix) in the tropical Paranoa Reservoir (Brasilia, Brazil): a mesocosm experiment. *Hydrobiologia*, **257**, 143–152.

Steinberg C. (1983). Effects of artificial destratification on the phytoplankton populations in a small lake. *Journal of Plankton Research*, **5**, 855–864.

Stratton G. W. and Giles J. (1990). Importance of bioassay volume in toxicity tests using algae and aquatic invertebrates. *Bulletin of Environmental Contamination and Toxicology*, **44**, 420–427.

Straus D. L. (1993). Acute toxicity of copper sulfate and chelated copper to channel catfish *Ictalurus punctatus*. *Journal of the World Aquaculture Society*, **24**, 390–395.

Sun D. L., Yu J. W., An W., Yang M., Chen G. G. and Zhang S. J. (2013). Identification of causative compounds and microorganisms for musty odor occurrence in the Huangpu River, China. *Journal of Environmental Sciences*, **25**, 460–465.

Sunesson A., Vaes W., Nilsson C., Blomquist G., Andersson B. and Carlson R. (1995). Identification of volatile metabolites from five fungal species cultivated on two media. *Applied and Environmental Microbiology*, **61**, 2911–2918.

Tanaka A., Oritani T., Uehara F., Saito A., Kishita H., Niizeki Y., Yokota H. and Fuchigami K. (1996). Biodegradation of a musty odour component, 2-methylisoborneol. *Water Research*, **30**, 759–761.

Tsai K. P. (2016). Management of target algae by using copper-based algaecides: effects of algal cell density and sensitivity to copper. *Water Air and Soil Pollution*, **227**, 238.

Tsao H. W., Michinaka A., Yen H. K., Giglio S., Hobson P., Monis P. and Lin T.-F. (2014). Monitoring of geosmin producing *Anabaena circinalis* using quantitative PCR. *Water Research*, **49**, 416–425.

Tsujimura S. (2004). Water management of Lake Yogo targeting internal phosphorus loading. *Lakes and Reservoirs: Research and Management*, **9**, 171–179.

Tung S. C., Lin T. F., Tseng I. C. and Lin H. M. (2006). Identification of 2-MIB and geosmin producers in Feng-Shen reservoir in south Taiwan. *Water Science and Technology: Water Supply*, **6**(2), 55–61.

Tung S. C., Lin T. F., Yang F. C. and Liu C. L. (2008). Seasonal change and correlation with environmental parameters for 2-MIB in Feng-Shen Reservoir, Taiwan. *Environmental Monitoring and Assessment*, **145**, 407–416.

Upadhyay S., Bierlein K. A., Little J. C., Burch M. D., Elam K. P. and Brookes J. D. (2013). Mixing potential of a surface-mounted solar-powered water mixer (SWM) for controlling cyanobacterial blooms. *Ecological Engineering*, **61A**, 245–250.

Van der Does J., Verstraelen P., Boers P., Van Roestel J., Roijackers R. and Moser G. (1992). Lake restoration with and without dredging of phosphorus-enriched upper sediment layers. *Hydrobiologia*, **233**, 197–210.

Vašek M., Prchalová M., Peterka J., Ketelaars H. A., Wagenvoort A. J., Čech M., Draštík V., Říha M., Jůza T. and Kratochvíl M. (2013). The utility of predatory fish in biomanipulation of deep reservoirs. *Ecological Engineering*, **52**, 104–111.

Visser P. M., Ibelings B. W., van der Veer B., Koedood J. and Mur L. R. (1996). Artificial mixing prevents nuisance blooms of the cyanobacterium *Microcystis aeruginosa* in Lake Nieuwe Meer, the Netherlands. *Freshwater Biology*, **36**, 435–450.

Vörös L., Oldal I., Présing M. and Katalin V. (1997). Size-selective filtration and taxon-specific digestion of plankton algae by silver carp (*Hypophthalmichthys molitrix* Val.).

In: Shallow Lakes'95 – 1997 Trophic Cascades in Shallow Freshwater and Brackish Lakes, L. Kufel, A. Prejs and JL. Rybak (eds), Springer, Dordrecht, the Netherlands.

Walsh K., Jones G. J. and Dunstan R. H. (1998). Effect of high irradiance and iron on volatile odour compounds in the cyanobacterium *Microcystis aeruginosa*. *Phytochemistry*, **49**, 1227–1239.

Wang D. H., Li L., Zhu C. W., Wang Z. Y. and Xie P. (2017). Combined application of natural sunlight and hydrogen peroxide on the removal of harmful cyanobacteria. *IOP Conference Series: Earth and Environmental Science*, **81**, 012068.

Warnecke F., Sommaruga R., Sekar R., Hofer J. S. and Pernthaler J. (2005). Abundances, identity, and growth state of actinobacteria in mountain lakes of different UV transparency. *Applied and Environmental Microbiology*, **71**, 5551–5559.

Wauer G., Gonsiorczyk T., Kretschmer K., Casper P. and Koschel R. (2005). Sediment treatment with a nitrate-storing compound to reduce phosphorus release. *Water Research*, **39**, 494–500.

Welch E. B. and Cooke G. D. (1995). Internal phosphorus loading in shallow lakes: Importance and control. *Lake and Reservoir Management*, **11**, 273–281.

Westerhoff P., Aiken G., Amy G. and Debroux J. (1999). Relationships between the structure of natural organic matter and its reactivity towards molecular ozone and hydroxyl radicals. *Water Research*, **33**, 2265–2276.

Westerhoff P., Rodriguez-Hernandez M., Baker L. and Sommerfeld M. (2005). Seasonal occurrence and degradation of 2-methylisoborneol in water supply reservoirs. *Water Research*, **39**, 4899–4912.

Whittington J., Sherman B., Green D. and Oliver R. L. (2000). Growth of *Ceratium hirundinella* in a subtropical Australian reservoir: the role of vertical migration. *Journal of Plankton Research*, **22**, 1025–1045.

WHO (2011). Guidelines for Drinking-Water Quality. World Health Organization, Geneva, Switzerland.

Windolf J., Jeppesen E., Jensen J. P. and Kristensen P. (1996). Modelling of seasonal variation in nitrogen retention and in-lake concentration: a four-year mass balance study in 16 shallow Danish lakes. *Biogeochemistry*, **33**, 25–44.

Wnorowski A. (1992). Tastes and odours in the aquatic environment: a review. *Water SA*, **18**, 203–214.

Wu J.-T. and Jüttner F. (1988). Differential partitioning of geosmin and 2-methylisoborneol between cellular constituents in *Oscillatoria tenuis*. *Archives of Microbiology*, **150**, 580–583.

Wu X., Joyce E. M. and Mason T. J. (2011). The effects of ultrasound on cyanobacteria. *Harmful Algae*, **10**, 738–743.

Xie Y., He J., Huang J., Zhang J. and Yu Z. (2007). Determination of 2-methylisoborneol and geosmin produced by streptomyces sp. and Anabaena PCC7120. *Journal of Agricultural and Food Chemistry*, **55**, 6823–6828.

Yagi M., Nakashima S. and Muramoto S. (1988). Biological degradation of musty odor compounds, 2-methylisoborneol and geosmin, in a bio-activated carbon filter. *Water Science & Technology*, **20**(8-9), 255–260.

Yang C. Y., Liu S. J., Ma X. X., Zhou J. and Xia C. H. (2012). Effects of ethyl 2-methyl acetoacetate (EMA) on the growth of *Phaeodactylum tricornutum* and *Skeletonema costatum*. *Chemistry and Ecology*, **28**, 524–534.

Yang J., Jiang H., Dong H., Wu G., Hou W., Zhao W., Sun Y. and Lai Z. (2013). Diversity of carbon monoxide-oxidizing bacteria in five lakes on the Qinghai-Tibet Plateau, China. *Geomicrobiology Journal*, **30**, 758–767.

Yen H., Lin T., Tseng I., Tung S. and Hsu M. (2007). Correlating 2-MIB and microcystin concentrations with environmental parameters in two reservoirs in south Taiwan. *Water Science & Technology*, **55**(5), 33–41.

Zaitlin B. and Watson S. B. (2006). Actinomycetes in relation to taste and odour in drinking water: myths, tenets and truths. *Water Research*, **40**, 1741–1753.

Zhang S., Zhou Q., Xu D., Lin J., Cheng S. and Wu Z. (2010). Effects of sediment dredging on water quality and zooplankton community structure in a shallow of eutrophic lake. *Journal of Environmental Sciences*, **22**, 218–224.

Zhao F., Xi S., Yang X., Yang W., Li J., Gu B. and He Z. (2012). Purifying eutrophic river waters with integrated floating island systems. *Ecological Engineering*, **40**, 53–60.

Zhong J.-C., You B.-S., Fan C.-X., Li B., Zhang L. and Ding S.-M. (2008). Influence of sediment dredging on chemical forms and release of phosphorus. *Pedosphere*, **18**, 34–44.

Zhou B., Yuan R., Shi C., Yu L., Gu J. and Zhang C. (2011). Biodegradation of geosmin in drinking water by novel bacteria isolated from biologically active carbon. *Journal of Environmental Sciences*, **23**, 816–823.

Zhou Q., Li L., Huang L., Guo L. and Song L. (2018). Combining hydrogen peroxide addition with sunlight regulation to control algal blooms. *Environmental Science and Pollution Research International*, **25**, 2239–2247.

Zhou S., Shao Y., Gao N., Deng Y., Qiao J., Ou H., Ou H. S. and Deng J. (2013). Effects of different algaecides on the photosynthetic capacity, cell integrity and microcystin-LR release of *Microcystis aeruginosa*. *Science of the Total Environment*, **463**, 111–119.

Chapter 8

Characterization and removal of minerals that cause taste

Andrea M. Dietrich[1] *and Ricard Devesa*[2]

[1]*Department of Civil and Environmental Engineering, Virginia Polytechnic Institute and State University, Blacksburg, VA, USA*
[2]*Aigües de Barcelona, Barcelona, Spain*

8.1 OVERVIEW

Consumers expect drinking water to be refreshing and possess an overall neutral taste that is free from off-tastes, off-odours, and unpleasant mouth feel (Burlingame *et al.*, 2007; Dietrich, 2006; Dietrich & Burlingame, 2015). For these reasons, it is important to understand the sensory perception of drinking water to determine what comprises a desirable taste versus an unacceptable taste.

Mineral content (Burlingame *et al.*, 2007; Marcussen *et al.*, 2013a; Platikanov *et al.*, 2013; Vingerhoeds *et al.*, 2016) and temperature (Gallagher & Dietrich, 2010; Pangborn & Bertolero, 1972) are major factors that affect the taste of water when no competing off-flavours are present. The sources of minerals in drinking water are weathering, erosion, disturbance of rock and soil, and also anthropogenic sources such as road salt or industrial discharges. Dissolved anions and cations contribute to the mineral content, with chloride, sulfate, and bicarbonate typically being the major anions, and sodium, calcium, and magnesium as the major cations. Trace metals, such as iron, copper, and zinc, are minor cations in drinking water but they can have a major impact on taste.

Editor(s): Tsair-Fuh Lin, Susan Watson, Andrea M. Dietrich and I. H. (Mel) Suffet
doi: 10.2166/9781780406664_0245

Collectively, anions and cations are measured by the aggregate parameter of total dissolved solids (TDS).

Consumers desire consistency in their foods and beverages, which is why changing water quality in terms of mineral content can be noticeable (Burlingame *et al.*, 2007; Devesa *et al.*, 2010; Dietrich & Gallagher, 2013; McGuire, 1995; Platikanov *et al.*, 2013). Changes to the mineral content of drinking water occur due to increased mineralization from anthropological activities such as industries or alteration of hydrogeology/stream flow due to changes in land-use patterns or climate. Historically, the mineral content of drinking water has been slightly altered from the source water to the tap by conventional treatment, but desalinization, remineralization, and blending are changing this paradigm. For desalinated water, aesthetics is listed among the top-ten post treatment challenges facing drinking water providers (Duranceau *et al.*, 2012). Researches in the last decade of the 20th century and first decade of the 21st century have drastically improved our understanding of the sensory qualities that allow water to be a tasty beverage, as documented in a recent critical review (Dietrich & Burlingame, 2015). This improved understanding empowers drinking water providers to continue to make available water that is safe, nutritious, delicious, and valued by consumers. For water to be sustainable, that last criterion – 'valued by consumers' – is critical. This chapter addresses the topics related to the taste of water, which include the sense of taste and factors that affect it including temperature, total mineral content, human preferences for minerals in waters, international aesthetic guidelines, taste of specific anions and cations, and treatment technologies for removal of tastants and remineralization to improve taste.

8.2 TASTING DRINKING WATER

8.2.1 The sense of taste

The sense of taste is also known as gustation. There are five known tastes – sweet, salty, sour, bitter, and umami. Research demonstrates that thresholds for human perception of the five basic tastes are genetic and do not vary with nationality, although liking and hedonic scaling of the five tastes do vary with culture (Laing *et al.*, 1993). The sense of taste is stimulated when molecules or ions interact with chemoreceptors in the taste buds (Fox, 2008; Smith & Margolskee, 2001). Taste buds are primarily located on the tongue, but are also present throughout the oral cavity. Substances that interact with the taste buds to generate a taste are called tastants. To generate a taste, the tastant must first dissolve in saliva, then interact with either an ion channel or protein receptor located in taste cells contained on the taste buds. These interactions generate electrical impulses that signal the brain to announce a sweet, salty, sour, bitter, or umami taste. Of these five tastes, four of them – sweet, salty, sour and bitter- have been used previously to describe the taste of drinking waters (Matia, 1995; Ömür-Özbek, 2012).

Flavour is the combination of the two chemical senses, taste and odour, and also mouth feel such as astringency, drying, and burning. There are two types of odour: orthonasal and retronasal. Orthonasal odours are those that enter the nose through the nostrils; retronasal odours enter the nose from the throat during swallowing. Retronasal perception dominates over orthonasal perception for providing flavour to ingested foods and beverages (Burdach & Doty, 1987).

Many aqueous minerals contribute to taste. Sour is the taste from hydronium ions (H^+ in the presence of water), salty is primarily due to sodium ions, and bitter tastes are contributed by many higher atomic and molecular weight ions, including calcium and magnesium (Smith & Margolskee, 2001). Sweet tastes are mostly caused by sugars and organic molecules with active hydroxyl groups, although lead is associated with a sweet taste (Smith & Margolskee, 2001). Umami is the savory or meaty taste that is caused by glutamate; umami has not often been reported as a taste in drinking water. Humans detect tastants relative to the composition of their saliva, which is composed of many of the same ions as found in drinking water, plus proteins and lipids (Tenovuo, 1989). For example, the salty taste of sodium is due to the sodium concentration in a food or beverage exceeding the sodium concentration of saliva (Matsou, 2000; Pedersen *et al.*, 2002).

8.2.2 Role of temperature in taste

Water temperature is an important factor that affects both the intensity of taste and also the liking and preference for water. Tap water's serving temperature generally ranges from ~0°C to about 30°C. North Americans generally prefer to drink cold water and better like the taste of minerals in water chilled to 0°C (Figure 8.1) (Pangborn & Bertolero, 1972). In Sweden, 48% of the water is consumed at very cold temperatures; this is a much greater percentage of cold-water consumption than many developed countries (Westrell *et al.*, 2006). The French and the Chinese tend not to chill their drinking water and prefer drinking it at room temperature.

While choosing the temperature of drinking water is a preference, the effect of temperature on the taste of mineral salts is a function of the taste buds. At 0°C, the taste intensity of mineral salts is low and similar, while at 22°C the taste intensity nearly doubles and is more variable for the same concentrations of the same salts (Figure 8.2) (Pangborn & Bertolero, 1972). At hot temperatures (about 70°C), the taste of minerals is very low and heat dominates. Water temperature is known to affect dissolved oxygen content through Henry's Law, but the dissolved oxygen content does not affect taste (Bruvold & Pangborn, 1970).

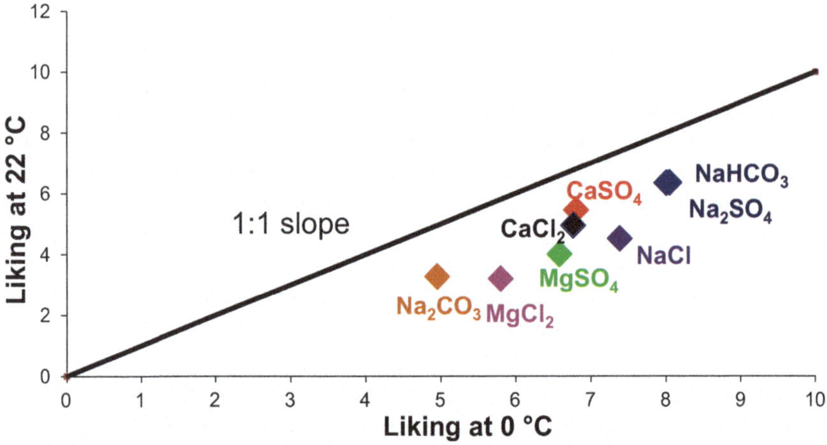

Figure 8.1. For North American consumers, liking of 750 mg/L TDS solutions of individual salts is higher at 0°C than 22°C. Liking scale ranged from 1 = extremely disliked to 17 = extremely liked. The salts tested are indicated on the graph (figure generated from data presented in: Pangborn & Bertolero, 1972).

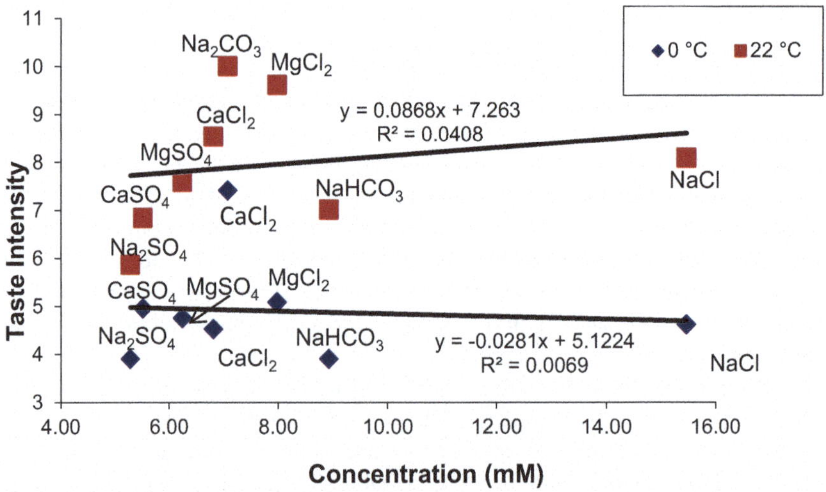

Figure 8.2. Relationship between molar concentration and taste intensity for 750 mg/L TDS solutions of individual salts at 0°C and 22°C. Taste intensity scale ranges from 1 = tasteless to 13 = extremely intense. *Note*: The salts tested are indicated on the graph and each salt had a different mM concentration so that related points are vertically aligned (*data from*: Pangborn & Bertolero, 1972).

8.2.3 Effect of water quality on taste thresholds

Dissimilar water qualities have a varying effect on taste sensitivity for different tastants. An extensive study by Hoehl *et al.* (2010) evaluated the effects of three waters: distilled water (0 mg/L hardness as $CaCO_3$, pH 5.1–6.4); tap water (162 mg/L hardness as $CaCO_3$; pH 7.6); and spring water (61.6 mg/L hardness as $CaCO_3$; pH 7.0) on identification of the five basic tastes and metallic flavour as caused by ferrous sulfate. These three water qualities had no effect on the ability of humans to identify and detect salty (sodium chloride) or sour (citric acid) tastes; there was only a slight effect on sweet (sucrose) and umami (monosodium glutamate). A significant and more variable effect was observed for bitter (caffeine) and metallic (ferrous sulfate). Human subjects were least able to identify bitter and metallic sensations in the high hardness tap water, which implies that the sensory thresholds for caffeine and ferrous are increased and sensitivity decreased in waters with more minerals and/or higher pHs. In another study, the metallic sensation of cupric sulfate was shown to decrease at pH values above 7.5 due to formation of insoluble complexes which reduced the concentration of dissolved copper in solution. When precipitation was avoided, the sensory threshold for copper was the same in distilled water and tap water (~75 mg/L TDS and 41 mg/L hardness as $CaCO_3$) (Cuppett *et al.*, 2006; Hong *et al.*, 2010a). While precipitation was shown to affect the sensory threshold of cupric ion at pH 7.5, this is unlike the case for the ferrous ion which is more soluble than the cupric ion and unlikely to precipitate in the pH 7–8 range at low mg/L concentrations (Hoehl *et al.*, 2010).

Chlorine and chloramine disinfectants are acknowledged to provide a chlorinous odour to drinking water, although their presence does not affect the human sensitivity to the salty taste of sodium chloride (Weisenthal *et al.*, 2007). Similarly, the presence of chlorine did not alter their perception and liking of TDS (Platikanov *et al.*, 2013). Thus, it appears that chlorine does not affect the taste of minerals, which is unlike its known masking effect on the odours of geosmin and 2-methylisoborneol (Worley *et al.*, 2003).

8.3 INTERNATIONAL AESTHETIC GUIDELINES

Drinking water regulatory organizations usually provide two types of standards: health-based standards designed to protect human health and aesthetic-based standards designed to minimize or prevent objectionable colour, taste, and odour when no health-threat is warranted. Table 8.1 provides example aesthetic guidelines for drinking water that are consumed across the globe. While many countries use the term aesthetic standards, the EU refers to the parameters in Table 8.1 as 'indicator parameters' and the USA refers to them as 'Secondary Maximum Contaminant Levels' (or secondary standards). The aesthetic standards may be treated as guidelines and may not be enforced.

Table 8.1. Comparison of international aesthetic guidelines for drinking water. Upper limits are listed unless noted.

Water Quality Parameter	Sensory Quality	Global/Regional Standards		Standards for Individual Countries				
		WHO[1]	EU[2]	Australia[3]	Canada[4]	China[5]	Taiwan[6]	United States[7]
Aggregate parameters								
Colour	Visual tint	15 Colour unit	Acceptable[8]	15 Hazen unit	15 True colour unit (CU)	15 CU	5 CU	15 CU
Odour	Odour	–	Acceptable[8]	Not offensive	Inoffen-sive	None	3, TON	<3, TON
pH (unit: -log [H^+])	Low: bitter, metallic taste; High: soda taste, slippery feel	6.5–9.5	6.5–9.5	6.5–8.5	6.5–8.5	6.5–8.5	6.0–8.5	6.5–8.5
Taste		–	Acceptable[8]	Not offensive	Inoffen-sive	None	–	–
Total dissolved solids, mg/L	Salty taste; scaling, staining	1,000	~1,600 mg/L (2,500 μS/cm 20°C)	600	≤500	1,000	500	500
Individual parameters								
Chloride mg/L	Salty taste	250	250	250	≤250	250	250	250
Copper mg/L	Metallic, bitter taste; astringent; green/blue colour	***2.0*** (health-based)	2.0	1	≤1.0	1.0	1.0	1.0 ***1.3*** (health-based)

Hardness mg/L as $CaCO_3$	Scaling	200	–	200	None required	450	300	–
Iron mg/L	Metallic taste; Red, brown, or yellow tint	0.3	0.2	0.3	≤0.3	0.3	0.3	0.3
Manganese mg/L	Metallic, bitter taste; Black/brown colour	0.1 (***0.4*** health-based)	0.05	0.1 (Desirable: <0.05)	≤0.05	0.1	0.05	0.05
Sodium mg/L	Salty taste	200	200	180	≤200	–	–	–
Sulfate mg/L	Salty taste	250	250	250	≤500	250	250	250
Zinc mg/L	Metallic taste	4.0	–	3.0	≤5.0	1.0	5.0	3.0

[1]WHO, Guidelines for drinking-water quality, 4th edition (2011a); values cited are non-health-based indicator of likely level at which human detection occurs. WHO Guidelines have not been established for constituents in water without direct adverse health impacts; [2]EU – COUNCIL DIRECTIVE 98/83/EC of 3 November 1998 on the quality of water intended for human consumption; COMMISSION DIRECTIVE (EU) 2015/1787 of 6 October 2015 amending Annexes II and III to Council Directive 98/83/EC; *NOTE*: mg/L TDS calculated based on conductivity; [3]Australian Drinking Water Guidelines 6 (2011); [4]Health Canada (2012); [5]China, Drinking Water Standard of China (GB5749-2006) (Zhang *et al.*, 2011); [6]Taiwan; Drinking Water Quality Standards-Revisions to Article 3 promulgated by Environmental Protection Administration Order Huan-Shu-Tu-Tzu No.0980106331E on Nov. 26, 2009; [7]USEPA (1979); [8]Acceptable and no abnormal changes.

8.4 TASTANT CHARACTERISTICS: AGGREGATE MEASURES OF WATER QUALITY

8.4.1 Measuring and characterizing TDS

Total dissolved solids (TDS) is a measure of all cations and anions in water and also organic compounds, although their concentration is usually minor compared to minerals. TDS is measured either by gravimetric methods or by conductivity. A gravimetric method requires that the sample be filtered to remove particulates, then the filtered sample is evaporated and the mass of the residual solids is measured by weight and usually reported as mg/L (APHA, 2012). This is the most accurate method although it can be time consuming. Electrical conductivity is a more straightforward measure of TDS in the water and can be measured with a conductivity meter. Specific conductivity is usually reported in units of microsiemens per centimeter (μS/cm). Conductivity provides an approximate value for the TDS concentration. According to Hem (1970), the TDS value can be calculated as: conductivity (μS/cm) * A = TDS (mg/L), where the value of A varies between 0.55 and 0.75. Due to the ease of measuring conductivity, this method is gaining popularity, although gravimetrically measured TDS is still used for many standards and guidelines.

Mineral content, as measured by TDS, is the major determinant in drinking water taste (Burlingame *et al.*, 2007; Platikanov *et al.*, 2013). As shown in Table 8.2, TDS provides a means to rate the overall mineral content of water from a low mineral water to a high mineral water. Concentrations of TDS from natural sources have been found to vary widely, from <30 to as much as 6,000 mg/L (WHO, 1996). Similarly, bottled waters can have the full range from almost no TDS to >1,000 mg/L TDS (Naddeo *et al.*, 2008).

Table 8.2. Classification of water based on TDS, or mineral content, according to the European Union (van der Aa, 2003).

Classifications Based on Minerals	TDS (mg/L)
Very low	<50
Low	50–500
Intermediate	500–1,500
High	>1,500

8.4.2 Consumer preferences for TDS in drinking water

People vary in their preference and liking for the amount of mineralization in their drinking water. The most important factor related to preference is the concentration of TDS in the drinking water (Platikanov *et al.*, 2013). Other factors that affect consumer preferences include prior exposure and being accustomed to a taste, which includes social, environmental, and cultural considerations (Laing *et al.*, 1993; WHO, 2011a).

The few studies that provide data for consumer preferences represent snap-shots of consumers from different locations across the globe. A study in France, where mineral waters are routinely consumed, found that consumers prefer waters with 300–350 mg/L TDS and describe these waters as tasteless and cooler (Teillet *et al.*, 2010). Californians in North America generally prefer low TDS waters with ~80 mg/L TDS, which they consider odourless, free from mineral tastes, and desirable (Bruvold, 1968; Bruvold & Daniel, 1990). The Japanese prefer waters with low calcium and magnesium ion concentrations and hardness values generally in the range of 50–80 mg/L as $CaCO_3$ (Koseki *et al.*, 2005). In Japan, consumer acceptability of drinking water could be predicted based on four parameters: total organic carbon, total iron, total hardness, and bicarbonate ion concentration (Sasaki *et al.*, 1996). An extensive study by Platikanov *et al.* (2013), with panelists from Barcelona (Catalonia, Spain), demonstrated that drinking waters with 200–400 mg/L TDS were preferred, while those with <30 mg/L TDS or >800 mg/L TDS had low acceptance. A river water with variable TDS, generally greater than 500 mg/L TDS, was reported to have a salty taste that increased as the TDS increased (Matia, 1995). Overall, consumers seem to prefer waters that are classified as 'low mineral' or 50–500 mg/L TDS according to Table 8.2. This preference is consistent with regulatory guidelines for aesthetic values of TDS in many countries (Table 8.1), although the WHO, the EU, and some countries allow for values ≥1000 mg/L TDS.

Insight into preference for TDS in water could be due to liking, as discussed above, but also due to habituation. Habituation is a psychological term referring to a settled behavior, such as repeatedly choosing the same beverage to drink because it is familiar and accessible. While no data are available for habituation for TDS, drinking water-related studies with chlorine can provide some understanding. In taste tests, French consumers were more sensitive to chlorinous taste in drinking water than American consumers, meaning that the French had a lower detection threshold for chlorine (Piriou *et al.*, 2004). The same effect was observed between French and Spanish consumers, when again the French were more sensitive to chlorine. Although this difference in sensitivity was significant for the French-Spanish study, the magnitude of difference was lower than for the French-American study (Piriou *et al.*, 2010). Chlorine levels and corresponding consumer thresholds for its taste followed the same trend: France < Spain < USA. Thus, regional chlorination practices appear to influence consumer perception. In contrast, differences were not noticed between French and Spanish consumers with respect to sensitivity to geosmin and 2-methylisoborneol, which are occasional taste-and-odor problems in drinking water (Piriou *et al.*, 2009). Therefore, habituation to chlorine appears to explain the lower threshold for chlorine for the French since they normally drink water with low chlorine residual (Piriou *et al.*, 2010). Like chlorine, TDS is a regional parameter of water quality, and unlike geosmin and MIB, TDS is not prone to be episodic. This comparison suggests that preference for a certain level of TDS could be habituation, although studies need to be performed to confirm this.

8.4.3 Ability to discriminate between TDS concentrations in drinking water

As stated previously, temperature affects the sense of taste so it is not surprising that water temperature affects the ability to discriminate between water samples of differing TDS. Human subjects were not able to distinguish between the tastes of two waters of 25 and 524 mg/L TDS at temperature 4°C, but the same waters were readily distinguished when warmed to 25°C (Gallagher & Dietrich, 2010).

While the level of TDS is the most important factor for drinking water preference, TDS levels are also important in the ability to discriminate between different waters. The magnitude of the TDS difference between two waters – or Δ TDS – is critical for identifying a difference in taste (Devesa & Dietrich, 2018; Dietrich & Gallagher, 2013). Comprehensive discrimination studies with consumer and trained panelists gathered for diverse water qualities and different countries demonstrate that $\approx\Delta$ TDS $>$ 150 mg/L is necessary for two waters to be discriminated based on the taste of TDS minerals (Devesa & Dietrich, 2018). The value of $\approx\Delta$ TDS $>$ 150 mg/L generally applies with some cautions that the presence of odourants, total organic carbon, temperature, and consumer variations for taste abilities and preferences will affect discrimination. Also, the direction of change in TDS, from high to low or low to high, also impacts discrimination as discussed below.

The ability to discriminate taste based on TDS is not simply based on the magnitude of Δ TDS, but also depends on whether the reference water being compared has high or low TDS. In a recent study, a high TDS water (524 mg/L) was diluted with taste-free water to produce waters of varying TDS levels and the same ratios of cations and anions (Dietrich & Gallagher, 2013). As shown in Figure 8.3, consumers who participated in the taste tests were better able to detect a taste difference when the reference water had low TDS (26 mg/L TDS) than when the reference water was high TDS (524 mg/L). When all the data were combined and statistically analyzed for values indicative of the general population, a Δ TDS $\approx$ 185 mg/L could be discriminated by taste when a low TDS water of 26 mg/L was the reference water to compare with waters of higher TDS levels. When a high TDS water of 524 mg/L TDS was the reference water, a Δ TDS $\approx$ 380 mg/L was necessary for consumers to be able to detect a difference based on taste when comparing waters with lower TDS values. Thus, if a lower TDS water increases in TDS and becomes 'saltier' it will be more readily noticed by consumers than if a high TDS water is altered to have a lower TDS.

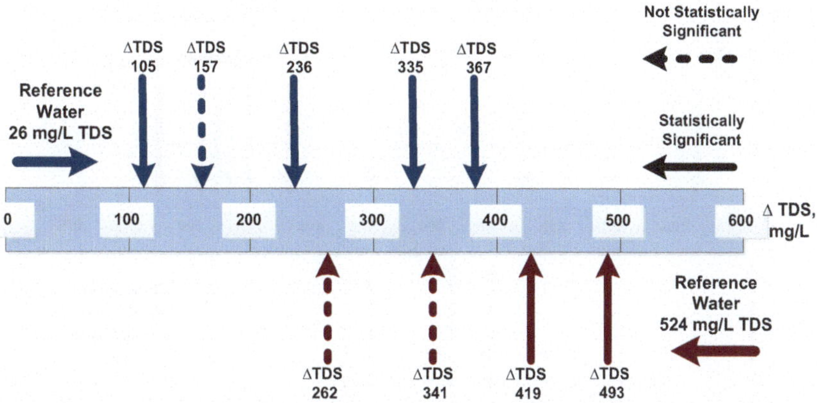

Figure 8.3. Results of untrained consumer panelists using taste to discriminate between two water samples with different TDS levels. A lesser Δ TDS level could be distinguished when the reference water was low (26 mg/L TDS) than when it was high (524 mg/L TDS). All waters were dilutions of the 524 mg/L reference water (*data from*: Dietrich & Gallagher, 2013).

The taste of water is a complex combination of the TDS and the concentrations of individual constituents that contribute varying amount of taste – for example, calcium, sodium, bicarbonate, chloride, etc. Thus, for actual drinking waters, the interpretation of Δ TDS is less clear than for one water that is diluted to alter TDS. As shown in Figure 8.4, studies from Europe and North America demonstrate that Δ TDS is often – but not always – correlated with the ability to detect a difference base on taste. A room temperature study with trained sensory panelists in Barcelona (Catalonia, Spain) found that the flavour improved when high TDS levels in conventionally treated surface water were reduced through blending with the same surface water treated by desalination; depending on the blend, the panelists were or were not able to detect a difference based in taste (Devesa *et al.*, 2010). Another room temperature study in Blacksburg (Virginia, USA) also found that consumers could discriminate taste based on Δ TDS (Gallagher & Dietrich, 2010), and that the greater the Δ TDS, the greater the ability of consumers to detect a difference in taste.

When evaluating alternative drinking water blends for distributed water, both consumers and sensory experts were able to discriminate between room temperature treated surface water of 480 mg/L TDS and desalinated seawater with 350 mg/L TDS (McGuire *et al.*, 2007a). These two treated waters varied in both TDS and the composition of individual anions and cations, and thus, both contributed to tastes. The same consumers and experts could also discriminate blends of 25, 50, or 75% desalinated water mixed with treated surface water.

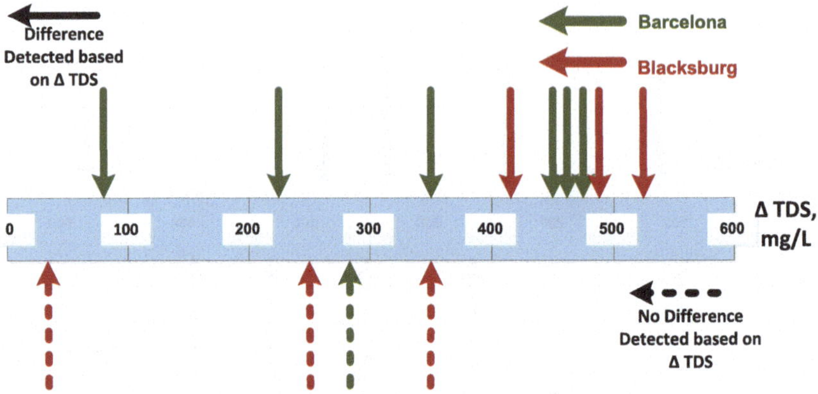

Figure 8.4. The application of Δ TDS as a variable associated with ability to discriminate drinking waters based on the taste of TDS. Solid arrows indicate the Δ TDS values of waters that could be discriminated, while dashed arrows indicate those that could not be discriminated based on taste (*data from*: Devesa *et al.*, 2010 (panelists from Barcelona, Catalonia, Spain); Gallagher & Dietrich, 2010 (panelists from Blacksburg, Virginia, USA)).

8.4.4 Taste characteristics of pH

pH is a measure of the H^+ concentration in water. The pH of drinking water should be between 6.5 and 8.5, which encompasses the typical pH range of saliva (Pedersen *et al.*, 2002). Drinking water regulatory agencies establish aesthetic guidelines of pH 6.5–9.5 (Table 8.1). Low pH water can produce a bitter, metallic taste, whereas high pH can have a slippery feel and soda taste (Burlingame *et al.*, 2007; Whelton *et al.*, 2007). Guidance for pH is based on both taste and also corrosion control as low pH water is aggressive toward metal pipes.

8.5 TASTE CHARACTERISTICS OF INDIVIDUAL MINERALS

This section summarizes known sensory characteristics of the minerals in natural and drinking water. The majority of the data available are responses from panels and groups of people; few data are available that portray the variability of individual human subjects. While data from both the food and the drinking water industry are provided in this section, it should be noted that the food industry, including the non-water beverage industry, usually performs taste testing at concentrations much higher than those found in drinking water. Also, food-related taste testing is often performed with compounds not typically found in drinking water, such as organic salts like lactates or gluconates.

8.5.1 Qualitative and quantitative analysis of minerals

8.5.1.1 Measuring metals and elements

Inductively coupled plasma (ICP) is the most common technique for the analysis of aqueous elements, including metals. The measurement is made in an argon plasma maintained at about 8,000 K to convert elements into their gaseous state. There are two main types: optical and mass-spectrometry. The optical method is referred as either ICP-optical emission spectrometry (ICP-OES) or ICP-atomic emission spectrometry (ICP-AES). The optical method measures specific emission lines from each elemental spectrum. In the mass-spectrometry technique (ICP-MS), elements are analyzed from the mass-to-charge ratio of each ion, which is the same principle used in organic analysis by mass spectrometry.

In general, the detection limits of ICP-optical emission methods are a few μg/L, which is generally sensitive enough for analyzing drinking water according to the requirements of international regulations. When lower detection limits are required, ICP-MS is the logical option as it is 10–100 times more sensitive than optical instruments. ICP-MS is also very useful for minor and trace components in seawater and brackish waters. In this case, the high concentration of sodium chloride requires the dilution of the sample before being nebulised in the instrument.

ICP-OES and ICP-MS are extremely useful for two main reasons. First, they are multi-element techniques; more than 40 relevant metals and some important metalloids and non-metals (e.g. boron, silica, and phosphorous) can be analysed simultaneously in water. And second, the linearity range is wide, covering several orders of magnitude. The technique allows the simultaneous determination of major components (sodium, potassium, calcium, magnesium, and strontium), minor components (iron, aluminium, zinc, copper, silicon, etc.), and trace components (chromium, nickel, lead, selenium, antimony, and other heavy metals). Mercury presents some difficulties due to memory and carry-over effects, which causes many laboratories to use fluorescence emission specifically for this element.

While ICP-OES and ICP-MS are the most used methods for elemental and metal analyses, other techniques can be used. These include atomic absorption spectrometry, emission or flame atomic spectrometry for alkaline and alkaline-earth metals, visible spectrophotometry, anodic stripping voltammetry, and cationic chromatography. Hardness, a crucial parameter to assess the corrosivity of waters, is based on a classical titrimetric method with a chelating agent (EDTA), where calcium, magnesium, and other cations are all included. In case of the analysis being performed by ICP, hardness is usually estimated from calcium and magnesium concentrations.

8.5.1.2 Measuring anions

Ion chromatography (IC) has simplified enormously the analysis of anions, which had been traditionally very tedious. Anion chromatography has unified anion analyses into a single technique to measure chloride, sulfate, nitrate, and minor components like fluoride, bromide, and nitrite. The technique replaced classical individual procedures such as: titrimetry with silver nitrate (argentometry) for chloride, nephelometry or gravimetry for sulfate, direct or induced UV spectrophotometry for nitrate, ion selective electrode for fluoride, just to name a few. The technique is also useful to analyse disinfection by-products at trace levels, like chlorite and chlorate when chlorine dioxide is used in the treatment or bromate in the case of ozonation of waters containing significant concentrations of bromide. Also, perchlorate, an emerging contaminant, is analyzed by IC. The technique has improved substantially in the last two decades, particularly in the area of sensitivity due to the conductimetric detection with electrolytic suppression and automated eluent generation. Mass spectrometry detectors have been recently introduced for trace analysis by IC. While analysis of bicarbonate (expressed generally in terms of alkalinity as mg/L as $CaCO_3$) continues to use potentiometric or colourimetric acid-base titrimetry, nearly all other titrimetric methods for anions have been discontinued for routine use.

8.5.1.3 Tastes of major metal cations

Calcium: Calcium, a divalent cation, is usually the major cation in drinking water. A study of surface and ground waters demonstrated that its concentration typically ranges from a few to ~100 mg/L Ca (Azoulay *et al.*, 2001). Drinking water regulatory agencies are not known to establish aesthetic guidelines for calcium (Table 8.1), although a guideline for hardness will relate to calcium. The presence of calcium is in drinking water is positively associated with a preference for taste (Platikanov *et al.*, 2013). This positive association of calcium with the taste of water is supportive of human health as drinking water is one of the three important sources of bioavailable calcium for humans; the other two are milk and dairy products (Azoulay *et al.*, 2001). The reported taste threshold for calcium is between 100 and 300 mg/L in water (WHO, 2011a). This is consistent with the findings of Zoeteman *et al.* (1978) who reported that panelist found the taste of calcium in water objectionable at concentration ≥100 mg/L Ca. Tordoff (1996) reported that human detection thresholds for calcium salts in water were similar, but that the taste descriptors were different for different calcium salts. Calcium chloride was described as more bitter than calcium lactate, which had sweeter and more sour tastes. Lawless *et al.* (2003) indicated that calcium chloride had a more intense taste than organic calcium salts. Although foods and beverages may have substantial concentrations of organic anions, drinking waters lack these constituents and therefore, inorganic calcium could be tasted easily. In foods, calcium is bound to proteins or insoluble as calcium phosphate, and this limits the ability of this cation to be tasted (Tordoff, 1996).

High levels of calcium may adversely affect the taste and acceptability of the water due to a bitter or salty taste and a slimy mouth feel (Lawless *et al.*, 2003).

Calcium detection thresholds are fairly close to salivary calcium concentrations, which are 200 mg/L (Neyraud & Dransfield, 2004). The relative concentration of calcium ions in water to saliva will determine calcium's taste (Burlingame *et al.*, 2007).

Magnesium: Magnesium, a divalent cation, is usually found in lower quantities in source waters than calcium. Typical concentrations in surface and ground waters vary from a few mg/L to ~30 mg/L Mg^{+2} (Azoulay *et al.*, 2001). Drinking water regulatory agencies do not establish separate aesthetic guidelines for magnesium (Table 8.1), although it can be captured in recommendations for hardness. In single metal taste tests, concentrations of above 10 mg/L Mg^{2+} in water cause a bitter taste that is perceived as offensive (Zoeteman *et al.*, 1978). The effect of magnesium on the likability of the taste of real drinking waters is unclear (Platikanov *et al.*, 2013) because the principle component analysis of concentrations of individual anions and cations in waters and their likeability by human subjects did not indicated a clear cut trend.

Hardness: Water hardness is the concentration of polyvalent cations. Hardness correlates highly with the concentration of calcium and magnesium cations, which are the most common divalent cations in natural waters and relate to the presence of limestone and dolomite in the geologic formations (Combs, 2005). Hardness is also controlled by the water quality parameters of pH and alkalinity. Water hardness greater than ~200 mg/L as $CaCO_3$ can cause scale formation (WHO, 2011a), thus, many drinking water regulatory agencies establish aesthetic standards for hardness at or above this concentration to avoid precipitation (Table 8.1).

The taste of combinations of cations, such as in calcium and magnesium, has proven to be a complex science. Therefore, the taste of divalent salts needs more descriptive studies to search for qualitative relationships with perceived sensation. Good tasting waters have been found to have hardness values between 10 to 100 mg/L as $CaCO_3$ due largely to calcium hardness (Burlingame *et al.*, 2007; Koseki *et al.*, 2005) as calcium is usually present at higher concentrations than magnesium. A common ratio of Mg:Ca is 1:4 found in ground and surface waters (Kozisek, 2003).

Sodium: The sodium ion is primarily responsible for salty taste (Bartoshuk, 1980). Sodium, a monovalent cation, frequently has an aesthetic standard of about 200 mg/L Na^+ due to its salty taste (Table 8.1; WHO, 2011a; Zoeteman *et al.*, 1978). Taste testing indicates that sodium in drinking waters is negatively associated with preference (Marcussen *et al.*, 2013b; Platikanov *et al.*, 2013). The taste of sodium is dependent on the counter anion, with chloride salts having a more intense taste than sulfate salts (Figure 8.2) (Pangborn & Bertolero, 1972). USEPA Drinking Water Advisory (USEPA, 2003) recommends 'that the sodium concentration in drinking water not exceed a range of 30 to 60 mg/L Na^+ because of possible adverse effects on taste at higher concentrations'.

8.5.1.4 Tastes of major anions

Chloride: Chloride, a monovalent anion, occurs naturally in fresh, estuarine or salt water. Anthropogenic sources include salting of roads in winter to melt ice or adding ferric chloride as a coagulant for water treatment. The taste threshold varies from 200–300 mg/L depending on counter cation (e.g. Na, K, and Ca; WHO, 2011a). Concentrations above 250 mg/L Cl^- are likely to be detected by consumers (WHO, 2011a), which is why most aesthetic standards are established at 250 mg/L Cl^- (Table 8.1). Overall, chloride is negatively associated with preference for the taste of drinking water (Platikanov *et al.*, 2013); thus, high chloride concentrations should be avoided.

Bicarbonate and carbonate: Bicarbonate (HCO_3^-), a monovalent anion, is likely to be the dominant anion in drinking waters, which are typically pH 6.5–8.5. At pH < 8.3, bicarbonate is predominant. At pH values greater than 8.3, the carbonate (CO_3^{2-}) concentration will begin to increase and it will exceed the bicarbonate concentration at pH values above 10.3. Both saliva and most drinking waters are buffered by bicarbonate, although lower levels of these anions occur in tap water compared with saliva. Bicarbonate shows a positive association with the good taste of water (Platikanov *et al.*, 2013) while carbonate can be associated with off-tasting water (Pangborn & Bertolero, 1972; Burlingame *et al.*, 2007; Whelton *et al.*, 2007). Waters with high bicarbonate concentrations are described as 'sweet' (Matia, 1995), further confirming that bicarbonate is a desirable anion for good tasting drinking water.

Sulfate: Off-tastes from this divalent anion typically occur at concentrations of 250 mg/L SO_4^{2-} or greater (WHO, 2011a) and thus, most aesthetic standards use 250 mg/L as a guideline for sulfate (Table 8.1). The associated cation affects the taste, with sodium sulfate having a taste threshold about 250 mg/L SO_4^{2-} and calcium sulfate at about 1,000 mg/L $CaSO_4$ (Pangborn & Bertolero, 1972; WHO, 2011a). Overall, sulfate has less perceived taste than chloride. Sulfate is positively associated with consumer taste preference for drinking waters (Platikanov *et al.*, 2013).

8.5.1.5 Minor metals can provide major sensory effects

Since the turn of the 21st century, there has been an increased interest in the presence, taste, and flavour of metals released from drinking water distribution materials, namely, iron, copper, and zinc. Renewed interest is a combination of health issues, metallic taste and odour dominating water quality complaints, and an ageing drinking water infrastructure with the associated increase in corrosion (Benson *et al.*, 2012; Dietrich *et al.*, 2004; Edwards, 2004; Kirmeyer, 2000). Metallic tastes were reported to be the third most commonly reported taste issues in a 2011 survey of water companies; the reported that metallic taste complaints had nearly doubled in the last 20 years (Ömür-Özbek, 2012). In this survey, chlorinous was reported as the most common aesthetic issue and earthy/musty was second. The result is that many of the aesthetic guidelines for metals in drinking water, which have been established for a number of decades, may not be effective at avoiding taste and odour issues. This is particularly true for redox-active metals that cause lipid oxidation and produce volatile and odourous organic chemicals in the oral cavity, such as copper and iron (Dietrich, 2009; Epke & Lawless, 2007; Hettinger *et al.*, 1990; Hong *et al.*, 2006; Ömür-Özbek & Dietrich, 2011; Ömür-Özbek *et al.*, 2012). These metals have both a taste and a retronasal odour that can be perceived by humans.

Most recent sensory studies for the thresholds of metals have been conducted primarily in distilled water. While distilled water is not considered tasteless by many consumers (Bruvold, 1968; Bruvold & Daniel, 1990; Burlingame *et al.*, 2007), its lack of anions and cations minimizes the formation of soluble and insoluble complexes with metal ions that could interfere with sensory perception. Insoluble complexes have less taste and flavour than soluble metals (Cuppett *et al.*, 2006; Hong *et al.*, 2009, 2010a). The acknowledgement that metals can cause oral lipid oxidation that contributes volatile compounds that cause 'flavour', or a combination of odour and taste, has dictated that sensory testing with metals be limited to one metal-containing sample per sensory session per morning or afternoon, as a second sample has less flavour because the oral lipids were recently oxidized. Additionally, unlike sweet, sour, and salty, it takes about 1–2 minutes for the metallic flavour to develop (Hong *et al.*, 2010b) so the time between tasting samples needs to be longer, even if comparing a metal sample with distilled water.

Copper: Copper is typically present as the divalent cupric ion in oxygenated waters. In natural waters, copper is usually present at low μg/L concentrations and can enter drinking water from corrosion of copper pipes used in premise plumbing that results in concentrations exceeding 1 mg/L Cu (Dietrich *et al.*, 2004). Copper has a slight taste and a significant retronasal odour, thus, perception of this metal in water is due to its flavour (Ömür-Özbek & Dietrich, 2011). Copper's aesthetic issues are a strong metallic flavour, slight bitter taste, and an astringent mouth feel; these sensory characteristics require around 1–2 minutes to develop (Cuppett *et al.*, 2006; Hong *et al.*, 2010b; Ömür-Özbek & Dietrich, 2011). Visual effects of blue-coloured water and particles can also occur (Booth & Brazos, 2004).

As shown in Table 8.1, copper is a metal that has both health-based standards of 1.3–2 mg/L Cu and aesthetic guidelines of 1.0 mg/L Cu. Recent investigations into the human threshold for copper reveal that soluble cupric ion has a threshold in the range of 0.4–0.8 mg/L in distilled water and a similar sensory threshold in tap water (Cuppett *et al.*, 2006). Thus, the human threshold appears to be slightly less than aesthetic guidelines (Table 8.1), although in either case, the human sensory perception of copper is generally protective of human health because consumers may detect its flavour before it reaches its health-based regulatory level. A concern is the variable solubility of copper as a function of pH. Copper may precipitate at concentrations and pH values normally found in drinking water; this precipitated copper would not be detected by taste or flavour (Dietrich *et al.*, 2008).

Iron: Iron is the most abundant element in the earth's crust and occurs in the reduced divalent ferrous form (Fe^{2+}), which is fairly water soluble, and the oxidized trivalent ferric (Fe^{3+}) form, which is highly insoluble in water. In groundwater and stratified lakes/reservoirs, the reduced ferrous dominates and if not removed, will be present in tap water. In oxygenated surface waters, the iron concentrations are low, but the addition of ferric chloride and corrosion of iron pipes allow iron to be present as soluble or particulates (rust) in tap water. A detailed description of iron corrosion and the occurrence of ferrous and ferric ions in tap water are available in a recent review article (Benson *et al.*, 2012).

Conventional aesthetic guidelines suggest 0.3 mg/L total Fe as adequate for preventing stain in laundry or reddish-brown water (WHO, 2011a; Table 8.1). While 0.3 mg Fe/L can protect against visual effects, it is not likely protective of the human chemical senses for ferrous ion. Recent literature reports indicate that the overall threshold in distilled water is 0.17 mg/L Fe^{2+} in distilled water, with an individual range of 0.007 to >14 mg/L Fe^{2+} for metallic flavour due to a strong retronasal metallic odour and slight bitter taste (Mirlohi *et al.*, 2011; Ömür-Özbek & Dietrich, 2011). Ferrous iron has a substantial retronasal odour from production of odourous lipid oxidation products in the oral cavity (Mirlohi *et al.*, 2011; Ömür-Özbek & Dietrich, 2011; Ömür-Özbek *et al.*, 2012). As for many tastants and odourants, the descriptors used to describe the flavour of the ferrous ion are variable (Figure 8.5). The ferric ion has very little taste or odour as it does not participate in lipid oxidation. Ferric ion can be reduced in the oral cavity to ferrous, and this would participate in lipid oxidation and have flavour (Ömür-Özbek & Dietrich, 2011).

Figure 8.5. Cloud tag of descriptors generated by North Americans who ingested a 10 mg/L ferrous ion solution. Note the consistency of 'metal-like' descriptors and the variability of the overall words selected.

Manganese: Soluble manganese occurs in drinking water as reduced Mn(II) in groundwater, stratified reservoirs, and lakes. When oxidized to Mn(IV), MnO_2 will form and it is highly insoluble (Stumm & Morgan, 1996). Because of the presence of Mn-oxidizing and -reducing bacteria in source water, treatment plants, and distribution networks (Cerrato *et al.*, 2006, 2010), Mn is in a constant state of redox cycling and can be either mobilized as the dissolved reduced form or precipitated as the oxidized species.

Manganese aesthetic standards are typically between 0.05–0.1 mg/L Mn (Table 8.1). The aesthetic effects of Mn are reportedly due to both colour and bitter taste, although recent research indicates that Mn(II) and Mn(IV) have very little taste. The taste threshold is about 75–100 mg/L Mn(II) as $MnSO_4$ while insoluble Mn(IV) particles have no detectable taste (Sain & Dietrich, 2015; Sain *et al.*, 2014). Mn(IV) as $MnO_{2(s)}$ has colour; Mn(II) does not, even at concentrations of 1,000 mg/L Mn(II). MnO_2 particles are visual as a yellow colour at about 0.05 mg/L Mn with formation of brown and black particles at higher concentrations (Booth & Brazos, 2004; Sain & Dietrich, 2015; Sain *et al.*, 2014; WHO, 2011a). Sloughing off of Mn precipitates is noticeable at about 0.2 mg/L Mn (WHO, 2011a) as black particles in drinking water.

Zinc: Zinc usually occurs in drinking water from dissolution of galvanized pipes (WHO, 2011a). When concentrations of 3–5 mg/L occur, water may appear cloudy or develop a greasy film when boiled (WHO, 2011a). The primary sensory quality of zinc is astringency but not taste; many zinc salts have a lingering astringency (Keast, 2006).

8.6 TREATMENT STRATEGIES THAT ALTER TASTE

Drinking water supply companies have traditionally focused their efforts on providing a product that is healthy, safe, clean, and meets enforceable regulatory standards. However, consumers primarily evaluate drinking water in terms of its aesthetic properties. For this reason, in the 21st century the water sector for human consumption has made noteworthy efforts to improve the odour and taste of drinking water by introducing new advanced processes, that is, membrane treatments. This section of the chapter was prepared using data available to the authors in mid-2012. Thus, the usage of treatment techniques and their effectiveness are representative of technology available and in use in mid-2012.

8.6.1 Removing tastants by traditional treatment techniques

A broad range of technologies based on totally different physical and chemical principles are available for the treatment of drinking waters. Some of them, such as carbon filtration, improve the taste of treated waters by removing or minimizing odourous organic compounds. Others can modify the flavour of water because of the formation of by-products (i.e. ozonation, which produces aldehydes) or the presence of odourous disinfectant residuals (the most common case being chlorination). But in most cases, conventional treatment techniques do not change the taste of water because they do not alter the TDS or the mineral composition.

There are, though, a few traditional techniques which change (in general, reduce) the mineral composition in water; consequently, they substantially modify the taste of the finished drinking water. These established traditional technologies were implemented for TDS removal and/or hardness removal to avoid scaling; taste alteration was usually a consequence rather than a goal. The technologies include ionic exchange columns, precipitative softening, and distillation techniques. Use of ionic exchange and precipitation is decreasing due to an increased use of membrane technologies, which includes primarily reverse osmosis (RO), nanofiltration (NF), and electrodialysis reversal (EDR), all of which have the ability to remove minerals and alter taste.

Ion exchange has been often used at industrial and domestic levels ('Point-of-use' and 'Point-of-entry' devices) and also in small treatment plants (Masters & Ela, 2007). The most common design for water treatment consists of using two sequential columns, one cationic and one anionic; one adsorbs cations and releases H^+, while the other removes anions and releases OH^-. The final result is water with significantly lower TDS and neutral pH. Depending on the purpose of treatment, broad range or species-specific columns have to be used. Ion exchange softeners are based on a simplified process where a column removes calcium and magnesium by exchanging them for sodium. In this situation, TDS is not reduced, and from the health and taste point of view, the treatment is not suitable.

Precipitative softening implements chemical precipitation by raising the pH to 10–11 through the addition of lime ($Ca(OH)_2$), soda ash (Na_2CO_3), sodium hydroxide (NaOH, also known as 'caustic soda'), or a combination of these chemicals. The high pH causes the formation of calcium carbonate and/or magnesium hydroxide precipitates that are removed by settling and/or filtration. The water is then recarbonated with carbon dioxide gas and the pH reduced to 7–8 before distribution to consumers (Masters & Ela, 2007).

8.6.2 Thermal and membrane desalination processes

Thermal and membrane processes are the two major techniques used for water desalination to remove TDS and improve water quality. Thermal, or evaporative, processes employ two basic techniques: Multiple-Effect-Distillation (MED) and Multi-Stage-Flash (MSF) distillation, both of which distill seawater in a series of stages. The vapor produced is condensed to produce demineralized water. The main difference between MED and MSF is found in the evaporation method and heat transfer. In MED, evaporation takes place from a seawater film in contact with the hot surface, while in MSF heating of the seawater is achieved by convection and steam is produced from brine by a 'flash' mechanism. Both types of installations are often constructed in cogeneration plants where power and water are produced at the same time. Although the systems are designed to recover the condensation heat of water, it is obvious that these processes are energy intensive.

Among the membrane techniques, reverse osmosis (RO), nanofiltration (NF), and electrodialysis reversal (EDR) are the most common. These techniques use semi-permeable membranes which permit the permeation of water but not the solutes. In the case of RO and NF, the driving mechanism is external pressure applied to overcome the osmotic pressure of the saline water. RO and NF have the same principle. The difference is in the pore sizes in the membranes, smaller for RO ($1–10 \times 10^{-10}$ m). In fact, NF (pore size = $10–100 \times 10^{-10}$ m) is sometimes referred as low pressure RO. NF requires a smaller operation pressure, but the rejection of salts is lower. NF is mainly applied to brackish and freshwater while RO can be applied to any type of water including seawater. For EDR, a magnetic field is applied to separate positive and negative charged species, which then circulate in opposite directions through the membrane. Although direct electrodialysis (ED) is technically available, EDR, where polarity of electrodes changes periodically, is mostly used due to operational advantages derived from easier maintenance of membranes.

Ultrafiltration (UF) and microfiltration (MF) are also membrane techniques, but the mechanism of actuation is different due to their large pore sizes. In MF, the membranes work as a conventional surface filter, with efficiency similar to conventional sand filtration. Particulate matter is removed, but dissolved aqueous constituents are not. On the other hand, the UF mechanism shares some characteristics with NF and MF. In UF, colloids are retained in addition to the suspension matter (i.e. iron, manganese, and aluminium from coagulation agents), but no significant removal of dissolved species is achieved. Therefore, MF and UF do not have any influence on the taste of water. In most cases, MF and UF are used as pre-treatments for NF, RO, and EDR.

8.6.3 Relevance of desalination for global treatment of water

The United Nations estimates that 1,200 million people currently live in regions facing water scarcity and this figure will increase to 1,800 million by 2025 (United Nations, 2012). Desalination of seawater and brackish water is currently a common process in some countries and regions of the globe, and it is rapidly growing as the main source of freshwater worldwide (WHO, 2011b).

The desalination industry began about 1950 and currently over 60 Mm^3/day (millions cubic meter per day) are treated. The installed capacity has risen by more than 50% over the past five years prior to 2012, although it seems that this huge rate of increase will not be maintained in the near future. It is estimated that about 60% of the current capacity corresponds to seawater, 21% to brackish water, 8% to freshwater, 6% for wastewater, and 5% for other uses.

At its beginning, the desalination industry was energy intensive and primarily limited to the freshwater-poor and oil-rich Arabian Gulf countries, which used thermal technologies (*vide infra*) to desalt seawater. The thermal capacity that was installed is still operational and will continue to be used. Membranes are increasingly installed and this technology (particularly RO) leads the market even in Middle Eastern countries (Dawoud & Al Mulla, 2012). As of 2012, it is estimated that for membrane technologies, about 60% of the installed capacity corresponds to RO, 27% to MSF, 8% to MED, 4% to EDR plus ED, and about 1% for hybrid systems and other techniques.

Traditionally, the desalination market was led by the Gulf Cooperation Council countries that desalted seawater through thermal processes, and in the USA, where mostly brackish waters were treated by membranes. Over the past decade, some Mediterranean countries and Australia have reached significant figures due to major programs instituted by public water agencies. The top countries by overall desalination capacity are Saudi Arabia (16% of market share), USA (14%), United Arab Emirates (12%), Spain (9%), Kuwait and China (4% each), and decreasing percentages below 3% for Japan, Qatar, Algeria, Australia, and others.

8.6.4 Rationalization for adding minerals to de-salted water

Permeates from membrane treatments and distillates from thermal desalination plants both possess very low TDS levels and are not suitable for human consumption due to health and taste considerations (Cotruvo, 2006; WHO, 2011b). Permeates and distillates are slightly acidic, usually in the pH range of 5.5–6.5. Pre-treatment steps in membrane processes require the addition of mineral acids to prevent scale formation on the membranes and maintain their performance. Acidification leads to the generation of CO_2, which passes through the membranes and acidifies the permeate. In thermal plants, the dissolution of atmospheric CO_2 in the distillates leads to acidification and low pH. The joint effect of low pH and low TDS produces highly aggressive and corrosive permeates and distillates, which are highly dangerous for the distribution network (WHO, 2011b).

Desalination plants differ in principle and practice from the traditional plants for freshwater for three main reasons: (1) saline source; (2) treatment techniques (thermal, RO); and (3) post-treatment including remineralisation. Consequently, the product water from desalination plants has very different properties in comparison with finished water from conventional plants. While the WHO Water Quality Guidelines (2011b) are fully applicable to desalinated water supply systems, a specific section in the guidelines is devoted to special water quality characteristics for desalinated water. These special characteristics deserve a brief discussion of their features and implications:

8.6.4.1 Minerals and health

Although drinking water and tap water are not globally considered a significant source of minerals in the daily diet, they can be relevant for special segments of the population. In addition to the important role of calcium and magnesium in bone health, many epidemiology studies over the last 50 years show the benefits of consumption of hard waters for cardiovascular disease (Cotruvo & Bartram, 2009; Monarca *et al.*, 2006; WHO, 2005). It has been suggested that the beneficial effects of calcium and magnesium could be an indirect consequence of the lack of aggressiveness against plumbing of hard waters in comparison with the highly corrosive soft waters. For this reason, whatever the actual biochemical mechanism of actuation is, it has been suggested that reintroduction of calcium and magnesium in desalinated water would be beneficial for the health of consumers.

Taste of water: Untreated desalinated waters are unpalatable due to the extremely low TDS. They are described as rough, bitter, and flat. Significant improvements are obtained by the addition of minerals, especially bicarbonate and calcium. This subject is treated in prior sections of this chapter.

Chemical instability: Demineralized waters are prone to pH shifts due to their lack of bicarbonate alkalinity and poor buffer capacity. This situation is very important as shown by Lahav and Birnhack (2007) because several relevant reactions that occur in the distribution network are pH dependent, such as chlorination/disinfection, fluoridation, or nitrification in chloraminated waters.

Aggressiveness/corrosivity: Desalination processes remove almost all of the minerals from the water source. Demineralized waters are corrosive for pipes in the distribution network and also for plumbing in buildings.

To summarize, for these four reasons - health, palatability, lack of stability, and corrosion risk – desalinated product waters must be properly conditioned. Two main treatment technologies are available: remineralization and blending. In remineralization, chemical salts, the same as those used for pH adjustment, are added to desalinated water. The usual chemicals are lime and limestone. In blending, the desalted water is blended with high mineral source water, preferably water with high alkalinity and mineral content. Blending is broadly used in thermal plants by addition of a small percentage of raw seawater.

There are other special characteristics of desalinated waters that have no relationship to remineralisation or blending, and therefore are out of the scope of the present chapter. Nonetheless, the authors will mention two of these: higher levels of boron and additional risk of microbial re-growth, as these are of concern to water producers and consumers (WHO, 2011b).

8.6.4.2 Remineralization techniques

Since the inception of desalination technologies, it was acknowledged that minerals needed to be added back to product waters in order to improve water quality. Several remineralization technologies have been developed and are currently implemented (Bloetscher *et al.*, 2013; Duranceau, 2009; Duranceau *et al.*, 2012; Whiters, 2005). These current technologies are discussed below.

Current remineralization methods have some limitation; therefore, it is necessary to note that post-treatment is an important and active area of research because there is a great need to improve this aspect of the desalination process.

The choice of remineralization method is case specific and depends on several parameters: characteristics and quality of the feed water; desalination technology and particular plant design; required water quality or specifications to comply with regulations; and logistics, availability and cost of the chemicals to be used. And most importantly, the taste of the water, because it is obvious that the remineralization process has a relevant influence on the aesthetics of tap water and customer's perception (Vingerhoeds *et al.*, 2016). Remineralization for fluoridation of drinking water for dental health reasons is not discussed here due to its lack of influence on taste.

The most common treatments for remineralization are addition of lime and limestone, in conjunction with the addition of carbon dioxide. In thermal plants, CO_2 can be recovered from the distillation process, although it is economically viable in MSF, but not in MED plants due to brine temperature considerations (Glade *et al.*, 2005). The addition of calcium to the finished water through the use of lime or limestone is beneficial to taste. The reported taste threshold for calcium is between 100 and 300 mg/L in water (WHO, 2011a).

The principle is based on dosing CO_2 gas and a slurry of hydrated or slaked lime (milk of lime, $Ca(OH)_2$) to the desalinated water. The use of quicklime (CaO) is very unusual due to operational reasons derived from the highly exothermic reaction for its dissolution in water. The desalinated water is re-carbonated according to the reaction:

$$2\,CO_2 + Ca(OH)_2 \rightarrow Ca(HCO_3)_2$$

This equation results in a mass ratio where 8.8 mg/L of CO_2 and 7.4 mg/L of lime are needed to increase total alkalinity by 10 mg/L as $CaCO_3$.

Remineralization facilities require short contact times and are relatively simple in comparison with calcite reactors. Lime is dosed in a mixer as a slurry, generally about 3–5%, which is prepared from a more concentrated solution coming from a saturator. The main drawback of this system is the risk of turbidity problems due to the low solubility of calcium hydroxide in water.

CO_2 and limestone: First, CO_2 gas is added and dissolved. The permeate is then forced to pass through a bed of limestone ($CaCO_3$), which is a sedimentary rock composed by three polymorphic phases of which calcite is the most common. In this case, the reaction is:

$$CO_2 + CaCO_3 + H_2O \rightarrow Ca(HCO_3)_2$$

It can be seen that this process requires 4.4 mg/L of CO_2 and 10 mg/L of $CaCO_3$ to increase total alkalinity by 10 mg/L as $CaCO_3$, which is one half of the CO_2 required for the lime process. In practice, due to the slow kinetics of calcite dissolution, an excess of the stoichiometric CO_2 is often necessary, and therefore, a smaller saving in this component is achieved.

The process can be accomplished in gravity or pressure contact chambers (commonly referred as 'bed filters' which is not strictly correct). These installations are large with continuous monitoring of operation conditions to comply with the finished water specifications. Turbidity problems are not usually present provided that properly washed calcite is used to remove fines.

The composition of commercially available limestone varies substantially depending on the geographical origin of the quarry. In addition to calcium carbonate, limestones can contain significant levels of dolomite ($CaMg(CO_3)_2$) and other components such as quartz and clays. The presence of magnesium is increasingly appreciated because of the beneficial effects of this metal on human health. However, the drawback is the additional difficulty in the operation of the installation owing to the slower kinetics of dolomite dissolution in comparison with calcite. Quartz and clays are undesirable due to their insolubility.

Ruggieri *et al.* (2008) performed a comparative study of the textural and dissolution properties of different samples of quarry limestone. They recommend avoiding dolomite and using an alternative source of magnesium. Birnhack *et al.* (2008) proposed a treatment consisting of calcite beds followed by ionic exchange with a cationic resin previously loaded with magnesium. In some countries, a product of marine origin (*lithothamnium*, a fossil calcareous alga) is used for remineralization because of its availability and high reactivity. Calcium carbonate is the major component, but contains a high quantity of magnesium carbonate (approximately 5–10%).

Other methods: When using limestone, an alternative to CO_2 is sulfuric acid. In this case, the pH of the permeate circulating through the limestone contactor can be adjusted to more acidic values, and this facilitates the dissolution of calcite. The main drawback is the risk of precipitation of calcium sulfate and that the ratio alkalinity/Ca is not very appropriate (Lahav & Birnhack, 2007). Other less common remineralisation processes include: lime plus sodium carbonate, sodium bicarbonate plus calcium sulphate, and sodium bicarbonate plus calcium chloride. The main drawback is the higher price of chemicals, and thus, they tend to be used in small plants. A final pH adjustment with CO_2 or mineral acid is often necessary.

8.6.4.3 Blending

Historically, blending waters has been commonly used to manage water quality in distribution networks served by two or more sources. One goal is to homogenize water quality from different sources. In 'time', homogenization is done to avoid consumer-perceived changes in quality from one day to another. In 'space', homogenization is done to avoid different qualities in different geographical areas of the network at the same time, to 'dilute' one water with a parameter above the permitted level with another water possessing a lower concentration in order to meet internal specifications or regulations, to 'dilute' a taste-and-odour event, for example, occurrence of geosmin or MIB in the drinking water, or to mitigate changes of temperature in the network (which can be dangerous in winter) by adding groundwater to surface sources.

For desalinated waters the concept of blending has acquired a new dimension. Blending is a useful tool to improve the physico-chemical quality of the distributed water, in particular quality parameters related to aggressivity/ corrosivity and aesthetics. It is important to pay attention to the quality of the water used for blending. For seawater facilities, it is important to take into account variable raw water quality, which may contain hydrocarbons, algae, or microorganisms. When groundwater is used for blending, it must be monitored and disinfected to avoid introducing pathogens into the finished water, even though the percentage of water added in blending is small. Reasons for and types of blending facilities are further described below.

Blending for corrosion control: Demineralized waters are highly aggressive toward cement materials and they are also highly corrosive for metallic materials used in the distribution network and storage. Corrosivity is due to very low TDS, acidic pH, extremely low alkalinity and calcium concentration, and in consequence, poorly buffer capacity. Corrosion is an enormous and complex concern for water companies, requiring that pipes be maintained or replaced periodically, independently of the construction materials. For iron pipes, 'red water' is acknowledged to be a global corrosion problem due to release of iron particulates or 'rust' (Imran *et al.*, 2005; Benson *et al.*, 2012). Metal plumbing is also not compatible with demineralized water, which has the potential to leach lead, copper, nickel and zinc, all of which have undesirable effects on health and taste. Additionally, distribution networks and plumbing systems can be disturbed when the distributed water quality changes.

In order to minimize corrosion, in addition to the usual recarbonation stages in post-treatment and dosing with anticorrosive agents (generally phosphates and polyphosphates), blending source waters has been a useful tool to minimize the corrosive behaviour of the distributed waters. Several studies can be found in the literature about the improvement achieved by blending demineralized waters with another source (Liu *et al.*, 2009; Plottu-Pecheux *et al.*, 2001; Tang *et al.*, 2006; Taylor *et al.*, 2005).

The water distribution industry often uses corrosion indices to predict the corrosivity of source waters and corrosion release models to predict iron release (Benson *et al.*, 2012; Imran *et al.*, 2005). The most commonly used indices are the Larson Ratio, Langelier Saturation Index, Ryznar Index, and Calcium Carbonate Precipitation Potential. These indicies consider pH, alkalinity, and concentrations of specific anions to account for the effects of scaling usually through formation of a calcium carbonate layer. Implementation of corrosion indices or corrosion release models are challenging and often require detailed investigations into the water quality and hydraulic characteristics of the distribution networks. This is an area of active reseach.

Post-treatment and blending facilities: Blending can be achieved in two different places or situations: as a stage of the desalination process before the finished water leaves the plant, or in a segregated mixing facility after the water leaves the treatment plant. The first practice of blending at the plant is normally used for seawater and brackish water treatment facilities when the source water has a good and consistent quality, as well as permeates and distillates having very low TDS and minimal concentrations of calcium, magnesium, and bicarbonate ions. In these situations, minerals can be added to the demineralised water by blending it with source water. For seawater plants, blending is generally limited to about 1% to achieve approximately 105 mg/L sodium, 4 mg/L calcium, 13 mg/L magnesium, and 190 mg/L chloride. It is clear that water enriched in sodium and chloride is not the best option for taste and aggressivity; additionally, blending with seawater provides no alkalinity or buffer capacity. Thus, a pH adjustment is usually necessary. Since the mixing is done before finished water leaves the desalination plant, sometimes this process is considered a special type of remineralisation rather than a blending process.

Conceptually different is the second situation; water that is circulated through the distribution network from a desalination plant is blended with one or more water sources in specially designed facilities located after the main treatment facility. In other words, mixing is done outside the perimeter of the desalting plant (although it can be located very close). In these situations, demineralized waters are usually blended with surface water or groundwater of elevated salinity.

All blending facilities have to be carefully designed to ensure that all waters reaching the blending facilities are properly treated, disinfected, and monitored. When these facilities are considered as part of the distribution network, then they are under the responsibility of the distribution department of the water company. On the contrary, sometimes in addition to the mixing stage some treatment stages, like filtration or disinfection, are needed. In this case, the process is 'returned' again to the responsibility of water production area. Thus, blending facilities can pose both administrative and technical challenges.

When the mixing process takes place outside the desalination plant, a particular circumstance must be taken into account: the possible existence of consumers situated between the desalination plant and the blending stage. In this case, depending on the distribution strategy, it is possible that these consumers receive unblended desalinated water. The water companies have to pay attention to the properties of this distributed water, in particular, the aggressivity and taste of the water.

A good example of the complexity of the blending process is presented by Kilmore (2009) through the challenge for Adelaide, Australia to blend desalinated seawater with surface water. The distributed water can have changing water quality as 'full flexibility' is required such that the blends can range from 0 to 100% desalinated seawater. An economic approach of desalting and blending is given by Rygaard *et al.* (2009, 2011). The authors developed a model, including the cost of desalination and remineralization, the health impacts on consumers, lifetime of distribution pipes and domestic appliances, detergent consumption, and purchasing of bottled water. Interesting results are obtained by application of the model to Perth, Australia and Copenhagen, Denmark.

8.6.4.4 Blending for taste assessment and improvement

The design of blending facilities and the proportions of water to be blended are generally determined based on physico-chemical considerations for distributing a product water that is stabilized, non-corrosive, and meets regulatory requirements. Aesthetic quality of the blended water is increasingly becoming an important driver for water companies and consumers (Devesa *et al.*, 2010; Duranceau *et al.*, 2012; Raich-Montiu *et al.*, 2014; Vingerhoeds *et al.*, 2016). Water companies are also in need of more information to make effective decisions about blending waters to satisfy their consumers. For this reason, several interesting articles concerning organoleptic studies of blended waters are summarized below.

Two cases from California, where desalination facilities have a long tradition, have been presented by McGuire *et al.* (2007a, b). The first case involves treated water for communities east of San Francisco (California, USA) and is a good example of a complex distribution network using different blends from surface water and groundwater that is partially treated by RO. The negative perception of water quality by consumers has historically been attributed to the high hardness of the groundwater. The results of the study suggested the negative relationship between liking and TDS, but in this particular case, the earthy/musty off-flavour produced by geosmin was the main factor for consumers' annoyance. The study recommended introducing activated carbon or ozonation to remove geosmin from the surface water and membrane treatment to reduce the mineralization of groundwater.

The second case was from San Diego, California, USA (McGuire *et al.*, 2007b). The main purpose of the study was to estimate if consumers could discern between the water currently delivered, which was freshwater from a conventional treatment plant, and desalinated seawater coming for a new plant. This information was crucial to decide if blending facilities would be needed. Two types of panels were summoned: professional tasters and untrained consumers. The trained panel followed the FPA method while the consumer panel participated in several preference taste test sessions. The results of the project showed good agreement between the types of panels. It was deduced that alkalinity adjustment did not significantly affect the perception of the desalinated water. Also, tasters could discern between the waters; their preference was for the surface water. The final result was that the discrimination between water types was not significant enough to recommend the construction of blending facilities to mitigate the differences in taste between the two waters.

Another relevant situation comes from the city of Tucson, Arizona, USA (Biggs *et al.*, 2007). In this study, historical blending had been used to mix Colourado River water (650 mg/L TDS) with local groundwater (300 mg/L TDS). Traditionally, blends supplied to the consumers had been up to 450 mg/L. Due to the need to preserve the aquifer and reduce groundwater consumption, there was a need to increase the proportion of surface water in the blends. Before implementing this change, a consumer tasting campaign was initiated by the Tucson water department. Two blends of 450 and 650 mg/L were prepared and about 14,000 customers were invited to taste and express their preference. The results were that 57% preferred the 450 blend, 28% preferred the 650 blend, and 15% had no preference. When consumers were informed about the additional cost associated to the 450 mg/L blend, the figures changed to 50, 35, and 15% respectively.

The same situation was again studied by McGuire *et al.* (2007c), but in this case, there was a consumer panel of 80 people who participated in a hedonic rating test. A 2×3 factorial design was used, with three levels of mineralization (450, 550, and 650 mg/L) and two levels of residual chlorine (0.4 and 0.8 mg/L). The differences in the mean liking scores were relatively small, although they permitted the deduction that subjects tended to give lower ratings as TDS increased.

Other studies with trained and untrained panels have been performed in the metropolitan area of Barcelona, Catalonia (Spain), where a complex distribution network is fed with several sources, including surface water, groundwater and treated seawater (Devesa *et al.*, 2010). For desalination, different technologies are used, including NF, RO and EDR. During a pilot plant study to evaluate treatment technologies, the taste changes were achieved by blending Llobregat River water, a high mineralization water source of 1,000–1,100 mg/L TDS, with different percentages of either RO or EDR membrane treated water. The results demonstrated a gradual improvement when the TDS of the blended water was decreased. Blends containing 70, 50, and 30% of membrane treated water (TDS values from 350 to 760 mg/L) were preferred and effectively discriminated from unblended water (970–1,060 mg/L). The results of the study confirmed the usefulness of membranes to improve the taste and taste testing to improve the aesthetic quality of the drinking waters (Devesa *et al.*, 2010). Raich-Montiu *et al.* (2014) confirmed the actual improvement in the aesthetic quality of the tap water in a study with samples taken directly from the distribution network. A later study also investigated taste quality of Barcelona's drinking water when treated with membranes for desalination (García *et al.*, 2014). Both trained and untrained panellists reported that water with lower salinity/TDS was preferred and that

TDS was the primary driver of flavour with pH and mineral composition also having influence.

8.7 SUMMARY

In the 21st century, consumers expect water that is chemically and microbiologically safe, as well as, good tasting. The major minerals that constitute TDS – calcium, sodium, magnesium, sulfate, chloride, and bicarbonate – are primarily responsible for giving waters their good taste, with calcium, bicarbonate, and sulfate the preferred ions for good-tasting water. Minor metals, such as copper, iron, and zinc, can cause water to have an off-flavour. Water providers, whether using conventional treatment or desalination/remineralization, have a responsibility to understand the effect that minerals have on water quality not only for control of corrosion and maintenance of water stability in distribution systems, but also for taste, which is the key criterion by which consumers judge water quality.

REFERENCES

American Public Health Association (APHA), American Water Works Association (AWWA), Water Environment Federation (WEF) (2012). Standard Methods for the Examination of Water and Wastewater, E. W. Rice, R. B. Baird, A. D. Eaton, L. S. Clesceri (eds), 22nd edn, Washington, DC.

Australian Drinking Water Guidelines 6 (2011). Volume 1, Natural Resource Management Ministerial Council, Commonwealth of Australia, Canberra.

Azoulay A., Garzon P. and Eisenberg M. J. (2001). Comparison of the mineral content of tap water and bottled waters. *Journal of General Internal Medicine*, **16**, 168–175.

Bartoshuk L. M. (1980). Sensory analysis of the taste of NaCl. In: Biological and Behavioral Aspects of Salt Intake, M. R. Kare, M. J. Fregly and R. A. Bernard (eds), Academic Press, New York, pp. 83–98.

Benson A. S., Dietrich A. M. and Gallagher D. L. (2012). Evaluation of iron corrosion release models for water distribution systems. *Critical Reviews in Environmental Science and Technology*, **42**, 44–97.

Biggs J., Thomure T., Arnol K. and Cotton C. (2007). Public outreach tool used in determining customer preferences of water aesthetics. Proceedings of the AWWA Water Quality Technology Conference, Charlotte, North Carolina, November 4th–8th.

Birnhack L., Penn R. and Lahav O. (2008). Quality criteria for desalinated water and introduction of a novel, cost effective and advantageous post-treatment process. *Desalination*, **221**, 70–83.

Bloetscher F., Stambaugh D., Hart J., Cooper J., Kennedy K., Sher L., Ruffini A. P., Cicala A. and Cimenello S. (2013). Use of lime, limestone and kiln dust to stabilize reverse osmosis treated water. *Journal of Water Reuse and Desalination*, **3**, 277–290.

Booth S. and Brazos B. (2004). Qualitative Procedures for Identifying Particulate Matter in Distribution and Tap Waters. AWWA Research Foundation, Denver, CO.

Bruvold W. H. (1968). Scales for rating the taste of water. *Journal of Applied Psychology*, **52**, 245–253.

Bruvold W. H. and Daniel J. I. (1990). Standards for mineral content in drinking water. *Journal American Water Works Association*, **82**, 59–65.

Bruvold W. H. and Pangborn R. M. (1970). Dissolved oxygen and the taste of water. *Journal American Water Works Association*, **62**, 721–727.

Burdach K. and Doty R. (1987). The effects of mouth movement, swallowing and spitting on retronasal perception. *Physiology and Behaviour*, **41**, 353–356.

Burlingame G. A., Dietrich A. M. and Whelton A. J. (2007). Understanding the taste of tap water. *Journal American Water Works Association*, **99**, 100–111.

Cerrato J. M., Reyes L. P., Alvarado C. N. and Dietrich A. M. (2006). Effect of PVC and iron materials on Mn(II) deposition in drinking water distribution systems. *Water Research*, **40**, 2720–2726.

Cerrato J. M., Hochella M., Knocke W., Dietrich A. M. and Cromer T. (2010). Use of XPS to identify the oxidation state of Mn in solid surfaces of filtration media oxide samples from drinking water treatment plants. *Environmental Science & Technology*, **44**, 5881–5886.

Combs G. F. (2005). Geological impacts on nutrition. In: Essentials of Medical Geology. In: Impacts of the Natural Environment on Public Health, O. Selenius, B. J. Alloway, J. A. Centeno, R. B. Finkleman, R. Fuge, U. Lindh and P. Smedley (eds), Elsevier Academic Press, New York, pp. 161–177.

Cotruvo J. (2006). Health aspects of calcium and magnesium in drinking water. *Water Conditioning and Purification (June).*

Cotruvo J. and Bartram J. (2009). Calcium and Magnesium in Drinking-water: Public Health Significance. World Health Organization, Geneva, Switzerland.

Cuppett J. D., Duncan S. E. and Dietrich A. M. (2006). Evaluation of copper speciation and water quality factors that affect aqueous copper tasting sensitivity. *Chemical Senses*, **31**, 689–697.

Dawoud M. A. and Al Mulla M. M. (2012). Environmental impacts of seawater desalination: Arabian Gulf case study. *International Journal of Environment and Sustainability*, **1**, 22–37.

Devesa R. and Dietrich A. M. (2018). Guidance for optimizing drinking water taste by adjusting mineralization as measured by total dissolved solids (TDS). *Desalination*, **49**, 147–154.

Devesa R., García V. and Matía L. (2010). Water flavour improvement by membrane (RO and EDR) treatment. *Desalination*, **250**, 113–117.

Dietrich A. M. (2006). Aesthetic issues for drinking water. *Journal of Water and Health*, **4** (Supplement 1), 1–11.

Dietrich A. M. (2009). The sense of smell: contributions of orthonasal and retronasal perception applied to metallic flavour of drinking water. *Journal of Water Supply: Research Technology – AQUA*, **5**, 562–570.

Dietrich A. M. and Burlingame G. A. (2015). Critical review and rethinking of USEPA secondary standards for maintaining consumer acceptability of organoleptic quality of drinking water. *Environmental Sciences & Technology*, **49**(2), 708–720.

Dietrich A. M. and Gallagher C. D. (2013). Consumer ability to detect the taste of total dissolved solids. *Journal American Water Works Association*, **105**(5), 255–263.

Dietrich A. M., Glindemann D., Pizarro F., Gidi V., Olivares M., Araya M., Camper A., Duncan S., Dwyer S., Whelton A. J., Younos T., Subramanian S., Burlingame G. A.,

Khiari D. and Edwards M. (2004). Health and aesthetic impacts of copper corrosion on drinking water. *Water Science and Technology*, **49**(2), 55–62.

Dietrich A. M., Cuppett J. D. and Duncan S. E. (2008). How much copper is too much? *OpFlow*, **34**, 28–30.

Duranceau S. J. (2009). Post-treatment of desalinated waters. Proceedings of the 2009 AWWA WQTC Conference, Seattle, Washington, November 15t–19th.

Duranceau S. J., Wilder R. J. and Douglas S. S. (2012). Guidance and recommendations for post-treatment of desalinated water. *Journal American Water Works Association*, **104**(9), E510–E520.

Edwards M. (2004). Controlling corrosion in drinking water distribution systems: A grand challenge for the 21st century. *Water Science and Technology*, **49**(2), 1–8.

Epke E. M. and Lawless H. T. (2007). Retronasal smell and detection thresholds of iron and copper salts. *Physiology and Behavior*, **92**, 487–491.

European Union (1998). EU COUNCIL DIRECTIVE 98/83/EC of 3 November 1998 on the quality of water intended for human consumption; NOTE: mg/L TDS calculated based on conductivity. EU, Strasbourg, France.

European Union (2015). COMMISSION DIRECTIVE (EU) 2015/1787 of 6 October 2015 amending Annexes II and III to Council Directive 98/83/EC on the quality of water intended for human consumption. EU, Strasbourg, France.

Fox S. I. (2008). Human Physiology. McGraw-Hill, New York, NY.

Gallagher C. D. and Dietrich A. M. (2010). TDS and temperature affect consumer taste preferences. *OpFlow*, **36**, 20–22.

García V., Fernández A., Medina M., Ferrer O, Cortina J., Valero F. and Devesa R. (2014). Flavour assessment of blends between desalinated and conventionally treated sources. *Desalination and Water Treatment*, **53**, 3466–3474.

Glade H., Meyer J.-H. and Will S. (2005). The release of CO_2 in MSF and ME distillers and its use for the recarbonation of the distillate: A comparison. *Desalination*, **182**, 99–110.

Health Canada (2012) Guidelines for Canadian Drinking Water Quality. http://www.hc-sc.gc.ca/ewh-semt/alt_formats/pdf/pubs/water-eau/2012-sum_guide-res_recom/2012-sum_guide-res_recom-eng.pdf (Accessed Dec 2012)

Hem J. D. (1970). Study and Interpretation of the Chemical Characteristics of Natural Water, 2nd ed. Geological Survey Water Supply Paper 1473, United States Government Printing Office, Washington, DC.

Hettinger T. P., Myers W. E. and Frank M. E. (1990). Role of olfaction in perception of non-traditional 'taste' stimuli. *Chemical Senses*, **16**, 755–760.

Hoehl K., Schoenberger G. U. and Busch-Stockfisch M. (2010). Water quality and taste sensitivity for basic tastes and metallic sensation. *Food Quality and Preference*, **21**, 243–249.

Hong J.-H., Duncan S., Dietrich A. M. and O'Keefe S. (2006). Effect of copper on volatile flavour compounds in a model mouth system. *Journal of Agricultural and Food Chemistry*, **54**, 9168–9175.

Hong J.-H., Duncan S. E., O'Keefe S. and Dietrich A. M. (2009). Ultrafiltration as a tool to study binding of copper to salivary proteins. *Food Chemistry*, **131**(1), 180–184.

Hong J.-H., Duncan S. E., Dietrich A. M. and O'Keefe S. (2010a). Evaluation of the operationally defined soluble, insoluble, and complexing copper consumed through drinking water in human saliva. *European Food Research and Technology*, **231**, 977–984.

Hong J.-H., Duncan S. E. and Dietrich A. M. (2010b). Effect of copper speciation at different pH on temporal sensory attributes of copper. *Food Quality and Preference*, **21**, 132–139.

Imran S. A., Dietz J. D., Mutoti G., Taylor J. S. and Randall A. A. (2005). Modified Larsons ratio incorporating temperature, water age, and electroneutrality effects on red water release. *Journal of Environmental Engineering*, **131**(11), 1514–1520.

Keast R. S. J. (2006). The effect of zinc on human taste perception. *Journal of Food Science*, **68**, 1871–1877.

Kilmore G. (2009). Challenges associated with mixing desalinated sea water and conventionally treated surface waters to minimise customer complaints. Proceedings of the 2009 AWWA WQTC Conference, Seattle, Washington, November 15th–19th.

Kirmeyer G. J. (2000). Guidance Manual for Maintaining Distribution System Water Quality. AWWA Research Foundation and American Water Works Association, Denver, CO, USA.

Koseki M., Fujiki S., Tanaka Y., Noguchi H. and Nishikawa T. (2005). Effect of water hardness on the taste of alkaline electrolyzed water. *Journal of Food Science*, **70**, S249–S253.

Kozisek F. (2003). Health Significance of Drinking Water Calcium and Magnesium. *National Institute of Public Health*. Feb 2003. 1–29. http://www.szu.cz/uploads/documents/chzp/voda/pdf/hardness.pdf (Accessed October 1, 2009)

Lahav O. and Birnhack L. (2007). Quality criteria for desalinted water following post-treatment. *Desalination*, **207**, 286–303.

Laing D. G., Prescott J., Bell G. A., Gillmore R., James C., Best J., Allen S., Yoshida M. and Yamazaki K. (1993). A cross-cultural study of taste discrimination with Australians and Japanese. *Chemical Senses*, **18**, 161–168.

Lawless H. T., Rapaki F., Horne J. and Hayes A. (2003). The taste of calcium and magnesium salts and anionic modifications. *Food Quality and Preference*, **14**, 319–325.

Liu H., Desormeaux E., Korshin G. V., Luckenbach H., Ferguson F. and Meyerhofer P. (2009). Effects of desalinated water and its blends with conventionally treated surface water on copper and lead release. Proceedings of the 2009 AWWA WQTC Conference, Seattle, Washington, November 15th-19th.

Marcussen H., Holm P. E. and Hansen H. C. B. (2013a). Composition, flavor, chemical food safety, and consumer preferences of bottled water. *Comprehensive Reviews in Food Science and Food Safety*, **12**, 333–352

Marcussen H., Bredie W. L. P., Stolzenbach S., Brüsh W., Holm P. E. and Hansen H. C. B. (2013b). Sensory properties of Danish municipal drinking water as a function of chemical composition. *Food Research International*, **54**, 389–396.

Masters G. M. and Ela W. P. (2007). Introduction to Environmental Engineering and Science, 3rd ed. Chapter 5, Prentice Hall, Upper Saddle River, New Jersey, USA.

Matia L. (1995). Treatment of tastes in drinking water: Causes and control. In: Advances in Taste-and-Odour Treatment and Control, J. H. Suffet, J. Mallevialle and F. Kawczynski (eds), American Water Works Association Research Foundation and Lyonnaise des Eaux, Denver, CO, pp. 247–253.

Matsou R. (2000). Role of saliva in the maintenance of taste sensitivity. *Critical Review of Oral Biology & Medicine*, **11**, 216–229.

McGuire M. J. (1995). Off-flavour as the consumer measure of drinking water safety. *Water Science and Technology*, **31**(11), 1–8.

McGuire M. J., Leserman J., Requa D., Stephenson S., Lang M. and Blute N. (2007a). Advantages of using a consumer panel to examine aesthetic challenges in a northern California water system. Proceedings of the AWWA Water Quality Technology Conference, Charlotte, North Carolina, November 4th–8th.

McGuire M. J., Loveland J., Means E. G. and Garvey J. (2007b). Use of flavour profile and consumer panels to determine differences between local water supplies and desalinated seawater. *Water Science and Technology*, **55**(5), 275–282.

McGuire M. J., Arnold K. F., Biggs J. and Pearthree M. S. (2007c). Using a consumer panel as a tool for making water resource blending decisions. Proceedings of the AWWA Water Quality Technology Conference, Charlotte, North Carolina, November 4th–8th.

Mirlohi S., Dietrich A. M. and Duncan S. E. (2011). Age-associated variation in sensory perception of iron in drinking water and the potential for overexposure in the human population. *Environmental Science & Technology*, **45**, 6575–83.

Monarca S., Donato F., Zerbini I., Calderon R. L. and Craun G. F. (2006). Review of epidemiological studies on drinking water hardness and cardiovascular diseases. *European Journal of Cardiovascular Prevention and Rehabilitation*, **13**, 495–506.

Naddeo V., Zarraa T. and Belgiornoa V. (2008). A comparative approach to the variation of natural elements in Italian bottled waters according to the national and international standard limits. *Journal of Food Composition and Analysis*, **21**(6), 505–514.

Neyraud E. and Dransfield E. (2004). Relating ionisation of calcium chloride in saliva to bitterness perception. *Physiology and Behavior*, **81**(3), 505–10.

Ömür-Özbek P. (2012). Global Taste and Odor Survey of Water Utilities: Final Report to the American Water Works Association from the Taste and Odor Committee. American Water Works Association, Denver, CO.

Ömür-Özbek P. and Dietrich A. M. (2011). Retronasal perception and flavour thresholds of iron and copper in drinking water. *Journal of Water and Health*, **9**(1), 1–9.

Ömür-Özbek P., Dietrich A. M., Duncan S. E. and Lee Y. W. (2012). Role of lipid oxidation, chelating agents, and antioxidants in metallic flavor development in the oral cavity. *Journal of Agricultural and Food Chemistry*, **60**(9), 2274–2280.

Pangborn R. M. and Bertolero L. (1972). Influence of temperature on taste intensity and degree of liking of drinking water. *Journal American Water Works Association*, **64**, 511–515.

Pedersen A. M., Bardow A., Jensen S. B. and Nauntofte B. (2002). Saliva and gastrointestinal functions of taste, mastication, swallowing and digestion. *Oral Diseases*, **8**, 117–129.

Piriou P., Mackey E. D., Suffet I. H. and Bruchet A. (2004). Chlorinous flavor perception in drinking water. *Water Science and Technology*, **49**(9), 321–328.

Piriou P., Devesa R., de Lalande M. and Glucina K. (2009). European reassessment of MIB and geosmin perception in drinking water. *Journal of Water Supply: Research and Technology – AQUA*, **58**, 532–538.

Piriou P., Devesa R., Puget S., Thomas-Danguin T., Roche V. and Glucina K. (2010). Evidences of regional differences in chlorine perception by consumers: Sensitivity differences or habituation? *Journal of Water Supply: Research and Technology—AQUA*, **64**, 783–792.

Platikanov S., Garcia V., Fonseca I., Rullán E., Devesa R. and Tauler R. (2013). Influence of minerals on the taste of bottled and tap water: A chemometric approach. *Water Research*, **47**, 693–704.

Plottu-Pecheux A., Democrate C., Houssais B., Gatel D. and Cavard J. (2001). Controlling the corrosiveness of blended waters. *Desalination*, **138**, 237–249.

Raich-Montiu J., Barios J., García V., Medina M. E., Valero F., Devesa R. and Cortina J. L. (2014). Integrating membrane technologies and blending options in water production and distribution systems to improve organoleptic properties. The case of the Barcelona Metropolitan Area. *Journal of Cleaner Production*, **69**, 250–259.

Ruggieri F., Fernandez-Turiel J. L., Gimeno D., Valero F., García J. C. and Medina M. D. (2008). Limestone selection criteria for EDR water remineralization. *Desalination*, **227**, 314–326.

Rygaard M., Arvin E. and Binning P. J. (2009). The valuation of water quality: effects of mixing different drinking water qualities. *Water Research*, **43**, 1207–1218.

Rygaard M., Arvin E., Bath A. and Binning P. J. (2011). Designing water supplies: optimizing drinking water composition for maximum economic benefit. *Water Research*, **45**, 3712–3722.

Sain A. E. and Dietrich A. M. (2015). Rethinking aesthetic guidelines for manganese and iron in drinking water. *Journal of Water Supply: Research and Technology-AQUA*. doi: 10.2166/aqua.2014.091.

Sain A. E., Griffin A. and Dietrich A. M. (2014). Assessing taste and visual perception of Mn(II) and Mn(IV) in drinking water. *Journal American Water Works Association*, **106**, 32–34.

Sasaki K., Iwanaga C., Wntanabe M., Suzuki K., Hamaoka T. and Kondo S. (1996). Correlation between good taste and chemical composition in natural and mineral water. *Nippon Nogeikagaku Kaishi*, **70**(10), 1103–1116.

Smith D. V. and Margolskee R. F. (2001). Making sense of taste. *Scientific American*, **284**, 32.

Stumm W. and Morgan J. (1996). Aquatic Chemistry. Wiley, New York.

Taiwan Government. Taiwan (2009). Drinking Water Quality Standards-Revisions to Article 3 promulgated by Environmental Protection Administration Order Huan-Shu-Tu-Tzu No.0980106331E on Nov. 26, 2009. (Accessed September 9, 2012)

Tang Z., Hong S., Xiao W. and Taylor J. (2006). Impacts of blending ground, surface, and saline waters on lead release in drinking water distribution systems. *Water Research*, **40**, 943–950.

Taylor J., Dietz J., Randall A. and Hong S. (2005). Impact of RO-desalted water on distribution water qualities. *Water Science and Technology*, **51**(6-7), 285–291.

Teillet E., Urbano C., Cordelle S. and Schlich P. (2010). Consumer perception and preference of bottled and tap water. *Journal of Sensory Studies*, **25**, 463–480.

Tenovuo J. O. (1989). Human Saliva: Clinical Chemistry and Microbiology, Vol. **I**. CRC Press, Boca Raton, FL.

Tordoff M. G. (1996). Some basic psychophysics of calcium salt solutions. *Chemical Senses*, **21**, 417–424.

United Nations (2012). *International Decade for Action: 'Water for Life' 2005– 2015*. Water Scarcity. http://www.un.org/waterforlifedecade/scarcity.shtml (Accessed November 2012)

USEPA (1979). National Secondary Drinking Water Regulations. Final Rule. *Federal Register*, **44**(140), 42195.

USEPA (2003). Drinking Water Advisory: Consumer Acceptability Advice and Health Effects Analysis on Sodium. EPA 822-R-03-006, United States Environmental Protection Agency, Office of Water, Washington DC. www.epa.gov/safewater/ccl/pdf/sodium.pdf.

van der Aa M. (2003). Classification of mineral water types and comparison with drinking water standards. *Environmental Geology*, **44**, 554–563.

Vingerhoeds M. H., Nijenhuis-de Vries M. A., Ruepert N., van der Laan H., Bredie W. L.P. and Kremer S. (2016). Sensory quality of drinking water produced by reverse osmosis membrane filtration followed by remineralization. *Water Research*, **94**, 42–51.

Weisenthal K. E., McGuire M. and Suffet I. H. (2007). Characteristics of salt taste and free chlorine or chloramines in drinking water. *Water Science and Technology*, **55**(5), 293–300.

Westrell T., Andersson Y. and Stenström T. A. (2006). Drinking water consumption patterns in Sweden. *Journal of Water and Health*, **4**, 511–522.

Whelton A. J., Dietrich A. M., Burlingame G. A., Schechs M. and Duncan S. E. (2007). Minerals in drinking water: impacts on taste and importance to consumer health. *Water Science and Technology*, **55**, 283–291.

Whiters A. (2005). Options for recarbonation, remineralisation and disinfection for desalination plants. *Desalination*, **179**, 11–24.

WHO (2005). Nutrients in Water. World Health Organization, Geneva, Switzerland.

WHO (2011a). Guidelines for Drinking Water Quality, 4th ed. World Health Organization, Geneva, Switzerland.

WHO (2011b). Safe Drinking Water from Desalination. WHO/HSE/WHS 11.03, Geneva, Switzerland.

World Health Organization (1996). Guidelines for Drinking-Water Quality, 2nd ed, Vol. **2**. Health criteria and other supporting information. WHO, Geneva, Switzerland.

Worley J. L., Dietrich A. M. and Hoehn R. C. (2003). Dechlorination techniques to improve sensory odour testing of geosmin and 2-MIB. *Journal American Water Works Association*, **95**, 109–117.

Zhang X., Chen C., Lin P., Hou A., Niu Z. and Wang J. (2011). Emergency drinking water treatment during source water pollution accidents in China: Origin analysis, framework and technologies. *Environmental Science & Technology, Supplemental Information*, **45**, 161–167.

Zoeteman B. C.J., de Grunt F. E., Köster E. P., Smit K. B.J. and Punter P. H. (1978). Taste assessment of individual salts in water – methodology and preliminary findings by a selected national panel. *Chemical Senses*, **3**, 127–139.

Chapter 9

Risk management of public perception

Gary Burlingame[1], Sarah Neiderer[2] and Lisa Ragain[3]

[1]*Philadelphia Water Department, Philadelphia, Pennsylvania, USA*
[2]*Arlington, Virginia, USA*
[3]*Metropolitan Washington Council of Governments, Washington, District of Columbia, USA*

9.1 THE VALUE OF COMMUNICATING WITH THE PUBLIC

Whether driven by political interests, the ability to increase rates for water service and usage, contracts for wholesale water provision, or the need to develop or improve public water supply, the attitudes and perceptions of the public being served by their water suppliers matter. Taste-and-odour (T&O) problems can be one source of customer complaints and public outrage. The news media and special interest groups can carry these concerns and complaints further. As more and more communities around the world develop safe and sufficient water supplies, the importance of an acceptable aesthetic quality to the drinking water will find its way into the public discourse. Water quality issues, including aesthetic issues, can erode public confidence in the suppliers of tap water.

Now more than ever, the public's relationship with their potable water and water supplier is facing major challenges. Media reports, the internet, bottled water marketing and a variety of emerging issues have persuaded the public that tap water may not be safe. These perceptions influence the public's interest and

Editor(s): Tsair-Fuh Lin, Susan Watson, Andrea M. Dietrich and I. H. (Mel) Suffet
doi: 10.2166/9781780406664_0281

responsibility in maintaining and renewing their public water systems. The public's willingness to pay for safe and reliable potable water is affected by the quality of the water, including its taste, smell and appearance.

Even in developing countries people will not support efforts to invest in safe and reliable water if they have a negative perception of it. The aesthetic characteristics of drinking water play an important role in the global goal of providing everyone with safe drinking water. Time and time again across the globe and throughout history, communities have rejected a safer supply because of their poor perceptions and lack of trust that are often shaped by undesirable T&O quality (whether due to chlorine in the water or the change from groundwater to surface water).

Water utilities too often maintain a distant relationship with their customers. Water professionals think about tap water 24 hours a day, seven days a week. However, they tend to only communicate with their customers during events that trigger public concern and require a utility response. Communication rarely involves strategic, consistent messaging. Consequently, the public is only thinking about water when they notice an increase in a water bill, changes in water quality or read a news report. Yet without continued public support, water systems cannot maintain essential infrastructure, meet drinking water standards, or maintain a desirable aesthetic quality.

It is time to get tap water on the minds of customers in their everyday life, when things are good and even when they are bad. For customers to support their public water system, they must understand the invaluable role tap water serves in our daily lives. Even more, it is important the public views their local water utility as a steward of public health and environmental sustainability, rather than a bill collector. The goal is for the public to view their water utility as the solution to resolving water quality issues. If utilities can establish a positive relationship with their customers, the two can be mutually supportive.

Communication strategies and approaches vary among utilities, but simple communication strategies can assist in leveraging public support for all water systems, large and small. The three pillars of municipal water are safety, affordability and reliability. Two mutual priorities exist for both: financial stability and health. The public is interested in affordable water bills and safe drinking water. Water utilities are interested in cost-effective measures to minimize customer bills and achieve safe drinking water and acceptable aesthetic standards. If the public understands that their priorities – financial stability and health – are shared with their local utility, building trust is possible, and trust impacts on acceptability. Because water utilities and customers define risks differently, it is important to understand priorities and interests of each party. The simplest way to achieve this is in identifying shared commonalities between the public and utilities and by identifying best practices for delivering a simple message, that investment and interest in safe water is a responsibility and valuable resource for all of us.

9.2 RISK PERCEPTIONS AND DRINKING WATER AESTHETICS

Aesthetics influence public perceptions and the public's trust in their water supply (Lazo *et al.*, 2004). Meeting the drinking water standards does not inherently result in trust. Poor aesthetic quality translates to perceptions of unsafe water and can influence customers choosing alternative drinking water sources such as private wells, springs, filtered water and bottled water (Mackey *et al.*, 2003). The literature regarding customers, perceptions and drinking water is replete with discussions of drinking water aesthetics: taste, odour, colour and clarity. This reflects the role of aesthetics as a drinking water quality indicator for water utilities and for customers. For water utilities, regulatory and public health agencies, taste, odour and colour serve as indicators of chemical and microbial water quality. T&O are the most frequent concerns expressed by the public, and a recurrent reason for customers opting for bottled water or home water treatment systems. Interest in drinking water and aesthetics is deeply associated with understandings of risk, safety and health.

Drinking water aesthetics that are well known to utilities are colour, odour and taste (Kelly & Pomfret, 1997; McGuire, 1995; Whelton *et al.*, 2007). Customer taste descriptions of high quality drinking water are often cited as 'freshness' or 'natural'. Customers with positive perceptions of drinking water characterized the taste as 'nothing' or 'fresh' (Mackey *et al.*, 2003). The absence of odour and colour was associated with natural drinking water, water free of additives or chemicals (Doria *et al.*, 2009). The residual level of chlorine in drinking water has been correlated with increased customer dissatisfaction with water quality and an increased perception of risk associated with drinking water (Turgeon *et al.*, 2004). In the final analysis, perception becomes reality. Experience with water quality and aesthetics have been demonstrated to influence drinking water perceptions. In a large scale Canadian study, prior negative experiences with either drinking water quality problems (such as boil water advisories) and tap water off-tastes diminished people's assessments of tap water acceptability and safety (Dupont *et al.*, 2010). Even positive changes in water quality (changing the level of hardness, for example or changing source waters) can result in negative perceptions if the customers are not explained, the reason and value of the change.

Some studies have provided an understanding of public perceptions of drinking water T&O quality as well as water safety (Mackey *et al.*, 2003, 2004). Taylor *et al.* (2006) stated, 'T&O episodes can not only lead to consumer dissatisfaction with water quality aesthetics but can also diminish consumer confidence in the safety of tap water and erode consumer trust in the utility.' The majority of customers in North America express a confidence in the quality of the drinking water and in the water utilities that provide the drinking water, believing the water is safe to drink (Taylor *et al.*, 2006). Perhaps this is because the majority of reported tastes or odours in tap water are chlorinous in nature, followed by earthy and musty. These types of odours, in North America, do not usually raise serious concerns. The chlorinous odour is readily linked to the chlorine residual found in almost all public tap waters (Mackey *et al.*, 2004). This may not be the same in other countries around the world as the use of chlorine in drinking water varies.

T&O compounds produced by algae (such as geosmin and 2-methylisoborneol or MIB) have not been shown to cause human health effects at the levels found in drinking water. As a result, water utilities have been eager to report that these offensive odours do not make the water unsafe for human exposure. For example, algae growth in raw water reservoirs caused earthy, fishy and dirty odours to break through treatment. The news report 'Algae causing unpleasant smell in Clyde water' (thenews-messenger, 2009) stated that, 'Even though the taste and smell may be objectionable to some, it is safe to drink'. A news headline for Shreveport stated, 'Shreveport: Dirt smell in water not toxic' (Shreveport Times, 2009). Topeka, Kansas had a news headline that read, 'Topeka drinking water experiencing strange taste, odour' (KTKA, 2010). This article went on to say that, 'Officials say it is safe to drink.' The Austin Water Utility, during an algal event, stated the following according to a news report: 'Recent testing has detected elevated blue algae levels, and while it may emit a musty odour, the water still meets all regulatory standards' (News8Austin, 2009).

The claim that algae-caused T&O have no associated health risks may, however, become a statement that water utilities can no longer release to the public. Graham *et al.* (2010) reported on the results of a survey of cyanotoxins associated with cyanobacterial blooms in lakes of the United States. They reported that, 'toxins and T&O compounds frequently co-occurred. However, toxins occurred more frequently than T&O compounds, so odour alone does not provide sufficient warning to ensure human-health protection.' The T&O compounds studied were geosmin and MIB. The press release issued by the United States Geological Survey (USGS) for this study was titled, 'If the Water Looks and Smells Bad, It May Be Toxic' (released 9/13/2010) and went on to say, 'Earthy or musty odours, along with visual evidence of blue-green algae, also known as cyanobacteria, may serve as a warning that harmful cyanotoxins are present in lakes or reservoirs.' *Water Environment and Technology* magazine (January, 2011 issue) covered the USGS release in this way: 'The presence of T&O compounds, along with the visual evidence of blue-green algae, known as cyanobacteria, could identify the presence of harmful cyanotoxins in lakes and reservoirs ... these compounds that result in earthy, musty odours, were found almost every time cyanotoxins were present.' The press release stated that cyanotoxins can be poisonous to people and pets, and exposure can result in skin rashes, severe stomach upsets, seizures and even death. These types of statements that exist in the public domain, especially from a scientific source, require that water utilities think twice before releasing a statement on the safety of drinking water.

9.3 STRATEGIC COMMUNICATION: PROACTIVE AND REACTIVE

Risk communication is an essential strategy to prevent or minimize the impacts of water quality issues on public perceptions and can establish the water utility as a credible source of information. Identifying, understanding and communicating potential risks help the public understand the vital role of water utilities in delivering safe and clean tap water.

It is paramount that water utilities begin to proactively communicate about drinking water issues. In reality, the majority of people do not understand the functions necessary for delivering clean, safe tap water. They simply expect it to be clean and safe, or assume it is never safe. Often when water utilities attempt to communicate about drinking water issues, the message is not effectively delivered and, therefore, people do not understand or value the message. The primary functions of water utilities that are most important to the public are delivering clean, safe water and addressing issues that affect water safety and aesthetics. Today, water utilities are faced with a variety of water quality issues that hinder a mutual relationship between the utility and its customers.

The most effective risk communication is a combination of proactive and reactive strategies. Ideally, engaging in proactive communication will minimize resource needs and potential loss of credibility when an event requires reactive communication. The exclusivity of proactive or reactive communication can potentially damage public confidence in a water supplier and the water it delivers. Proactive communication occurs 24 hours per day, 365 days per year and reactive communication occurs in response to a water quality issue or emergency, such as a taste or odour event.

Often, as a result of limited resources, the water sector relies upon reactive communication when communicating to the public. Minimal to no communication occurs when operations and services are operating efficiently. Reactive communication occurs when a water quality issue emerges or a customer calls in a complaint. However, when a water quality event or emergency occurs, haphazard, reactive communication requires significant resources and causes damage – sometimes permanently – to public perceptions of tap water and the water supplier. As is often the case with T&O problems, these issues occur sporadically, such as during a seasonal algal bloom, extreme weather event, or a backflow event in the distribution system. This typically is managed by reactive and temporary communication.

An effective complement to reactive communication is proactive communication. This type of communication is continuous and consistent during normal operations. Proactive communication involves staying in contact with customers when water quality is optimal and customer complaints are minimal or feedback positive. A proactive, transparent and collaborative water utility establishes itself as the primary source and expert for credible information on tap water at all times.

The participation in proactive and reactive communication creates a continuum of communication with the public, offering day-to-day customer information. In this approach, the water utility manages its risks during both normal operations and emergency events. Routine messaging also enables a utility to manage risks in everyday communications. It is valuable to understand the current perceptions of customers and knowledge of their water system. If there is a specific issue that the public is concerned about, communicate proactively so that when the water bill increases or an issue hits the media, the customers instinctively contact the water utility for more information.

Utility-driven communication is key. It is always the better option for a water utility to communicate information to the public, rather than a third party source. This proactive approach can prevent misinformation and establish utility credibility. A few major components are important for water utility communications, including diversity in staff expertise (technical and communications) and a risk management plan (Pollard *et al.*, 2009).

Internal communication is the first step in developing a risk communication strategy that is both proactive and reactive. Proactive communication relies upon continuity in internal communications to ensure information is exchanged and consistent at all levels of the water utility. Communication about current and emerging water quality issues must be a shared, two-way conversation between field staff and management. Information must also be shared across various sections within the utility, including field staff, laboratory and customer service. Internal communication includes information exchange about water treatment, distribution and quality.

The first step in prioritizing proactive communication is to create a utility culture of collaboration. Routine interdepartmental meetings are simple opportunities for setting an agenda that focuses on internal and external issues that my impact the public. These meetings ensure information is consistent wherein staff begins to establish a rapport with one another. Special briefings can also serve as a tool to ensure consistent and proactive communication. One example is to have water quality staff conduct briefings with customer service staff on various water quality issues, including T&O. Briefings are opportunities to share technical information and identify consistent, simple messaging for sharing this information with customers.

Routine internal meetings and briefings also establish essential relationships for addressing routine customer complaints as well as response during an emergency or water quality event. When a customer reports a common water quality issue, a utility can efficiently manage the risk with consistent messaging to the customer because it has already been discussed internally within the water utility.

When customers contact a utility to report a water quality complaint, it is also important that this information is tracked and shared internally. Customer contacts are an effective tool for tracking T&O trends in the distribution system and at geographic locations. This information is important to share among customer service staff and operational staff to ensure an expedited response and investigation. On the other hand, operational staff must also notify customer service staff when a water quality issue is identified in the system to ensure staff are prepared to respond to customer calls. If T&O issues are common or seasonal in a water system, it is helpful to provide customer service staff with prepared factsheets and talking points about the issue, including important information to share with customers during these specific periods. This often requires a brief explanation of any technical issues, how it impacts the customers, and what the customers and utility can do to resolve the issue.

External, proactive communication can begin with engaging the media on drinking water issues. Whether it is negative or positive news coverage on a water quality issue, a water utility should always contact a news source to offer corrected information or additional information, and to express appreciation if the news is positive. A water utility can also offer exclusive facility tours and a roundtable discussion for members of the media. This type of forum allows the media to inquire about current or emerging issues affecting public tap water and establishes the water utility as the single, credible source of information on water quality. When newsworthy events occur, the media will learn to directly contact the water utility for accurate information. The utility should also have media contacts for releasing public information during water quality emergencies.

Community meetings are another effective source for exchanging information with the public. Discussing the importance of water quality with local groups, including T&O issues, can improve public satisfaction and confidence in their local water utility. A utility can proactively seek out local and neighborhood meetings and request to be a guest speaker to discuss current water quality (when it is optimal) and respond to issues (when it is poor). Local environmental and health groups are specific groups that serve as effective tools for collaboration and sharing information with the public.

If T&O issues are highlighted in the news or community forums it is important to respond and highlight the shared responsibility model, what you know, what you may not know, your action plan and what the public can do. Responding to newsworthy events does not end when the media coverage ends. It is important to followup with the public because issues always resurface. Proactive communication includes continuously providing water quality monitoring updates, policy updates, research updates and simple messaging or reassurances that the water is safe.

9.4 RISK COMMUNICATION PLANNING

Risk communication is a set of strategies and approaches to effective communication. It is an interactive process of information exchange about risks involving individuals, groups and institutions. Risk management decisions and actions flow from this exchange. The public is likely to only identify a drinking water issue when the taste, odour or colour of their tap water changes or does not meet their personal standard of 'safe'. Conversely, water utilities think about their operations and essential role in delivering clean, safe drinking water, 24 hours a day, 7 days per week. Strategic risk communication aims to get tap water more often on the minds of the public, in addition to associating their local tap water with their local water utility. Risk communication can be effectively shared if customers understand the role, responsibility and functions of the water utility in delivering tap water. The ultimate goal of strategic risk communication is that the public identifies the water utility as the primary, trusted source for all drinking water information, positive or negative.

When establishing a risk communication framework, it is important to acknowledge that no single approach may work for all utilities. The priority is to establish a strategy that is effective for an individual utility audience and targets issues specifically affecting a system. Most importantly, communication should be consistent, reliable and understandable. A utility should consider the current public perceptions and knowledge of their tap water. This understanding helps a utility identify the learning needs of their public, preferences for risk communication messages and delivery methods, ability to uptake information and willingness for behavior or attitude change.

Risk communication uptake relates to previous experience of the public with drinking water risks and an ability to accept the water utility as a trusted source of information. A person viewing a water utility as credible and trustworthy is more likely to receive and utilize information about their drinking water. In addition, people that are well informed have higher levels of acceptance or information uptake, while the majority of people tend to simplify risk information. Message delivery method and tone are also important factors in public information uptake.

Risk perceptions can be influenced by voluntary versus involuntary risks, familiar versus unfamiliar risks, level of knowledge about the risk and how the risk is described and communicated. Water quality issues that cause changes in T&O are challenging concerns for the public and utilities. These issues are complex and involve serious consequences for the reputation of the utility, triggering changes in public perception of its tap water. Communicating about and responding to the public on aesthetic issues is essential to building public trust and credibility. When communicating these risks, it is important to understand the needs of the public relative to information. What do they want to know? How do they want to know it? When do they want to know it? In general, it is important to always provide basic information for the least informed groups and additional information for well informed groups.

Effective risk communication practices are essential for gaining public confidence in a utility and the water it delivers. This requires consistent communication with targeted health and environmental messaging. Utilities should practice a commitment to risk communication, develop a strategic process rather than reactive measures, identify specific water quality issues specific to their system (e.g. T&O), and engage a full range of employees. Figure 9.1 shows a possible decision-making process involved with addressing an emerging water quality concern and how internal and external communication might fit into the process. Therefore, a risk communication plan would benefit the water utility.

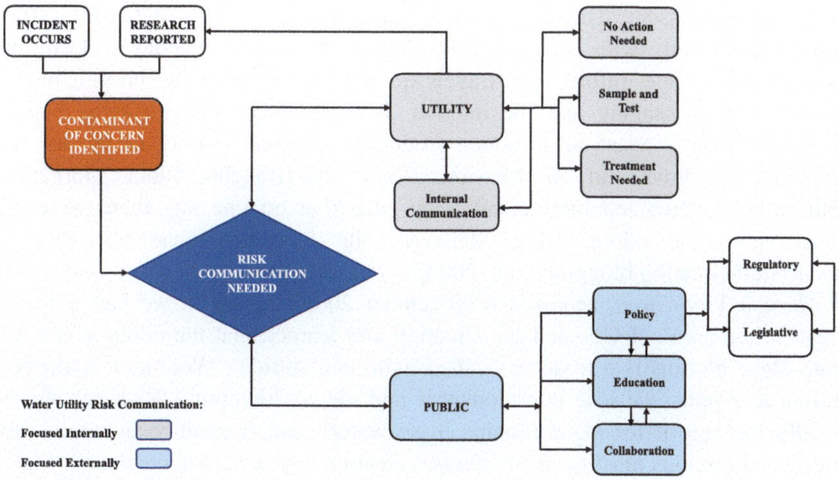

Figure 9.1. Decision making within a water utility concerning water-quality contaminants.

A risk communication plan identifies current and potential issues that can affect a water system and trigger public concern, particularly T&O issues. A list of public concerns can be developed by examining customer complaints for the last three years, or by reviewing local news reports for the past five years. The water utility should identify who, within the water utility, are the experts on the technical issues of each concern to develop white papers, or technical papers, on specific subject areas. From these technical papers, public communication experts can develop question-and-answer (Q&A) sheets in non-technical language for dissemination to the general public. This Q&A format is most helpful for people within and outwith the water utility. The water utility should also share and request feedback from key stakeholders for technical papers and Q&A sheets, including regional experts and agencies that would be involved in water quality issues. The water utility should also identify special or vulnerable populations that may need specific information during a water quality event. This includes identifying contacts for these groups (the people who provide them with trusted information) and determine if translation into other languages is needed.

9.5 UTILITIES MUST BECOME THE TRUSTED SOURCE FOR INFORMATION

Taylor *et al.* (2006) summarized some of the major sources of information that customers in the U.S. will go to when experiencing a T&O problem: health-care providers; health departments; United States Environmental Protection Agency (EPA) or governmental agencies; environmental groups; academic experts; water utilities; and the news media. Customers will frequently use the internet to gather information on issues. There are many organizations, as well as individuals, that communicate daily with the public by blogs and Twitter. In today's electronic communication age, information travels quickly and extensively through blogs, Twitter, text messaging and the internet. For example, when an algae bloom combined with a reservoir turnover produced persistent odours in the air for communities downwind in Mission Viejo and Rancho Santa Margarita, California, it generated internet attention such that on one day there were 82 blogs about the odour. Blogs delivered the following messages: (http://sciencedude.freedomblogging.com/2009/11/01, *Algae bloom causes bad smell in Mission Viejo area*, accessed 6 November 2009): 'Thought we had a sewer break in the area', 'I Googled the situation and learned that the odour given off from algae blooms is a toxic gas called hydrogen sulfide', 'We smell hydrogen sulfide at 5 ppm, we lose consciousness and die at 700 ppm', 'Raw sewage is usually the culprit for algae blooms in seawater', and 'Certainly an odour this strong and noxious must have an adverse effect on any breathing creature'.

One way to help predict misinformation and the need for clarification about an issue is to conduct an internet search using key words. Using the key word 'cyanobacteria', the following information was found on Wikipedia (www.wikipedia.com, accessed on 24 December 2010): 'Some algae, particularly blue-green algae (cyanobacteria) such as *Anabaena*, produce MIB together with other odourous chemicals such as geosmin' and 'Cyanobacteria can also produce chemical toxins that have been a problem in drinking water in some cases'.

An internet search for 'bad smells' and 'water' resulted in the following information (http://www.wisegeek.com/what-can-cause-tap-water-to-smell-bad.htm, accessed on 24 December 2010): 'A number of factors can cause tap water to smell bad, including concentrations of metal from the pipes used to deliver it, organisms living in the water, chemicals used to remove organisms living in the water, and environmental contaminants', 'If the tap water smells extremely unpleasant or comes from a well, have the department of public health test it for potentially harmful contamination', 'Copper, iron, and lead have all been found in tap water, and lend unique smells and tastes', and 'Lead is a more serious concern, because it can have negative impacts on human health in even small amounts.'

T&O issues in a water distribution system may also receive media coverage. One example is customer complaints about a localized smell in the water distribution system. A news headline called it, 'Nasty-Smelling Water.' The cause was reported as naphthalene, a primary ingredient in mothballs, from a stagnant section of a water main that had a bitumastic or coal tar lining. A Google search of 'naphthalene' would trigger public concern or confusion and highlights the importance of a utility communication strategy during water quality events (http://www.eco-usa.net/toxics/chemicals/naphthalene.shtml, accessed on 28 December 2010): 'Naphthalene is a white solid that evaporates easily. It is also called mothballs, moth flakes, white tar, and tar camphor', 'Naphthalene has a strong, but not unpleasant smell. You can smell it in water when 21 ppb are present', and 'Hemolytic anemia (a condition involving the breakdown of red blood cells) is the primary health concern for humans exposed to naphthalene for either short or long periods of time. These effects can occur from either breathing or eating naphthalene. In pregnant women, naphthalene and its breakdown products in blood can reach the fetus.'

All of these events trigger the common question, 'Is my water safe to drink?' The public often poses this question, particularly during water quality events or changes in aesthetic quality. Responding to this question consistently throughout the year – even when it is not asked – can establish the utility as a reliable source of information. It is important for customers to hear and learn information directly from the utility, rather than the media or other third parties. If the water utility does not build a proactive and positive relationship with the public, then they will lose control of where the public seeks and finds information. Good and bad information gets out quickly in today's social networking and internet culture. As soon as negative or misinformation is publicly released, it is difficult to retract and counteract without investing significant resources.

Communication with customers is a two-way conversation. A water utility can communicate with its customers using phone, website, text, email, social media, print media, outreach events, field investigations, billing, newsletters, door hangers, letters and community meetings. Customers will often communicate to their utility using phone, email, social media, websites, mail and community forums. Community groups are a powerful, cost-effective tool for two-way dialogue between utilities and the public. Utilities can annually contact local community groups and request that they share any interests or concerns about their drinking water. Utilities can also ask members of active community groups to serve as liaisons between the utility and their members.

9.6 INTERNAL COMMUNICATION TOOLS

Tools exist that can assist in managing customer complaints and water quality issues. First, a water utility should consider tracking all customer complaints received via phone, email or website. Complaints can be logged by issue type, such as taste, odour or colour (Figure 9.2). Reports that are a public health concern or affecting more than one customer in a similar geographic area can assist in identifying a problem. Customers should also be able to contact the utility and speak with a knowledgeable utility staff member to discuss water quality issues and concerns, and receive guidance to improve their water quality. This requires trained, skilled staff that understand water quality issues and know how to effectively communicate their knowledge with the public. Figure 9.3 is an example of a guide that a utility can develop for its staff to effectively respond to customer complaints.

Microsoft Excel - Complaint Breakdown-Daily Tracking

	A	B	C	D	E	F	G	H	I	J	K	L	M	N	O	P	Q	R	S	T	U	V	W	X	Y	Z	AA	AB	AC	AD	AE	AF	AG
1	March-10	Date																															Month Total
2	Type of Complaint	1	2	3	4	5	6	7	8	9	10	11	12	13	14	15	16	17	18	19	20	21	22	23	24	25	26	27	28	29	30	31	
3	Foul Odor / Bad Taste	1		3	2	1			1									1						1	1	1				2	2		16
4	Chlorine Odor/ Taste	5		2	2					4	2	2	1			1	1		1					1						1			23
5	Cloudy Water								1			1														1	2						5
6	Discolored Water	1	4	5	4	4			4	3	1	5	5			6	1	1	5	7			2	6	4					2	5	2	77
7	Chemical Odor / Taste																																0
8	Particles																																0
9	Air Bubbles																																0
10	Oily Water																																0
11	Skin Irritation											1																					1
12	Possible Medical Issue																																0
13	High Temp of Cold Water																																0
14	Possible Intentional Contamination				1																												1
15	DBP Calls																																0
16	Vacant Home Test																																0
17	Other'																																0
18	Total number of calls	7	4	10	9	5	0	0	6	7	3	9	6	0	0	7	2	2	6	7	0	0	2	8	5	2	2	0	0	5	7	2	123
19	Number scheduled visits	0			3	1			3	0	0					1	1	2	1					2		1	1			2	1		19
20	Retest															1																	1

Figure 9.2. Example of a customer complaint database.

	IDENTIFY THE PROBLEM	POTENTIAL CAUSES	RECOMMENDATIONS	ACTION
DISCOLORED WATER	• How long has the problem occurred? • Color of water? • Cold and/or hot water? • Throughout building or localized? • Have you run the cold water for 15 minutes or more?	• Brown/yellow - typically iron caused by old pipes; not a health risk • Black - carbon filter or manganese • Green/blue - copper	• Flush cold water for 15 minutes. • Do not wash clothes when water is discolored. If customer is in the middle of doing laundry, keep clothes wet until discoloration is no longer present & re-wash. • Black - replace carbon filter (if applicable)	• Brown/yellow - FLSH 6 • If FLSH 6 does not resolve issue, schedule appointment • Black - test for manganese • Green/blue - test for copper
CLOUDY WATER	• How long has the problem occurred? • Cold and/or hot water? • Throughout building or localized?	• Typically caused by air that enters pipes and leaves tap as air bubbles • Can occur more often in colder weather • Not a health risk • Can also be calcium or phosphate	• Fill a glass with water, if the cloudiness disappears from bottom to top in a few minutes, it is air bubbles. • If cloudiness settles to the bottom or does not clear, likely calcium or phosphate.	• Schedule appointment if cloudiness does not clear
PARTICLES IN WATER	• How long has the problem occurred? • Color of particle? • Cold and/or hot water? • Throughout building or localized? • Water treatment installed?	• White particles - can be caused by calcium that becomes solid as it travels through the hot water heater or white plastic material from the hot water heater (dip tube) • Black particles - can caused by disintegrated rubber gaskets in plumbing fixtures (usually slimy), iron scales from pipes after a large main break/major construction, carbon filter cartridges, or manganese.	• White particles - place particles in distilled vinegar. If particles bubble or dissolve, it is calcium, otherwise it is most likely plastic from the hot water heater. Clean aerators; flush hot water heater. • Black particles - flush cold taps; replace filter cartridges; replace gaskets/o-rings with chloramine-resistant materials.	• White particles - if necessary, schedule appointment, ask customer to save particles • Black particles - if necessary, note on appt sheet to test for manganese
ODOR	• How long has the problem occurred? • Type of odor? • Cold and/or hot water? • Throughout building or localized? • Water treatment installed?	• Sewer odor - commonly from sink drain, not water. • H2S (Hydrogen Sulfide) - rotton eggs • Chlorine odor - drinking water is disinfected with chlorine and chloramine to ensure protection against contaminants. • Musty odor - earthy	• Sewer odor - Fill a glass halfway with tap water and smell the water in a separate room. If the odor is no longer present, the odor is likely from the sink plumbing. Pour 1/2 cup of bleach down drain to remove debris and odor. • Chlorine Odor - Flush cold water taps for 5 - 10 min for 3 days.	• Chlorine odor - schedule an appointment • Musty odor - shedule an appointment & contact supervisor, could be MIB or geosmin
TASTE	• How long has the problem occurred? • Type of taste? • Throughout building or localized?	• Metallic - new pipes/fixtures installed. • Chlorine - drinking water is disinfected with chlorine and chloramine to ensure protection against contaminants. • Burning taste - schedule crew ASAP, notify supervisor.	• Chlorine taste - Flush cold water taps for 5 - 10 min for 3 days AND routinely run cold tap for at least 2 minutes, then collect and refrigerate cold water in an open pitcher. Within a few hours, the chlorine taste and odor will disappear.	• Chlorine taste - schedule appointment • Metallic taste - schedule appointment • Burning taste - schedule crew ASAP
WHITE RESIDUE	• How long has the problem occurred? • Throughout building or localized?	• The result of dissolved minerals found in water, such as calcium and magnesium, also known as water hardness. • Typically water hardness is higher during warmer months. • Not a health risk.	• Commercial products are available to remove white residue caused by minerals in water.	• If necessary, schedule appointment
ADDITIONAL INFO	Note if customer mentions the following: • Pin hole leaks • Skin irritation - visited a doctor? • Other medical issues			• Pin hole leaks - contact supervisor • Medical issue related to water - schedule an appointment and notify supervisor.

Figure 9.3. Example of a customer complaint communication guide.

Internal strategic communications are essential for effective risk communication, particularly in preparation and response to emergency events. This can involve sharing a daily tracking log of customer complaints among utility staff, either daily or weekly. Water quality, communications personnel and field staff should meet routinely to discuss trends in customer complaints and potential distribution system issues impacting water quality. Annual and biannual communications workshops are another opportunity to train field staff and customer service personnel on consistent messaging relative to current or potential water quality issues. When water quality events emerge, it is easy to reconvene this group and develop talking points or fact sheets to share internally and with the public.

The development of a water quality committee or team is an effective method to create an organizational culture for proactive thinking and action on water quality and public perceptions of risk (Burlingame *et al.*, 2010). Philadelphia's Water Quality Committee has been functioning since the early 1990s and is chaired by its manager of Public Affairs. The committee's original purpose was to help Philadelphia speak with one voice on water quality issues and concerns. The Philadelphia Department of Public Health participates in the activities of the committee. The committee has developed risk communication capabilities that build upon Philadelphia's existing understanding of its customers' perceptions of, and experiences with, drinking water T&O (Burlingame & Mackey, 2007). The water quality committee is also a source of trained and informed spokespersons.

9.7 EXTERNAL RISK COMMUNICATION TOOLS

Defining communication methods specific to a system's water quality and audience is important. First, identify sources that the public generally uses for routine information uptake, for example, internet, newspaper, word-of-mouth, etc. Depending on the size and demographics of a water system, communication can occur in a variety of ways, including bill inserts, social media, newspapers, newsletters, water-fountain signage, and so much more. For utilities with minimal resources, identify a cost-effective communication method – website banner, Facebook or a bill message – and reuse basic tap water messages periodically throughout the year.

Routine, proactive communication should include simple messages informing the public that water quality continues to be clean and high quality. This simple message can be coupled with additional information about monitoring and detection activity. It is important to always provide an explanation about each issue or contaminant, including definitions, potential health effects and impacts on aesthetic water quality.

Taylor *et al.* (2006) stated that, 'Effective communication about T&O events can depend as much on *who* delivers the message and *how* the message is delivered as on the content'. A water utility must develop its capacity to develop messages to the audiences who will be affected by an incident. Proactive messages are needed when issues are expected to arise, such as for seasonal T&O problems. Messages are needed when an event occurs as well as to re-notify people when the problem is corrected. A message may require translation into different languages, which can be difficult with T&O and its associated technical language. The basic content of most public messages includes: a description of the problem (What happened?); alternative water supply information (Where can I get safe water?); what actions customers can take (What should I do for now?); and telephone, website, and contact information (How can I stay updated?).

The USEPA has specified the information that must be included in a public notification, such as for the violation of a drinking water quality standard: Who is the population at risk?; When did the event occur?; What is the description of the event?; What are the potential adverse or health effects?; Whether alternative water supplies should be used?; What actions consumers should take?; What is being done to correct the problem?; When will the water return to normal?; and Who is the contact for more information?

Utilities communicate with their customers using a variety of methods, including: web sites, call centres, press releases, news media interviews, newspaper, direct mailings, and consumer confidence reports (Taylor *et al.*, 2006). Public messages must be crafted to fit the requirements and limitations using a variety of outreach tools, such as: web sites; door hangers; press releases; call-centre script; online message-centre script; blogging and texting; automated phone-messaging script; television news, cable TV channels, news radio, print and online news; social networking sites such as Myspace, Facebook, Blogspot, Twitter and Youtube; and finally direct messaging, including text, email and voice messaging.

Developing talking points for traditional and social media is a simple, effective tool for ensuring that consistent, credible information is communicated. The water utility can develop talking points and trained spokespersons are essential to effectively deliver the talking points. Message mapping is one way to develop talking points. When used consistently, message maps promote multiple partners speaking with one voice or in a consistent and clear manner. Message mapping (Covello *et al.*, 2007) is a seven-step process by which users can: *Predict* 95% of questions likely to be asked by the media and others following an incident; *Prepare* clear and concise answers to the questions along with supporting information ahead of time; and *Practice* effective message delivery before a crisis occurs.

Message mapping can begin with a group of water utility experts preparing a technical paper with help from outside experts. This can serve as the scientific information behind the issue and is not intended as public information. Q&A and message mapping can be developed from a technical paper and shared with the public and other groups who need to understand a particular issue. The talking points can also be developed using facts from the technical paper as needed to emphasize key points.

Figure 9.4 shows an example of utility message mapping. Each water utility will need to understand its core message, the single most important message the public should hear. The spokespersons need to practice bridging back to this message when speaking to the public, such as through the news media. Figure 9.5 provides an example of this message map filled out for response to an algal bloom.

CORE STATEMENT: What is the main point that the public needs to hear?
PROBLEM Statement: What is it and why should I care? How does it impact my world, environment and health?
Talking Points
1)
2)
3)
SAFETY Statement: Is the water safe? How do we know?
Talking Points
1)
2)
3)
WHAT YOU CAN DO Statement: What can the public do to help protect the water supply?
Talking Points
1)
2)
3)

Figure 9.4. Message mapping template.

On a regional basis, spokespersons for organizations that represent public health should share notes and agree on the main messages, although their core statements may differ according to their own missions. In addition, practice is important and outside experts exist to provide this discipline (of risk communication) (Friedman, 2010; Ragain, 2010). A good process in risk communication to address T&O events can increase public confidence, result in a more-informed public, and improve support for controlling T&O problems when they arise.

CORE STATEMENT: This water utility's job is to provide clean, safe, high quality drinking water and we view that responsibility as a matter of public trust. The safety of our drinking water supply is important to all of us.
PROBLEM Statement: The drinking water for the city has an earthy-musty smell due to algae growth in the river
Talking Points
1) When algae grow in excess they create what is called a bloom that can colour the water green because the algae are green in colour.
2) Certain types of algae naturally produce an earthy-musty odour in the water.
3) This happens in the Spring when the weather warms up and it happens all across the country.
SAFETY Statement: The earthy-musty odour is not a health concern
Talking Points
1) The water treatment plant is adding carbon to the water before the filters to remove much of the earthy-musty odour.
2) The water laboratory is testing the water daily to monitor the effects of the algae.
3) The algae bloom will end when we get a good rain to flush and refresh the river.
WHAT YOU CAN DO Statement: All of us share in the responsibility to protect our water resources
Talking Points
1) Stay in touch with us regarding the algae bloom via our website.
2) Use phosphate-free detergents to help reduce the nutrients in the river that provide food for the algae to bloom.
3) Use fertilizer-free practices in maintaining your grass lawns since fertilizer can also provide nutrients for algae to grow.

Figure 9.5. Example of message mapping for an algae bloom.

9.7.1 Trends in communication

Over the last 30 years, customer complaints that come in by phone or opinions expressed at community meetings, followed by news media reports have been the driving force that water utilities had to reckon with when a T&O event occurred. Today, that is changing. Social media and apps designed to link the public to their politicians and community leaders are changing the way public opinion is expressed.

The internet and social media are global. In places where people may not own a car for personal transportation, they may have access to social media. Apps are linking people to the government for open data and transparency, and for expressing public opinion. One concept or practice is to make information, even data, openly available via the internet (known as 'open data'). The public will demand, more and more, that water utility data be accessible and open.

Another concept is that consensus can be established via the internet through the wisdom of the crowd. The public, in large or small groups, does not need a politician to speak for it. The opinion of a large group expressed by social media is being validated as reliable. The 'crowd' could be any group of trained or untrained people, and does not have to represent any particular demographic or organization.

This is happening irrespective of country status or level of development. If a water utility is caught in a serious and extended T&O episode, the public will use internet means to express their concerns and opinions. This will not be controllable, as it was in the past, with press releases and community meetings. Therefore, water utilities need to enhance their own experience and practice using available skills and tools for the risk management of public perception.

REFERENCES

Burlingame G. A. and Mackey E. (2007). Philadelphia obtains useful information from its customers about taste and odour quality. *Water Science and Technology*, **55**(5), 257–263.

Burlingame G. A., Estes-Smargiassi S. and Akagi Y. (2010). Committees streamline, coordinate water quality control efforts. *Opflow*, **36**, 22–23.

Covello V., Minamyer S. and Clayton K. (2007). Effective Risk and Crisis Communication during Water Security Emergencies. Summary Report of EPA Sponsored Message Mapping Workshops, EPA/600/R-07/027, National Homeland Security Research Center, Cincinnati, OH, USA

Doria M. De F., Pidgeon N. and Hunter P. R. (2009). Perceptions of drinking water quality and risk and its effect on behavior: A cross-national study. *The Science of the Total Environment*, **407**, 271–276.

Dupont D., Adamowicz W. L. and Krupnick A. (2010). Differences in water consumption choices in Canada: The role of socio-demographics, experiences, and perceptions of health risks. *Journal of Water and Health*, **8**, 671–686.

Friedman K. (2010). Shut Up and Say Something: Business Communication Strategies to Overcome Challenges and Influence Listeners. Praeger, Santa Barbara, CA, USA

Graham J. L., Loftin K. A., Meyer M. T. and Ziegler A. C. (2010). Cyanotoxin mixtures and taste-and-odour compounds in cyanobacterial blooms from the Midwestern United States. *Environmental Science & Technology*, **44**, 7361–7368.

Kelly M. G. and Pomfret J. R. (1997). Tastes and odours in potable water: perception versus reality. In: The Microbiological Quality of Water, D. W. Sutcliffe (ed.), The Freshwater Biological Association, Ambleside, UK, pp. 71–80.

KTKA (2010). www.ktka.com/news/2010/aug/19/ (Accessed on 20 August 2010)

Lazo J. K., Pratt J. L., Herrick C. N., Hagestand M. L., Raucher R. S., Hurd R. E. and Rambo E. H. (2004). Understanding and enhancing the impact of consumer confidence reports. In: Efficient and Customer-Responsive Organization, The Water Research Foundation (formerly AWWA Research Foundation), Denver, CO, USA.

Mackey E. D., Davis J., Boulos L., Brown J. C. and Crozes G. F. (2003). Customer Perceptions of Tap Water, Bottled Water, and Filtration Devices. The Water Research Foundation (formerly AWWA Research Foundation), Denver, CO, USA.

Mackey E. D., Baribeau H., Fonseca A. C., Davis J., Boulos L., Brown J. C., Crozes G. F., Piriou P., Rodrigues J. M., Fouret M., Bruchet A. and Hiltebrand D. J. (2004). Public Perception of Tap Water Chlorinous Flavor. The Water Research Foundation (formerly AWWA Research Foundation), Denver, CO, USA.

McGuire M. J. (1995). Off-flavour as the consumer's measure of drinking water safety. *Water Science and Technology*, **31**, 1–8.

News8Austin (2009). www.news8austin.com 11th September 2009, "Despite odd smell, Austin Water Utility says water still safe". (Accessed on 9th October 2009)

Pollard S., Bradshaw R., Tranfield D., Charrois J. W. A., Cromar N., Jalba D., Hrudey S. E., Abell P. and Lloyd R. (2009). Developing a Risk Management Culture – 'Mindfulness' in the International Water Utility Sector. The Water Research Foundation (formerly AWWA Research Foundation), Denver, CO, USA.

Ragain L. (2010). Drinking water advisories: Invest time and resources to plan ahead. *Opflow*, **36**, 10–13.

Shreveport Times (2009). www.shreveporttimes.com 4 August 2009. (Accessed on 18 August 2009).

Taylor W. D., Losee R. F., Torobin M., Izaguirre G., Sass D., Khiar D. and Atasi K. (2006). Early Warning and Management of Surface Water Taste-and-Odour Events. The Water Research Foundation (formerly AWWA Research Foundation), Denver, CO, USA.

Thenews-messenger (2009). Algae causing unpleasant smell in Clyde water. www.thenews-messenger.com, July 11, 2009. (Accessed on 13 July 2009)

Turgeon S., Rodriquez M. J. and Levallois P. (2004). Perception of drinking water in the Quebec City region (Canada): The influence of water quality and consumer location in the distribution system. *Journal of Environmental Management*, **70**, 363–373.

Whelton A. J., Dietrich A. M., Gallaghe D. L. and Roberson J. A. (2007). Using customer feedback for improved water quality and infrastructure monitoring. *Journal American Water Works Association*, **99**, 62–76.

Index

IWA Publishing's authorised EU representative for General Product Safety Regulations is Diane D'Arras, 15 rue Duret, 75116 Paris, France, e-mail: safety@iwap.co.uk.

Printed and bound by CPI Group (UK) Ltd, Croydon, CR0 4YY
12/05/2026
02108869-0006